NEW GCSE SCIENCE

Science A

For Specification Units B1, C1 and P1

AQA

Series Editor: Ken Gadd

Authors: Mary Jones, Louise Petheram and Mike Tingle

Student Book

William Collins' dream of knowledge for all began with the publication of his first book in 1819. A self-educated mill worker, he not only enriched millions of lives, but also founded a flourishing publishing house. Today, staying true to this spirit, Collins books are packed with inspiration, innovation and practical expertise. They place you at the centre of a world of possibility and give you exactly what you need to explore it.

Collins. Freedom to teach

Published by Collins
An imprint of HarperCollins*Publishers*
77 – 85 Fulham Palace Road
Hammersmith
London
W6 8JB

Browse the complete Collins catalogue at
www.collinseducation.com

10 9 8 7 6 5 4 3 2 1

ISBN-13 978 0 00 741462 8

British Library Cataloguing in Publication Data
A Catalogue record for this publication is available from the British Library

Commissioned by Letitia Luff
Project managed by Hanneke Remsing and 4science
Edited by 4science
Proofread by Life Lines Editorial Services
Indexed by Jane Henley
Designed by Hart McLeod and Peter Simmonett
New illustrations by 4science
Picture research by Caroline Green and Thelma Gilbert
Concept design by Anna Plucinska
Cover design by Julie Martin
Production by Kerry Howie
Contributing authors John Beeby, Mike Willmott and Ed Walsh
'Bad Science' pages based on the work of Ben Goldacre

Printed and bound by L.E.G.O. S.p.A. Italy

Acknowledgements – see page 320

Contents

Biology

Chemistry

Physics

How to use this book

Welcome to Collins New GCSE Science for AQA!

The main content

Each two-page lesson has three levels:

> The first part outlines a basic scientific idea

> The second part builds on the basics and develops the concept

> The third part extends the concept or challenges you to apply it in a new way.

Information that is only relevant to the Higher tier is indicated with 'Higher tier'.

Each section contains a set of level-appropriate questions that allow you to check and apply your knowledge.

Look for:

> 'You will find out' boxes

> Internet search terms (at the bottom of every page)

> 'Did you know' and 'Remember' boxes

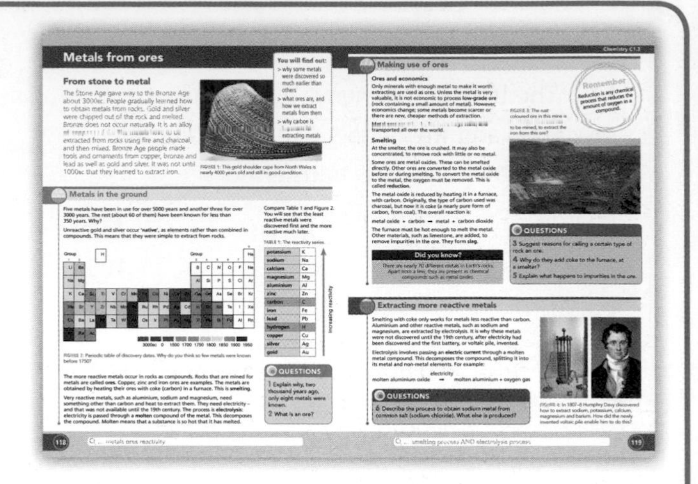

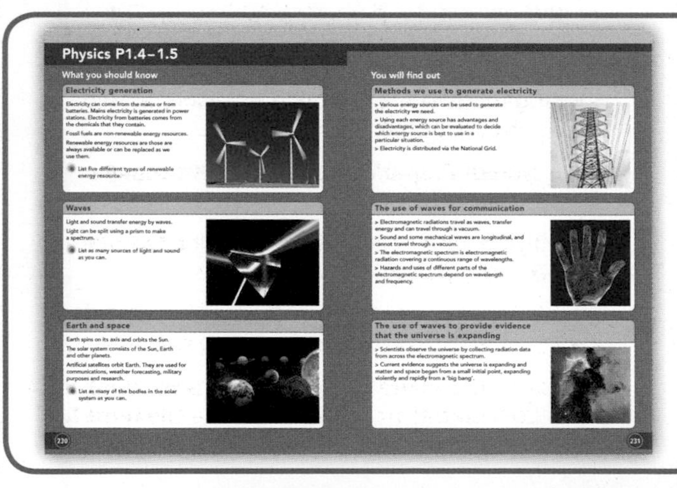

Units and sections

Each Unit is divided into two sections, allowing you to easily prepare for Route 1 or Route 2 assessment.

Link the science you will learn with your existing scientific knowledge at the start of each section.

Checklists

Each section contains a checklist.

Summarise the key ideas that you have learned so far and see what you need to know to progress.

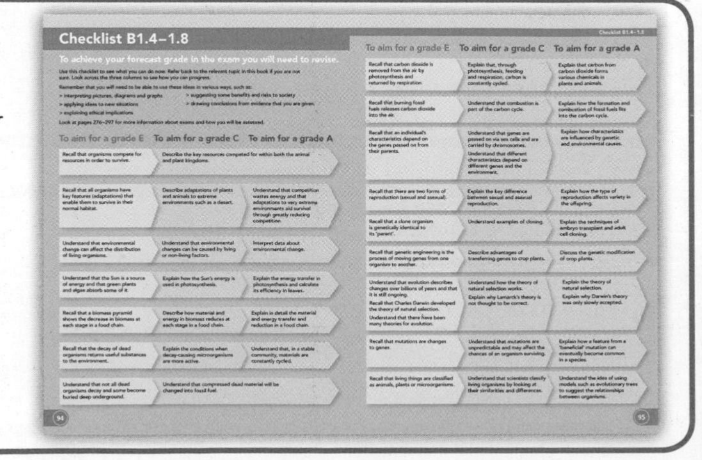

Exam-style questions

Every section contains practice exam-style questions for both Foundation and Higher tiers, labelled with the Assessment Objectives that they address.

Familiarise yourself with all the types of question that you might be asked.

Worked examples

Detailed worked examples with examiner comments show you how you can raise your grade. Here you will find tips on how to use accurate scientific vocabulary, avoid common exam errors, improve your Quality of Written Communication (QWC), and more.

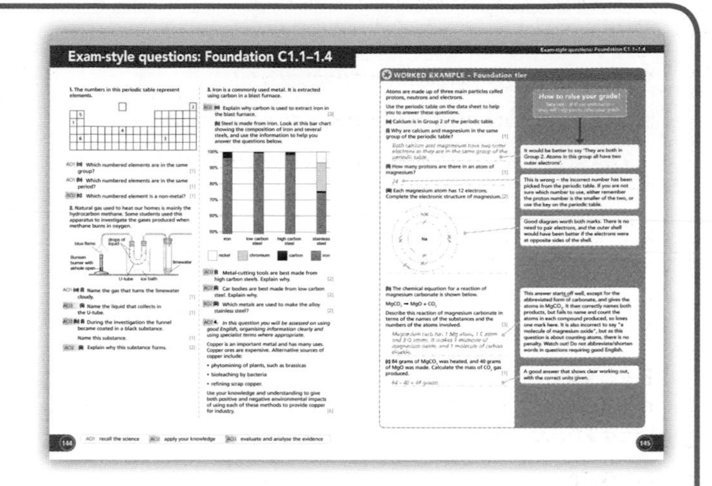

Preparing for assessment

Each Unit contains Preparing for assessment activities. These will help build the essential skills that you will need to succeed in your practical investigations and Controlled Assessment, and tackle the Assessment Objectives.

Each type of Preparing for assessment activity builds different skills.

> Applying your knowledge: Look at a familiar scientific concept in a new context.

> Planning an investigation: Plan an investigation using handy tips to guide you along the way.

> Analysing and evaluating data: Process data and draw conclusions from evidence. Use the hints to help you to achieve top marks.

Bad Science

Based on *Bad Science* by Ben Goldacre, these activities give you the chance to be a 'science detective' and evaluate the scientific claims that you hear everyday in the media.

Assessment skills

A dedicated section at the end of the book will guide you through your practical and written exams with advice on: the language used in exam papers; how best to approach a written exam; how to plan, carry out and evaluate an experiment; how to use maths to evaluate data, and much more.

Biology B1.1–1.3

What you should know

Diet

Food is an energy source.

Different people require different amounts of energy, depending on their age, gender and level of activity.

We also need food so that the body can grow, and to protect it from disease.

What does a 'balanced diet' mean?

Cells, organisms and life processes

All living things are made of cells.

Different cells are adapted for different purposes.

Cells are arranged into tissues, organs and organ systems.

Name two organ systems in the human body, and outline the function of each of them.

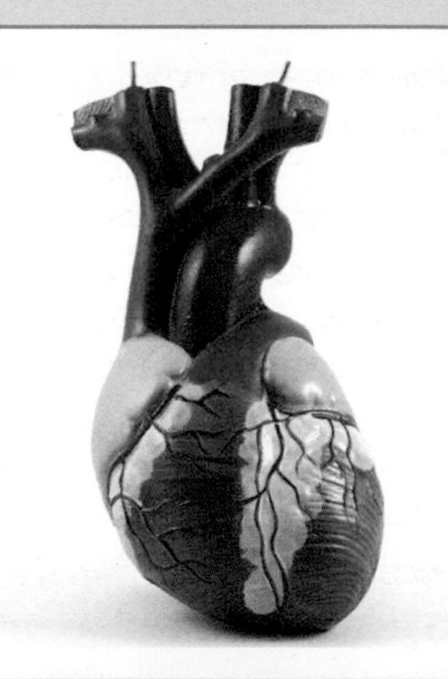

Disease and drugs

Doctors prescribe medicines, such as antibiotics, to help recovery from illness.

Diseases caused by microorganisms can be prevented through immunisation.

Medicines are useful drugs but, if misused, can cause harm to the body.

Drugs, such as alcohol and nicotine, can be harmful to health. Illegal drugs can damage vital organs in the body and affect the way that people behave.

Describe some ways of preventing the spread of infectious diseases.

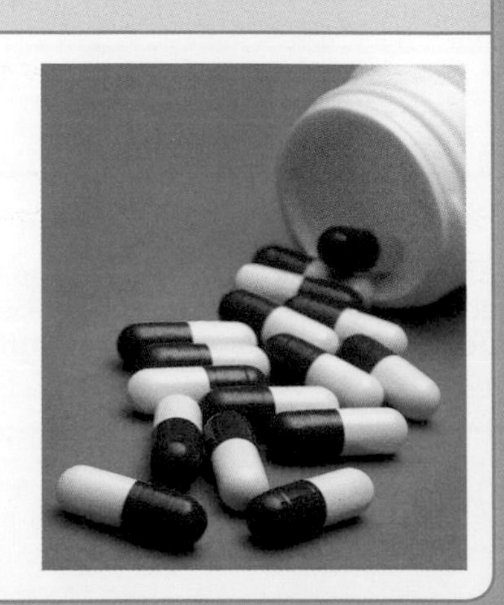

You will find out

Keeping healthy

> Eating a balanced diet and doing plenty of exercise helps you to stay healthy.

> Microorganisms that cause disease are called pathogens. White blood cells defend us against them, for example by producing antibodies.

> Antibiotics are drugs that kill bacteria inside the body.

Nerves and hormones

> The human nervous system includes nerves, brain and spinal cord. It transfers information quickly around the body, as electrical impulses. Receptors receive information about the environment and effectors respond to it.

> Many conditions inside the body are controlled by nerves and hormones. These include water and ion content, temperature and the sugar concentration in the blood.

> Hormones control the menstrual cycle in women. Hormones are also used to prevent pregnancy, or to help an infertile couple to have a baby.

> Plants produce hormones, such as auxin, which help them to respond to their environment.

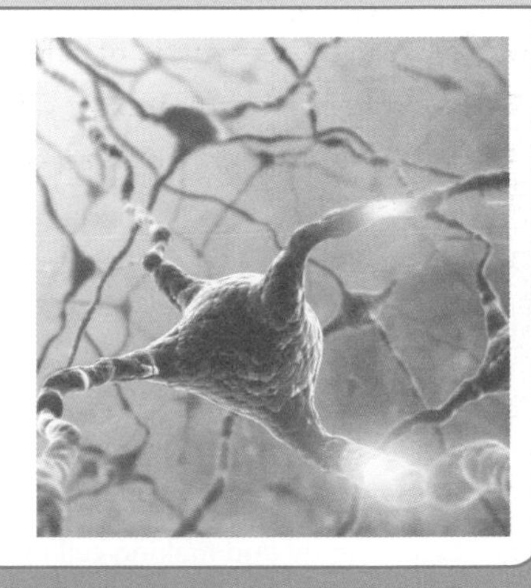

The use and abuse of drugs

> Drugs affect the working of the body. All drugs have side effects.

> Both legal and illegal drugs can cause harm if misused.

> Scientists test new drugs extensively, before they are given to patients.

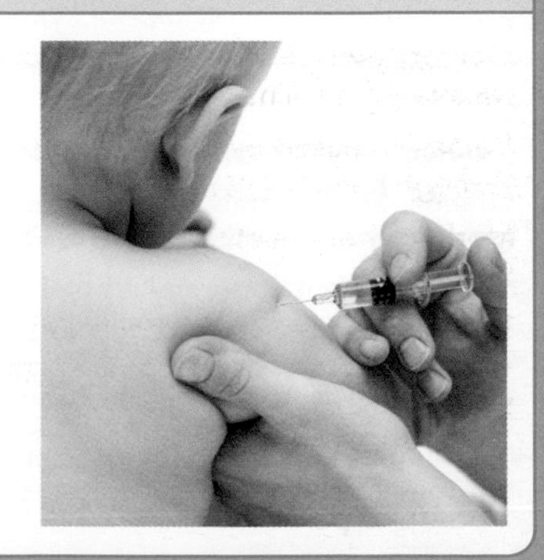

Diet and energy

You will find out:

> what a 'balanced diet' means

> energy needs differ between people

Body heat

Ten people dancing generate as much heat as a gas fire. The energy comes from the food they eat. The dancers lose some of this body heat through their skin.

FIGURE 1: An infrared camera detects hot things. Which colours show 'hot' and 'cool'?

Getting the right balance

Energy balance

Each day, you use energy. Every cell in your body uses food to supply the energy it needs to stay alive and function properly.

Everybody needs to eat food containing the right amount of energy to balance their needs. If you take in more energy than you use, your body stores it as fat. If you take in less energy than you use, your body raids its fat stores to provide your cells with the extra energy that they need.

Balanced diet

The food you eat each day is your diet. Most people eat a **balanced diet**, containing some of all these different kinds of nutrients:

> **carbohydrates** for energy

> **fats** for energy and making cell membranes

> **proteins** for growth (making new cells), repair and energy

> **vitamins** and mineral ions for keeping healthy

> **roughage** (fibre) to keep the digestive system working well

> **water** to transport other nutrients.

Vitamins and minerals are needed in much smaller amounts than the other nutrients.

Most foods contain several different nutrients. For example, milk contains carbohydrates, fats, proteins, calcium, vitamins and water. If you eat plenty of different foods, you will probably get all the nutrients that you need. If your diet is not balanced, you may become **malnourished**. You may become too fat or too thin, or suffer from deficiency diseases.

Dieting

The best way to lose body mass ('weight') is to change your diet so that you eat less overall while still having a balanced diet. Taking extra exercise also helps. The idea is simple: use up more energy than you take in. Then your body has to use some of its energy stores.

Foods and nutrients

When writing about balanced diets, you need to check whether you mean **nutrient** or food. For example, bread is a food. Bread contains the nutrients carbohydrates, proteins and calcium (a mineral ion).

One kind of nutrient cannot contain another. For example, vitamins do not contain proteins – they are different chemicals.

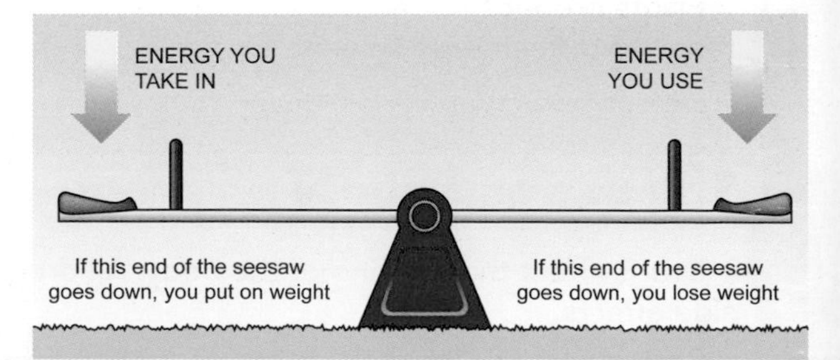

FIGURE 2: Energy balance.

QUESTIONS

1 Which three types of nutrients provide energy?

2 What is a balanced diet?

3 What happens if you take in more energy than you use?

Diet and exercise

Slimming diets

Although people generally lose body mass when following a slimming diet, they usually put it back on again once they go back to their normal diet. People tend to be more successful at keeping their new, lower body mass permanently, if the slimming diet helps them to change their eating habits permanently.

Energy needs

People doing manual work, or a lot of exercise, need to eat more energy-containing foods than people who do not exercise much.

Different people have different **metabolic rates**. This is the rate at which chemical reactions take place in your cells. Two people, just sitting and resting, use different amounts of energy.

Generally:

> Men have faster metabolic rates than women.

> Young people have faster metabolic rates than older people.

> The greater the proportion of muscle to fat in the body, the higher the metabolic rate is likely to be.

Metabolic rate can be affected by your genes, too. These are inherited from your parents.

FIGURE 3: What makes this meal a good one for a person who is on a slimming diet?

FIGURE 4: What happens to metabolic rate during exercise?

QUESTIONS

4 What does metabolic rate mean?

5 Two people eat the same quantity of food each day, but one of them has a much higher body mass than the other. Suggest three reasons that could explain this difference.

Did you know?

Many snakes detect their prey not by using sight, sound or smell, but by sensing the infra red radiation the prey gives off.

Fatty foods

One gram of fat releases almost twice as much energy as one gram of carbohydrate, or one gram of protein. Proteins are not usually a major source of energy for the body because they have other, more important, functions.

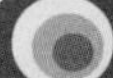

QUESTIONS

6 Explain why eating a lot of foods high in fat is more likely to make a person overweight than eating foods containing mostly carbohydrate and protein.

Diet, exercise and health

You will find out:
- > exercise increases your metabolic rate
- > diet, exercise and your genes can affect your health

Fast food or fat food?

Many people do not want to spend time cooking or shopping. Fast food is easy – it's cheap, can taste good and fills you up. But take care. Some fast foods contain more saturated fat and salt in one serving than you should be eating in a whole day.

FIGURE 1: A typical burger and fries meal contains about 36 g of saturated fat.

Cholesterol

Exercise

In general, people who exercise regularly stay healthier and fitter than people who do not. Exercise increases the use of energy from the food you eat and increases your metabolic rate. The good news is that metabolic rate tends to stay up for quite a long time afterwards. Research shows that 20 minutes exercise, two or three times a week, has a really noticeable, positive effect on fitness and health.

Diet and cholesterol

Your liver makes **cholesterol**. Cholesterol is needed to make cell membranes, but too much is bad for you. It can form blockages in blood vessels. If the blockages restrict blood flow, this may lead to **heart disease**.

What you eat can increase your cholesterol level. **Saturated fats** are especially to blame. These are the fats found in animal products – foods like eggs, dairy products and meat. Eating other kinds of fats is not so bad. Some kinds actually seem to lower your blood cholesterol level. These are called **unsaturated fats**. They are found in plant oils, such as sunflower oil.

Inherited factors and cholesterol

Some people's bodies are better than others at keeping low levels of cholesterol in their blood. It just depends on your genes. A few people have inherited genes that allow them to eat quite a lot of saturated fats and still keep their blood cholesterol levels low.

Did you know?

For the film *Supersize Me*, Morgan Spurlock ate nothing but fast food for a month. He gained 11 kg and his body fat went up from 11% to 18%.

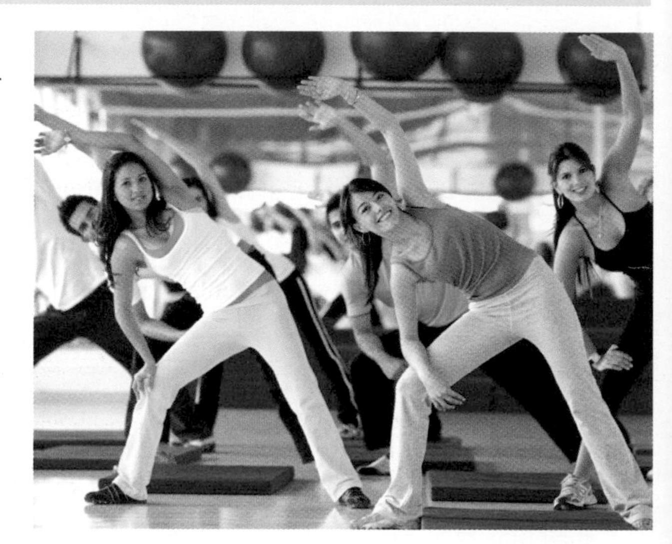

FIGURE 2: Why does exercise help you to stay healthy?

QUESTIONS

1 Why does the body need cholesterol?

2 Why is having a high blood cholesterol level bad for your health?

3 What kind of foods can raise your blood cholesterol level?

4 Explain why, even if two people eat the same diet, one may have a higher blood cholesterol level than the other.

... cholesterol saturated fat

Diet and health

Cholesterol and heart disease

A high level of cholesterol in the blood increases the risk of developing **plaques** in the walls of the arteries. The diagrams below show how this can happen.

The plaque reduces the space that the blood can flow through. It can slow down the blood so that it clots. The clot, or part of the plaque if it breaks away, can get carried along in the blood. It then gets stuck in a smaller blood vessel, blocking blood flow.

Sometimes, a clot blocks one of the arteries that take oxygenated blood to the heart muscle. This causes a heart attack – the muscle cannot work, so the heart cannot beat properly.

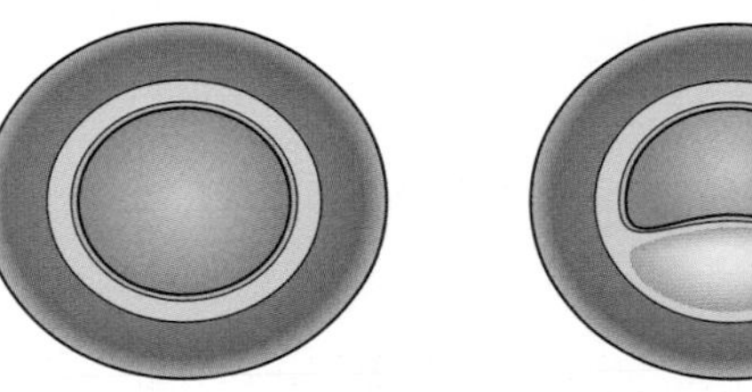

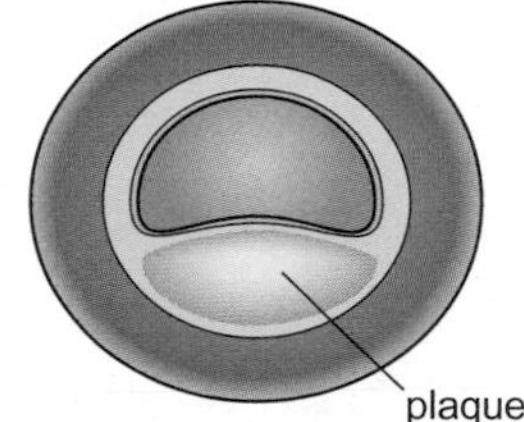

A healthy artery has a stretchy wall and a space in the middle for blood to pass through.

Sometimes, a substance called plaque builds up in the wall. This is more likely to happen if you have a lot of cholesterol in your blood.

The plaque slows down the blood and a clot may form. A part of the plaque may break away.

FIGURE 3: How a plaque develops in an artery.

Diabetes

Diabetes can increase the risk of heart attacks. In Type 2 diabetes, blood sugar levels rise high because certain cells in the body do not respond properly to the hormone **insulin**. You are much more likely to develop Type 2 diabetes if you have close relatives who have diabetes, or if you are very overweight. A healthy lifestyle, with plenty of exercise and eating a balanced diet, greatly reduces the risk of getting diabetes.

QUESTIONS

5 Describe how cholesterol is involved in the development of plaques in arteries.

6 Explain how plaques can cause heart attacks.

7 Suggest factors that are likely to increase someone's risk of contracting diabetes.

Good and bad cholesterol

Cholesterol is carried in your blood in two ways. Low-density lipoprotein **(LDL)** cholesterol is 'bad' and can cause heart disease. High-density lipoprotein **(HDL)** cholesterol is 'good'. To stay healthy, it is best to have at least twice as much HDL as LDL in your blood. HDLs can actually protect against heart disease. They help to remove cholesterol from the walls of blood vessels.

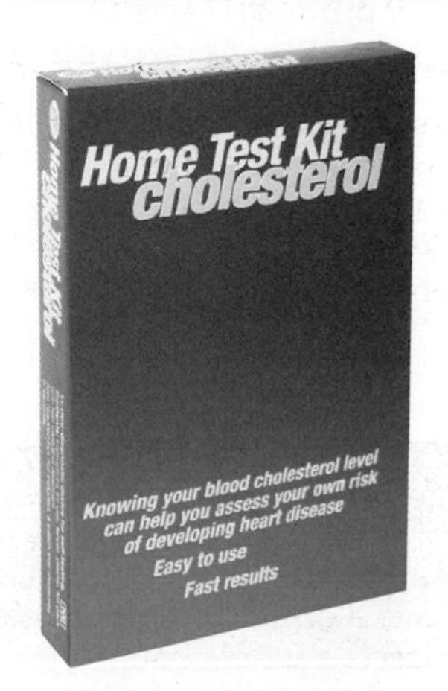

QUESTIONS

8 If someone wants to test their cholesterol levels, why should they choose a kit that measures LDL and HDL cholesterol, not just overall cholesterol levels?

Pathogens and infections

You will find out:

> pathogens are microorganisms that cause disease

> toxins produced by pathogens make us feel ill

> how Semmelweiss began the control of infection

Food for thought

The green blobs on this beef are *Salmonella* bacteria. They probably arrived on the feet of a visiting housefly. If the beef is left in a warm place, the bacteria will multiply. Eating this beef could make you very ill.

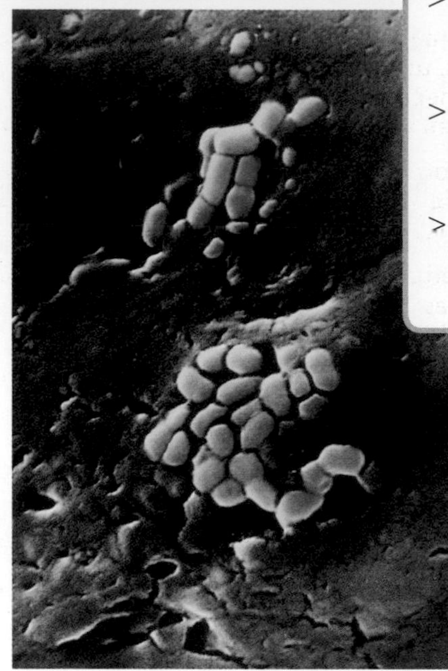

FIGURE 1: A piece of roast beef, magnified 2000 times.

Did you know?

You may have up to fifty million bacteria on one square centimetre of your skin. Almost all of the bacteria are harmless, but a few could be ones that cause disease.

Microorganisms

Microorganisms are living things that are too small for us to see without a microscope. They include **bacteria** and **viruses**.

Look at the millimetre markings on a ruler. You could fit 1000 bacteria, lined up side by side, into one millimetre. Viruses are even smaller – 1000 times smaller than a bacterium. Some viruses can actually get inside bacteria, so even bacteria become ill.

Microorganisms and disease

Some bacteria and viruses cause disease. Microorganisms that cause disease are **pathogens**.

Bacteria can reproduce rapidly inside the body. They may produce waste substances called **toxins** (poisons) that make you feel dreadful. The toxins get carried round the body in the blood. So, even if the bacteria are only inside your intestines, you can feel ill all over.

Viruses are reproduced inside body cells. When lots of viruses have been produced, they burst out of the cell, destroying it. The viruses then invade other cells. Next time you have a sore throat, imagine the viruses bursting out of the cells in the lining of your throat. No wonder it feels sore.

Disease spread

An **epidemic** occurs when a wide spread of people have a disease. A **pandemic** is when the disease affects a whole country or even spreads worldwide.

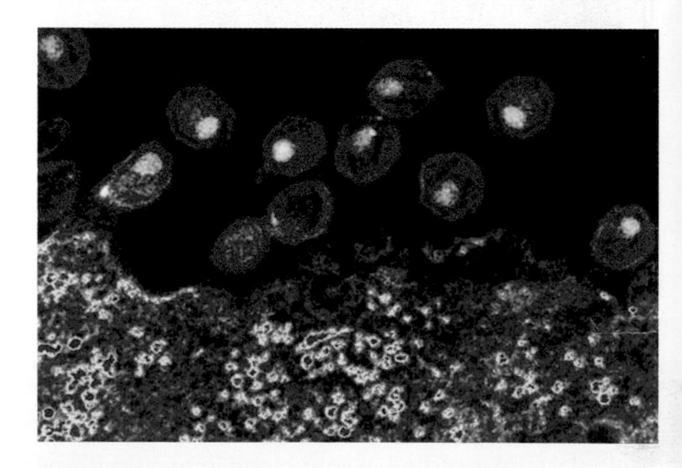

FIGURE 2: The red and purple blobs are HIV, the viruses that cause AIDS. They are bursting out of a white blood cell (blue).

QUESTIONS

1 What is a pathogen?

2 How do bacteria make you feel ill?

3 How do viruses damage the body?

4 Arrange these in order, smallest first: human cell, virus, bacterium.

Semmelweiss

Ignaz Semmelweiss, a Hungarian doctor, worked in an Austrian hospital in the 1840s. In two wards, there were women due to give birth. One ward was run by midwives. In the other, medical students were trained. The doctors rarely washed their hands and often wore dirty coats. Both wards were cleaned just once a month.

Semmelweiss was horrified by how many women died after trouble-free births. They died within a few days, from puerperal fever. No one had any idea what caused it. No one knew about bacteria or viruses then.

Semmelweiss realised that three times as many women died from puerperal fever in the teaching ward, as in the midwives' ward. However, nothing he thought of seemed to explain this.

Then, a professor carrying out a postmortem was accidentally cut with a knife. He died within days, with symptoms similar to the women. Semmelweiss thought that something might be transferred to the doctors' hands, from bodies in the mortuary. The doctors' hands perhaps carried disease to the women in the wards.

He made all the doctors wash their hands in chlorine water, before examining the women. Within a very short time, the death rate plummeted.

Today, we know about bacteria and how important it is to keep everything clean in hospitals. Yet, there is still the problem of people, in hospital, dying from infections.

Remember

Do not use 'germ' for a microorganism that causes disease. The correct word is pathogen.

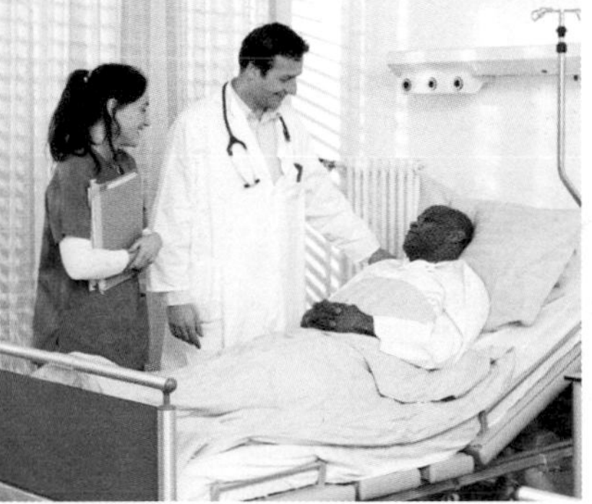

FIGURE 3: 19th and 21st century hospital wards. What might explain why so many fewer people die from infections, in hospitals, today?

QUESTIONS

5 Explain how Semmelweiss tracked down the cause of the high death rate in the teaching ward.

6 Suggest why so little attention was paid to hygiene in labour wards, in the 1840s.

A new hypothesis

Scientists are still learning about pathogens.

A stomach ulcer is a sore, raw patch in the stomach wall. Doctors were taught that it was caused by stress or over-secretion of acid in the stomach.

In 1982, two Australian researchers, Marshall and Warren, found bacteria in the stomachs of people with ulcers. The bacteria were given the name *Helicobacter pylori* (meaning spiral bacterium of the stomach). Marshall and Warren put forward the hypothesis that these bacteria caused the ulcers.

Doctors were not ready to believe them, so Warren swallowed some of the bacteria. As he expected, he developed a stomach ulcer. Marshall and Warren were awarded a Nobel Prize for their work.

Nowadays, stomach ulcers can be treated with antibiotics, which kill the *H. pylori* bacteria.

QUESTIONS

7 Suggest less drastic methods that Marshall and Warren could have used, to collect evidence for their hypothesis.

Fighting infection

Battling the bacteria

This 18-year-old's body piercing has become infected. Bacteria have got in through the wound. Now it is swollen, red and painful. This shows that her white blood cells are trying to destroy the bacteria causing the infection. The wound probably needs to be treated with something to kill the bacteria, before they spread through her body, in her blood, and make her very ill.

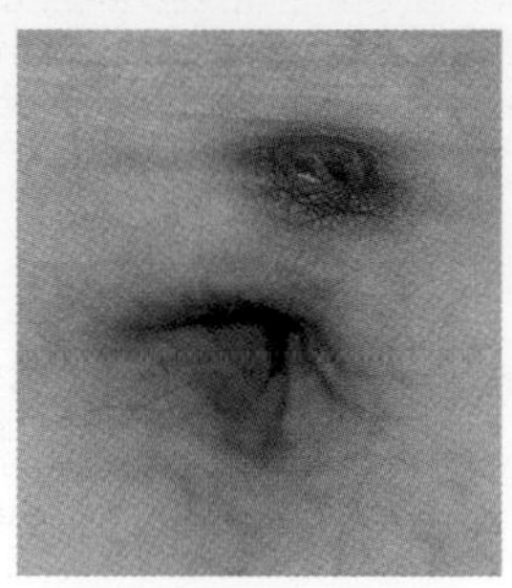

FIGURE 1: An infected navel piercing.

You will find out:

- > white blood cells fight against pathogens in the body
- > some white blood cells ingest pathogens and kill them
- > some white blood cells make antibodies and antitoxins

White blood cells

White blood cells are like defence forces. They attack and destroy **pathogens** that have found their way inside the body.

White blood cells are part of your **immune system**. Some white blood cells surround bacteria and ingest them. They take the bacteria into their cytoplasm and kill them.

Other white blood cells make chemicals called **antibodies**, which destroy the bacteria. Or, they may make **antitoxins**, which counteract the toxins that the bacteria are making.

The antibodies and antitoxins have to be exactly the right kind to match a particular pathogen or toxin. White blood cells can produce the right ones much more quickly, if they have met that particular pathogen or toxin before.

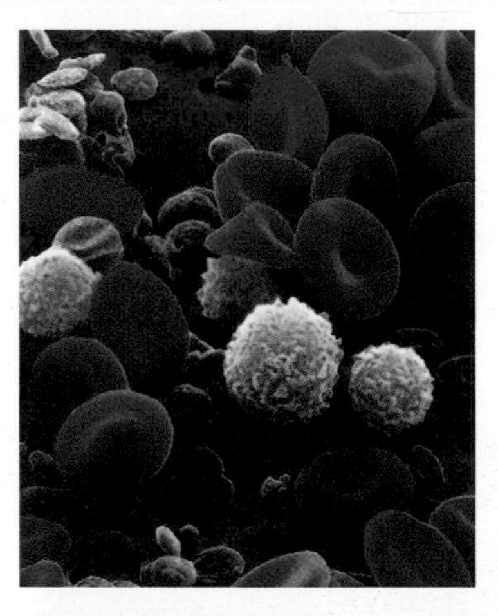

FIGURE 2: This is a microscopic drop of someone's blood, showing some of their blood cells. The white blood cells are coloured yellow. How many more red blood cells are there, than white blood cells?

Did you know?

The pus in an infected wound is a mixture of some of your own dead and dying skin cells, dead white blood cells and dead bacteria – the aftermath of a battle between the bacteria and your body defences.

QUESTIONS

1 Outline two ways in which white blood cells can kill pathogens.

2 What is the difference between an antibody and an antitoxin?

Phagocytosis and lymphocytes

Phagocytosis

Figure 3 shows how a type of white blood cell, called a **phagocyte**, can surround and ingest bacteria. This activity is called phagocytosis.

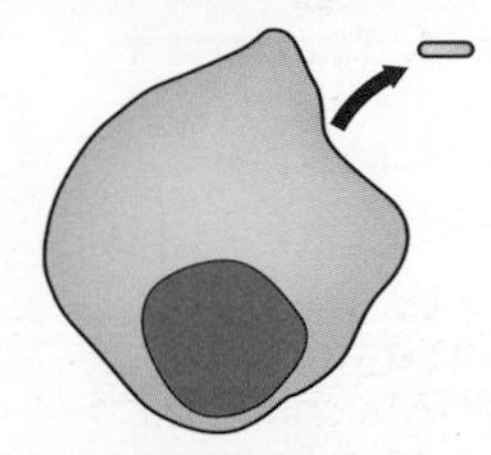

1 A phagocyte moves towards a bacterium

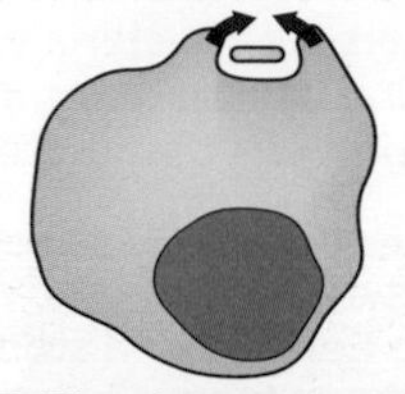

2 The phagocyte pushes a sleeve of cytoplasm outwards to surround the bacterium

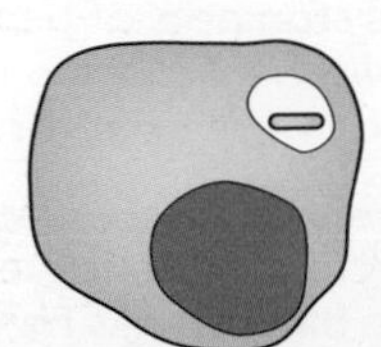

3 The bacterium is now enclosed in a vacuole inside the cell. It is then killed and digested by enzymes

FIGURE 3: Phagocytosis.

Q ... white blood cells for students AND phagocytes

Around an infected wound, some of the cells produce chemicals that signal to the phagocytes that they are needed. Extra blood flows to the infected site, bringing more phagocytes with it. The wound becomes inflamed and red. Underneath the skin, phagocytes are doing their best to kill the invading pathogens before they do you too much harm.

Remember
Do not confuse lymphocytes and antibodies. Lymphocytes are cells. Antibodies are molecules.

Antibodies

Other white blood cells, called **lymphocytes**, attack pathogens in a completely different way. They produce chemicals called **antibodies**.

People have millions of different lymphocytes. Each lymphocyte can only make antibodies against one particular sort of pathogen. Between them, they should be able to make antibodies that match almost any pathogen that gets into us.

The antibodies group round and stick to the pathogen. Sometimes they kill it directly. Sometimes they stick the pathogens together into clumps, so that phagocytes can gather and destroy them more easily.

Some lymphocytes make antitoxins that can stick to the dangerous toxins made and given off by bacteria, and destroy them. Like antibodies, these are very specific – each kind of antitoxin only works against a particular kind of toxin.

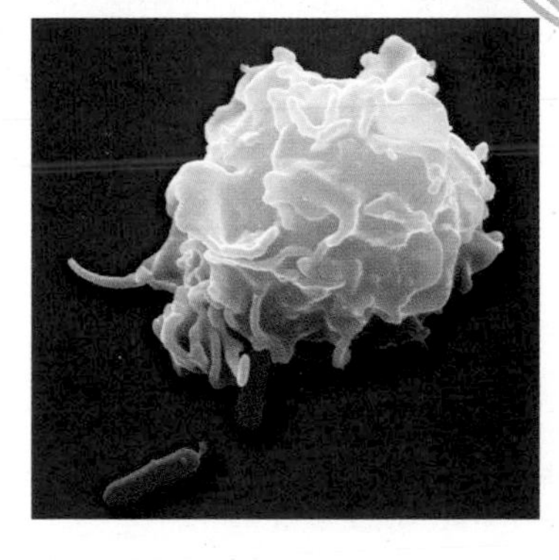

FIGURE 4: This white blood cell is ingesting several rod-shaped bacteria.

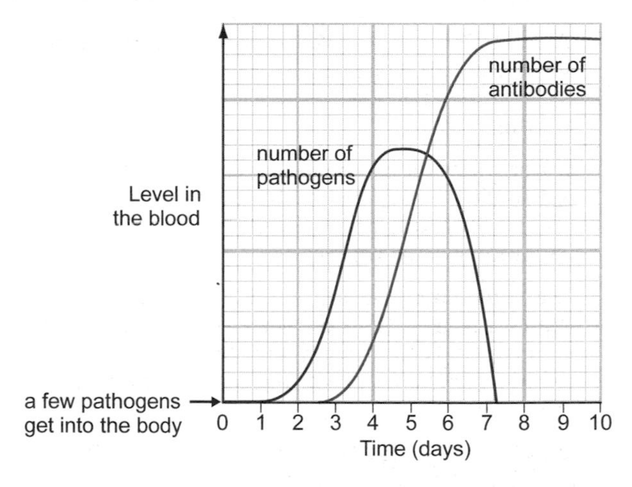

FIGURE 5: What happens to pathogen and antibody numbers when you get an infectious disease? When do you think you might begin to feel ill? When might you start to feel better?

QUESTIONS

3 Explain what these cells do: (a) phagocytes (b) lymphocytes.

4 Why does the body need to produce many different kinds of antibodies?

5 How do antibodies help to defend you against invading pathogens?

More about antibodies

Figure 6 shows an antibody molecule. The bits on the end of the Y arms can come in millions of different shapes. Each lymphocyte can make just one kind. The end bits fit onto **molecules** on the pathogen. Each shape only fits one kind of pathogen.

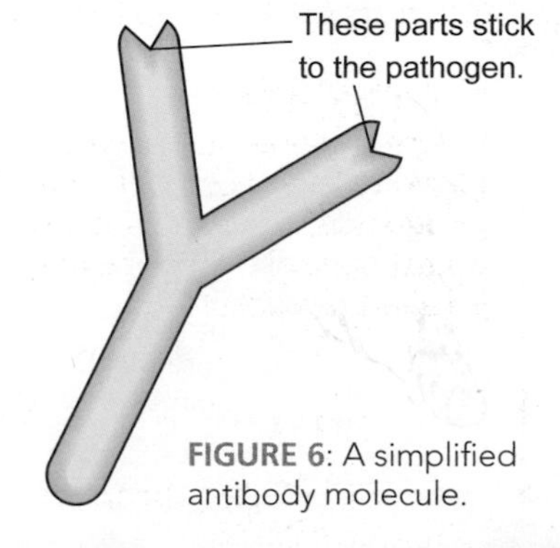

FIGURE 6: A simplified antibody molecule.

QUESTIONS

6 Which of these are cells, and which are molecules: lymphocyte, pathogen, antibody, toxin, antitoxin, phagocyte?

Drugs against disease

You will find out:

- > painkillers do not cure diseases
- > antibiotics kill bacteria but not viruses

Kill or cure?

Until the 1940s, there were really no weapons at all against bacteria. Thousands of people died from infections in wounds and bacterial diseases. Then antibiotics were discovered. Within a few years, doctors could cure illnesses that previously would have been fatal.

FIGURE 1: Even in the late 19th century, doctors in Europe treated many illnesses by blood letting. It probably made many patients more ill than they already were.

Treating pain and disease

Painkillers

You would be a bit unusual if you never had a headache. Unfortunately, thousands of people are in pain almost all the time – for example, pain in their joints because of arthritis, or pain in their back.

Most people take some kind of **drug** to get rid of pain. Painkillers that you can buy 'over the counter' include aspirin, paracetamol and ibuprofen. You do not need a doctor's prescription for over the counter (OTC) medicines. For really serious pain, doctors might prescribe powerful painkillers.

Painkillers can reduce symptoms. They cannot cure the underlying problem.

Antibiotics

Antibiotics are drugs that kill bacteria inside your body, without killing your own cells.

Antibiotics do not destroy viruses. They are not poisonous to viruses. Most antibiotics harm bacteria only, not any other kind of cell.

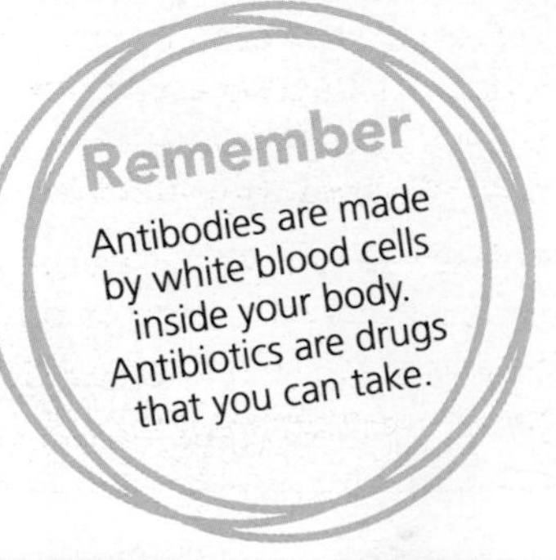

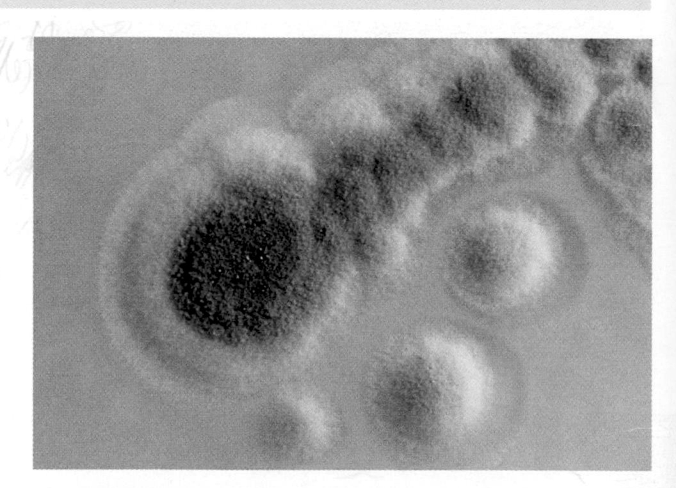

FIGURE 2: Many antibiotics come from fungi. This blue mould fungus makes the antibiotic penicillin.

Did you know?

It is estimated that six million children in developing countries die from preventable infectious diseases, each year.

QUESTIONS

1. Give two examples of painkillers.
2. Flu is caused by a virus. Explain why antibiotics will not help you to get over a bout of flu.

Choosing the right drug

Antibiotics

There are many different antibiotics. They include penicillin and streptomycin. There are different ones because they do not all work equally well against all the kinds of bacteria.

Figure 3 shows a test to find the best antibiotic to kill a bacterium called *E. coli*. The dish contains a clear jelly that was wiped over with a liquid containing the bacteria. Then little paper discs, each soaked in a different antibiotic, were placed on the jelly. The antibiotic diffused (seeped) out of the discs and through the jelly. If the antibiotic kills the bacteria, they do not grow around the disc. The bigger the no-growth area around the disc, the more effective the antibiotic.

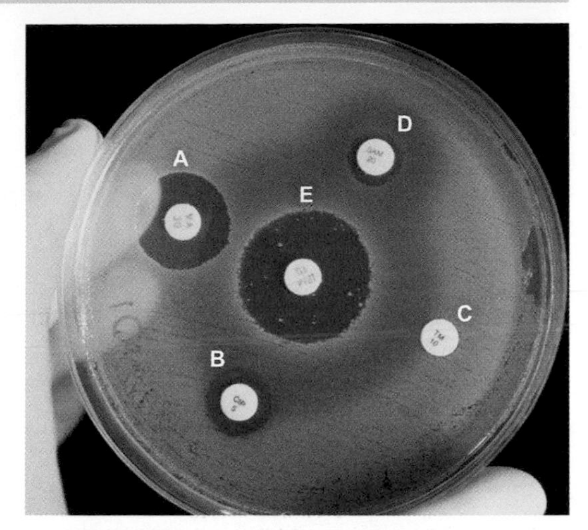

FIGURE 3: Testing antibiotics.

Antivirals

Viruses are more difficult to destroy than bacteria. Viruses get right inside our cells. It is practically impossible to destroy the viruses without killing the cells.

Drugs that destroy viruses are called **antivirals**.

AIDS is caused by the virus HIV. As yet, no cure has been developed. An antiviral drug called AZT does help to slow down the development of AIDS. Some people experience severe side effects when they take AZT.

QUESTIONS

3 Which antibiotic, shown in Figure 3, is not having any effect on the bacteria?

4 Looking at Figure 3, explain which antibiotic you would choose to treat an illness caused by *E. coli*.

5 How do the antibiotics spread out of the discs and through the jelly? Choose from: diffusion, radiation, secretion, infection.

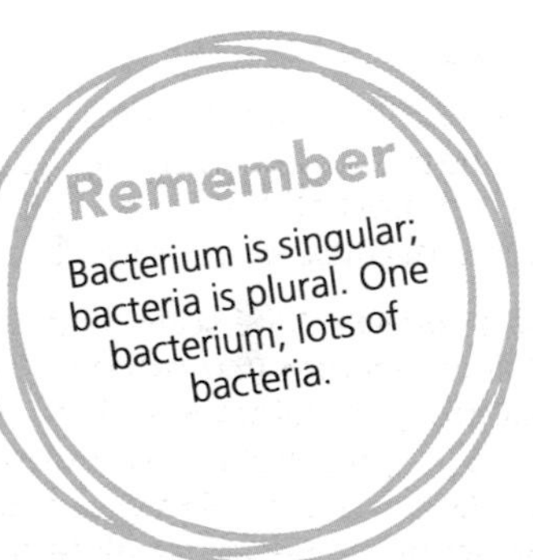

Prescribing antibiotics

Scientists now know that people must not use antibiotics unnecessarily. Overuse makes it more likely that bacteria will become resistant to them.

In Scotland, there was a campaign to persuade doctors not to prescribe antibiotics for everyone who wanted them. The table below shows average number of prescriptions per 100 patients, before and during the campaign.

Year	Prescriptions for all antibiotics	Prescriptions for penicillin
1992	95.6	51.4
1995	105.6	59.8
1999	86.1	47.5
2002	82.0	44.9
2003	81.9	44.7
2004	79.2	41.7

QUESTIONS

6 Display the data, in the table, as a graph.

7 Describe and compare the trends in (a) total antibiotic prescriptions and (b) prescriptions for penicillin.

8 Explain why this is not true: in 1992, only 4.4 people out of every 100 were not prescribed an antibiotic.

Q ... antivirals AND HIV ... AIDS facts

Antibiotic resistance

You will find out:

> bacteria can develop resistance to antibiotics
> resistance to antibiotics happens by natural selection
> about the need to avoid overusing antibiotics

Superbugs

The photo shows a person with skin ulcers caused by a type of bacterium. This is no ordinary bacterium – it is one that is really difficult to kill with antibiotics, and it's called methicillin-resistant *Staphylococcus aureus*, or MRSA. The media like to call these antibiotic-resistant bacteria 'superbugs'.

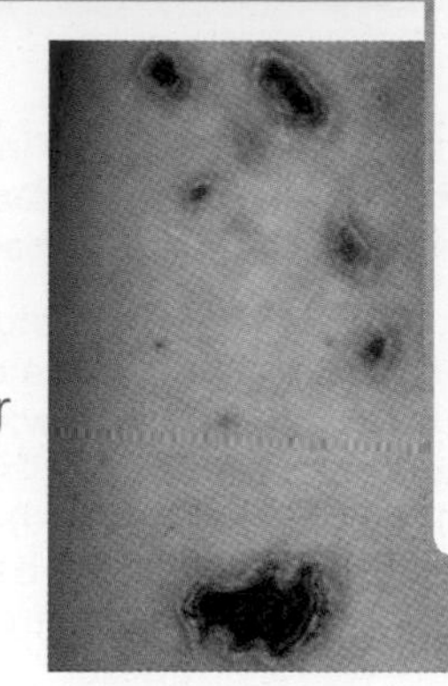

FIGURE 1: Bacteria are destroying this person's skin.

Resistance to antibiotics

When antibiotics were discovered in the 1940s, people thought that was the end of deaths from illnesses caused by bacteria. Since then, many bacteria have become **resistant** to the antibiotics. This means that the bacteria are not affected by the antibiotics that used to kill them.

One dangerous bacterium that is resistant to antibiotics is **MRSA**. MRSA infection can be difficult to treat, because most antibiotics do not work against the bacterium. Hospital patients, who are already ill, may not have strong defences against disease. They can pick up infections easily.

Bacteria do not become resistant to antibiotics on purpose. By chance, a sudden change may happen, in the DNA of a bacterium, that makes it different. This is called **mutation**. Just occasionally, one mutated bacterium may be resistant to antibiotics. This bacterium will reproduce to form a large number of resistant bacteria just like itself. This is called a **resistant strain** of bacteria.

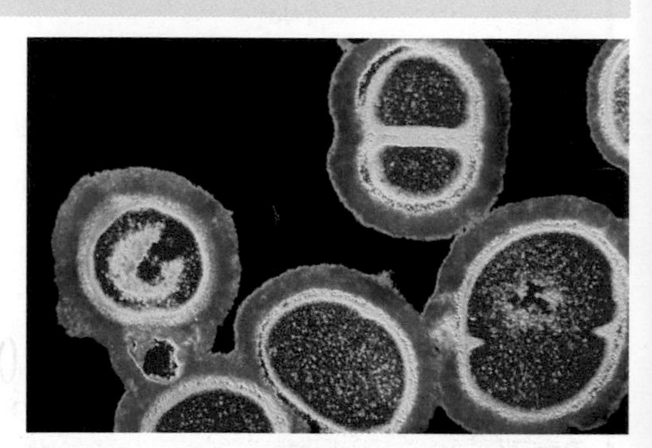

FIGURE 2: MRSA bacteria are most commonly found in hospitals.

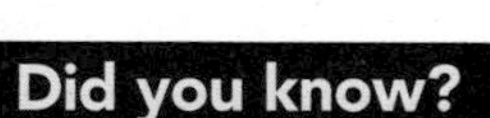

Did you know?

One strain of MRSA can kill a person only 24 hours after they have been infected.

QUESTIONS

1 Explain why an infection with MRSA is very difficult to cure.

2 What is a mutation?

3 Outline how an antibiotic-resistant strain of bacteria may develop.

Reducing the risk (Higher tier)

How antibiotic resistance arises

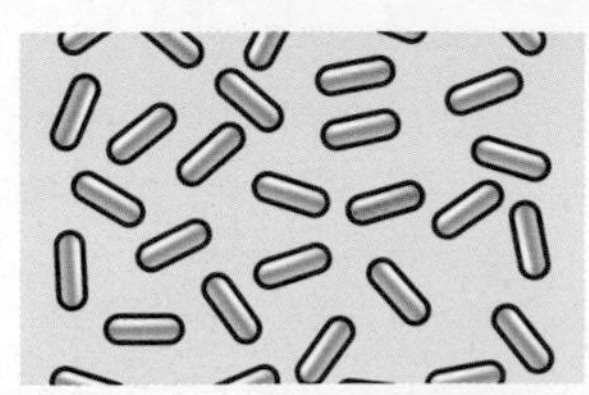

This is a population of bacteria in someone's body. By chance, one of them has mutated and is slightly different from the others.

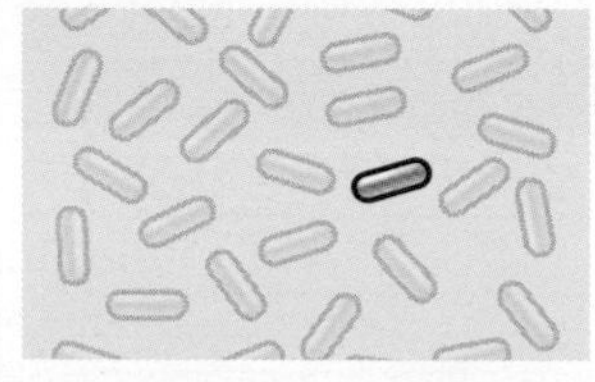

The person takes antibiotics to kill the bacteria. It works – except on the single odd bacterium. This one is resistant to the antibiotic.

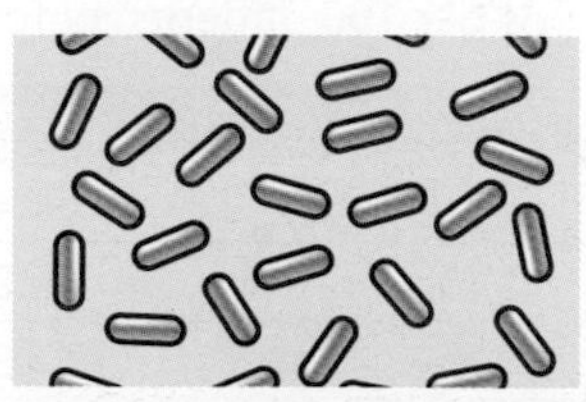

The bacterium now has no competitors and grows rapidly. It divides and makes lots of identical copies of itself. There is now a population of bacteria that the antibiotic cannot kill. This process is an example of **natural selection**.

... MRSA statistics ... bacterial mutation

As new antibiotic-resistant strains of bacteria emerge, we have to find new antibiotics to kill them. There is a constant race between the bacteria mutating and scientists finding new antibiotics.

Reducing the risk of antibiotic resistance

Increasing numbers of antibiotic-resistant bacteria are worrying. If antibiotics cannot kill bacteria, people have no defence other than their own white blood cells.

The most important tactic is to reduce the use of antibiotics. Whenever antibiotics are used, it gives an advantage to any mutant bacterium that happens to be resistant to them. If they are not used, then a mutant bacterium does not have any advantage: it is no more likely to reproduce than any other bacterium.

Doctors therefore avoid prescribing antibiotics unless it is really necessary. You may think it is tough, if your doctor will not give you antibiotics for a painful throat infection. However, a throat infection is not life threatening and your white blood cells will sort it out in a couple of days. You can see that there is very good reason for the refusal.

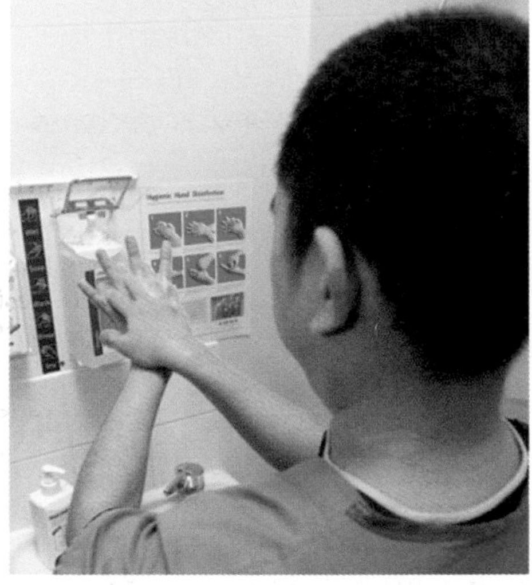

FIGURE 3: The number of MRSA infections in many hospitals has been greatly reduced in the last few years. How can thorough hand washes, by nurses and doctors, help?

QUESTIONS

4 Suggest why MRSA seems to be more common in hospitals than elsewhere.

5 Explain why there is always a need to develop new antibiotics.

MRSA deaths

Staphylococcus aureus is a common, usually harmless, bacterium. However, it is a risk to someone who has a weak immune system, or is very young or very old. It comes in different strains. One type is MRSA. The methicillin referred to in methicillin-resistant *Staphylococcus aureus* is an antibiotic. The media like to call MRSA a superbug because it is resistant to most antibiotics.

QUESTIONS

6 Describe and explain the trend in total deaths from all types of *S. aureus*, between 1993 and 2005.

7 Suggest a reason for the change in trend from 2006 to 2008.

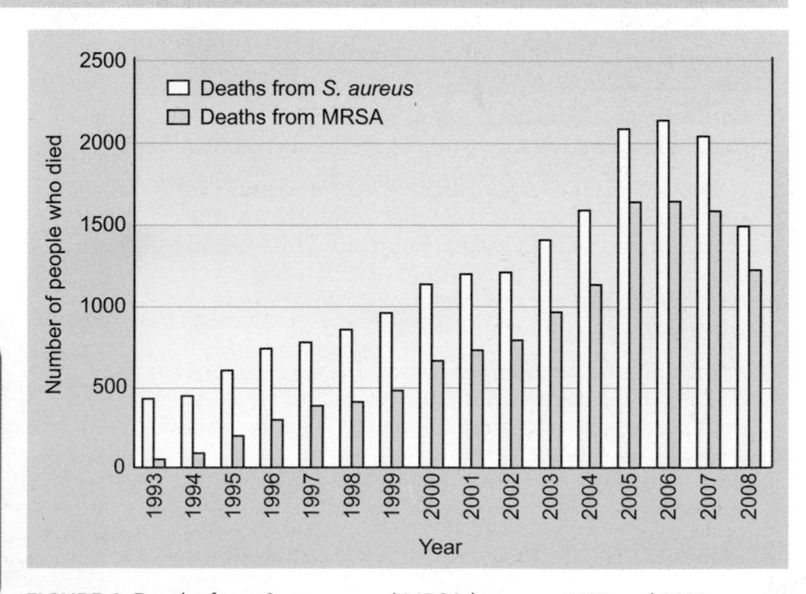

FIGURE 4: Deaths from *S. aureus* and MRSA between 1993 and 2008.

Vaccination

You will find out:

> how injection of a harmless pathogen produces immunity

> the MMR vaccine protects against measles, mumps and rubella

Smallpox is history

Smallpox was a dreadful, often fatal, disease caused by a virus. In 1956, the World Health Organization began a programme to wipe it out. They succeeded by vaccinating over 80% of the people who were at risk. In 1980 the world was declared free from smallpox.

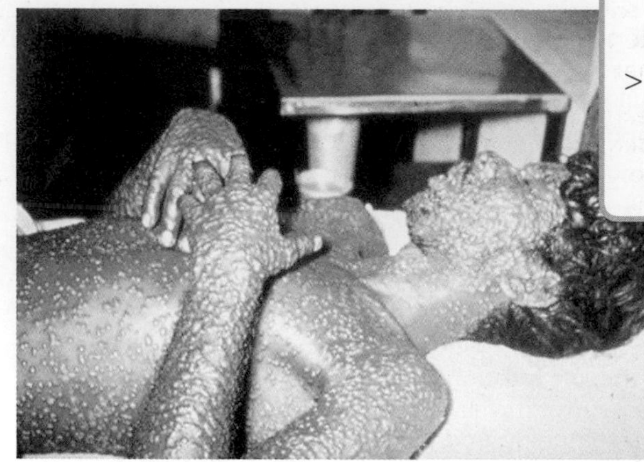

FIGURE 1: This Bangladeshi man was one of the last people to get smallpox.

Immunisation

Immunity

Certain white blood cells can make antibodies that destroy pathogens. Each pathogen needs a specific antibody to destroy it.

> If you get infected, the correct white blood cells multiply and make the antibody needed to destroy the pathogen. This takes time. You may become ill, or even die, before enough of the antibody is made to wipe out the pathogen.

> If you survive, then the next time you are infected, your white blood cells remember how to make the antibody. The antibodies can be made much faster and the pathogen destroyed before it causes harm.

> You had developed **immunity** to the disease, which prevented the disease affecting you a second time.

Immunisation

You have probably been immunised against diseases such as mumps, measles, rubella, polio and diphtheria.

A small amount of the dead or inactive viruses or the bacteria that cause a disease are injected into your blood. This is called **vaccination**. Your white blood cells attack them, just as they would attack living pathogens. You are now immune to this disease without having to suffer it first.

In the UK, one of the most important vaccinations that children have is the **MMR** jab. This cocktail makes them immune to three potentially deadly diseases – measles, mumps and rubella.

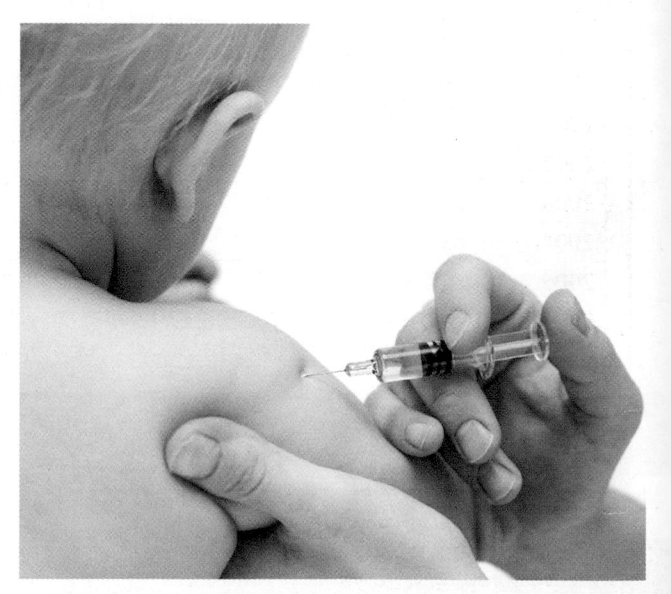

FIGURE 2: Injections are better than infections.

QUESTIONS

1 What is injected into someone when they have a vaccination?

2 Explain how a measles vaccination makes you immune to measles.

3 Explain why a measles vaccination does not make you immune to flu.

4 Suggest why the MMR vaccination is sometimes called the triple vaccine.

The MMR controversy

In 1998, a group of scientists published an article about autism. An autistic person cannot relate to other people in the same way as most people. The article suggested that the MMR vaccination might cause autism.

The article made headlines. Parents were frightened for their children. Many parents decided not to let their child have the MMR vaccination. The government tried to reassure them that no link had been proved, but not everyone trusted the advice.

Later, scientists who had published the original article published another one. They said that there had never been any scientific evidence for a link and that the first article should not have been published.

Many other studies have been carried out since. None has found any link between the MMR vaccination and autism.

Year	Percentage of children having MMR vaccination	Number of mumps cases
1996	92	94
1997	92	180
1998	91	119
1999	88	372
2000	88	703
2001	87	777
2002	84	502
2003	82	1549
2004	83	8104
2005	85	over 30 000
Since 2005, the percentage of children having the MMR vaccination has remained steady at about 85%; the number of mumps cases has fallen.		

TABLE 1: MMR vaccination and cases of mumps.

Did you know?

Researchers are working on genetically modifying banana trees so that they produce vaccines in their fruit.

QUESTIONS

5 Draw a graph to display the data in Table 1.

6 Describe the trend for MMR jabs over these years and suggest reasons for it.

7 Describe the trend for the number of mumps cases and suggest reasons for it.

New diseases

New infectious diseases appear from time to time. They happen when mutations occur in bacteria or viruses.

In 2009, a new kind of flu virus, called swine flu, appeared in Mexico. It spread quickly to all parts of the world. Governments were worried because they knew that current flu **vaccines** would not work against it. They paid drug companies to develop new vaccines and to manufacture antiviral drugs as quickly as they could. By the time the vaccines and drugs were available in large quantities, it appeared that the new disease was not as deadly as feared.

QUESTIONS

8 Explain the difference between an antiviral drug (such as *Tamiflu*®) and a vaccine.

9 Explain why the vaccines we already had against flu would not work against the 2009 swine flu.

10 The swine flu virus had similarities with the flu virus of the 1950s and 1960s. Suggest how this could explain the fact that young people were more likely to get swine flu than people over 50 years old.

… swine flu outbreak Mexico

Growing bacteria

> **You will find out:**
> - how microorganisms can be grown safely on agar jelly
> - how to avoid contamination with unwanted microorganisms

Take care!

The door into your school laboratory probably does not have quite as many safety notices as the one in the photograph. Through this door, people work with dangerous microorganisms. They must take precautions against infecting themselves, or anyone else, with a serious disease.

FIGURE 1: What do these warning signs tell you?

Growing microorganisms

How to grow microorganisms safely

Anything that you are growing in your school laboratory is not likely to be dangerous. However, you must treat microorganisms as though they might be nasty pathogens – just in case.

It's quite easy to grow microorganisms. Bacteria and **fungi** can be grown in a liquid or a jelly containing all the nutrients that they need. The liquid or jelly is called a **nutrient medium**.

A nutrient medium is a food source for bacteria. The bacteria (or other microorganisms) that grow on it is called a **culture**.

The nutrient medium must be **sterile** – it must not contain any microorganisms. It is also important to try not to let any unknown microorganisms get in, accidentally, while you are setting up your experiment.

Growing microorganisms on agar

The following pictures show you how to prepare for growing microorganisms, safely, in a Petri dish. The jelly in the dish is called **agar**.

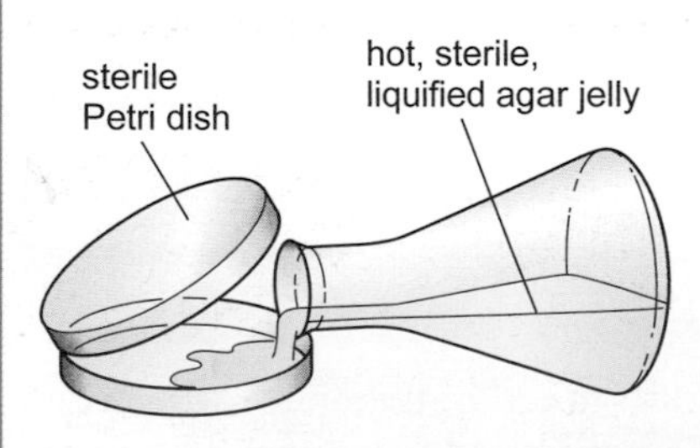

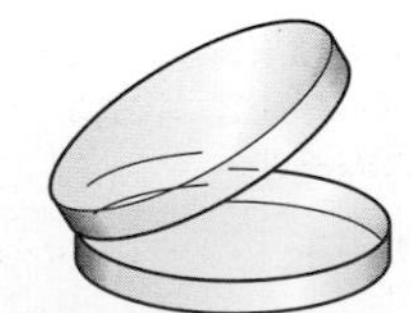

FIGURE 2: Preparing a Petri dish with sterile agar jelly.

> **Did you know?**
>
> There could be more than one million bacteria in each little blob that grows on the agar jelly in a Petri dish – all produced from just one bacterium.

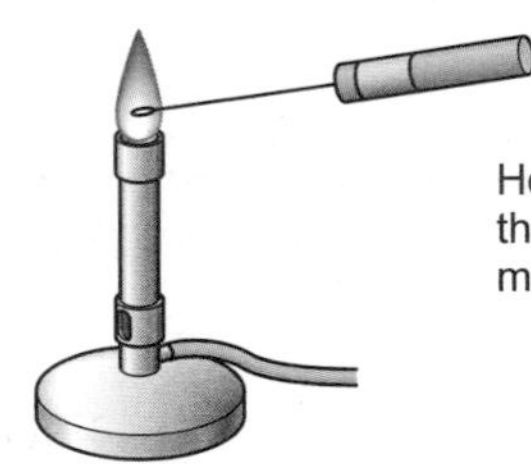

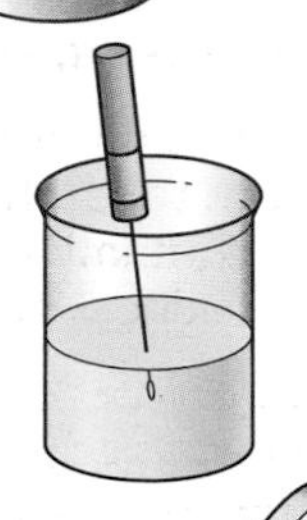

FIGURE 3: Putting microorganisms onto the jelly.

QUESTIONS

1 What is a 'nutrient medium' and what should it contain?

2 When you put the liquid onto the jelly, why will you not see any bacteria?

3 After a day or so, little blobs, called colonies, will appear on the agar, each containing lots of bacteria. Where have they come from?

Avoiding contamination

The microorganisms growing on the nutrient medium are called a culture. There are several things that can prevent unwanted microorganisms getting into a culture. These procedures are called **sterile technique**.

All equipment should be sterilised before use. Sterilising means killing all microorganisms that might be on the equipment.

> Metal equipment, such as a wire **inoculating loop**, can be held in a flame.

> Glass Petri dishes can be heated to a high temperature.

> New plastic Petri dishes are sterile and are packed in sterile bags.

The nutrient medium should be sterilised.

> Liquids can be boiled thoroughly, sealed and allowed to cool.

> Agar jelly is made by dissolving agar powder in liquid, heating the mixture to a high temperature and leaving it to cool until it is still just runny. It is poured quickly into the sterile dish.

> You must not touch the agar jelly with your fingers, because there will be bacteria on them, however clean you think they are! Do not breathe over it, either.

After putting the microorganisms onto the agar jelly in the Petri dish, seal the dish with tape. This prevents microorganisms, from the air, contaminating the culture. It lessens the chance of anyone taking off the top accidentally.

It also helps to keep the cultures at a temperature of about 25 °C. If you keep them at human body temperature (37 °C), then this might encourage the growth of microorganisms that live and breed in the body. These might be pathogens.

Remember
Keeping the lid on a Petri dish stops unwanted bacteria getting in and any harmful ones getting out.

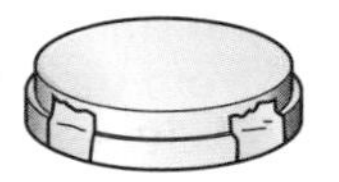

FIGURE 4: How to avoid contamination with unwanted bacteria in your Petri dish.

QUESTIONS

4 What does 'sterilised' mean?

5 Outline two reasons why it is important to stop people taking the top off a Petri dish containing a culture.

Pathology and industrial laboratories

People working in a hospital pathology lab are often asked to grow some of the bacteria that are causing an illness in a patient. The Petri dishes, on which the bacteria are growing, will be put into an incubator.

In industrial labs, harmless bacteria are grown commercially, for example, to make **enzymes**. They are often grown at about 40 °C.

QUESTIONS

6 Suggest a suitable temperature for the incubator in a pathology lab. Explain your suggestion.

7 Suggest an advantage of growing bacteria at a relatively high temperature, in enzyme manufacture.

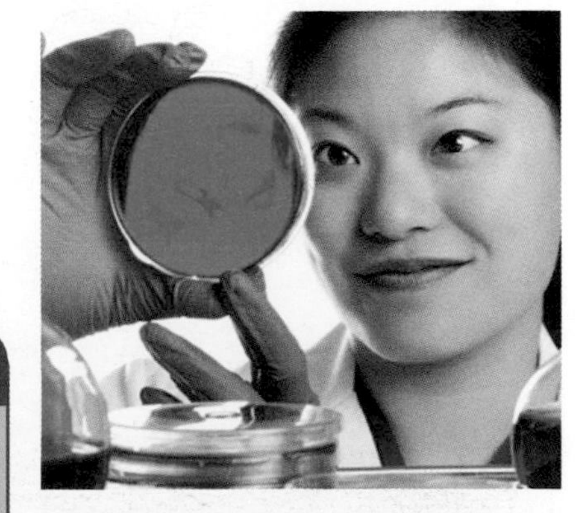

FIGURE 5: This pathologist is inspecting a Petri dish that contains a culture of bacteria from a patient. Why is she wearing gloves?

… working as a medical pathologist

Preparing for assessment: Applying your knowledge

To achieve a good grade in science, you not only have to know and understand scientific ideas, but you need to be able to apply them to other situations and investigations. These tasks will support you in developing these skills.

Bouncing off the walls

Andy is a lively, confident eleven-year-old boy, keen to have a go at anything. His mother worries about his diet though, and she has good reason. Andy is fussy about what he eats and the idea of healthy eating isn't one that troubles him.

He eats meat of various sorts and chips. He also eats fish but not vegetables (apart from cucumber). He will eat pizza or pasta (as long as it only has butter or cheese on it) and ice cream or some kinds of yogurt. He eats chocolate in almost any form. He will drink milk occasionally, but much prefers fruit juice (as long as it doesn't have 'bits' in it) and loves cola. Breakfast is a chocolate flavoured cereal with no milk on it.

Andy is certainly not fat. He is one of the shortest in his class. He is active and alert. When he starts at secondary school, he wants to join the school cross-country running team, though in a few of the practice sessions he seems to lack stamina. He goes to the local playing field with his friends sometimes and they kick a ball around. His mother is worried that his poor eating habits may put him at a disadvantage when he is older. In the winter, he spends a lot of time in front of the TV, though he does like reading and doing puzzles as well.

Task 1

Read the information about Andy. List the strengths and weaknesses of his current diet. Discuss whether his mother is right to be concerned about his diet.

Task 2

Andy says he is doing fine and that his mother should 'lay off'. He points out that he is not fat, that he is active and gets energy from his food to do loads of energetic things. To what extent is he right?

Task 3

His mother says he only eats what he fancies: sometimes he will eat a whole packet of biscuits and barely any tea. She knows that he is slim and active at the moment, but thinks that sport may not be a big part of his life in the future and that his body shape may then change. Is she right to be concerned?

Task 4

What might the future hold for Andy? Suggest two different outcomes. For each one, refer to diet, activity levels and physical health.

Maximise your grade

Grade	Answer includes showing that you...
	know one way in which excess body mass can cause health problems.
	can describe several ways in which excess body mass can cause health problems.
E	know one way in which someone may try to control their body mass.
	understand one way in which someone may try to control their body mass.
	understand how different people have different levels of energy requirement.
C	can explain how, even if a younger person has a large intake of energy from foods but uses it up through an active lifestyle, problems may arise from poor eating habits when linked to a more sedentary lifestyle, later in life.
A	can explain how inherited factors in combination with diet and exercise can affect a person's health.

Coordination, nerves and hormones

You will find out:
> nerves and hormones coordinate body activities
> the brain, spinal cord and nerves make up the nervous system
> hormones are secreted by glands and affect target organs

Watch the ball

Good tennis players watch the ball right to the point at which it hits the racket. To excel at tennis, you need excellent hand–eye coordination as well as technique, strength, fitness, agility and speed.

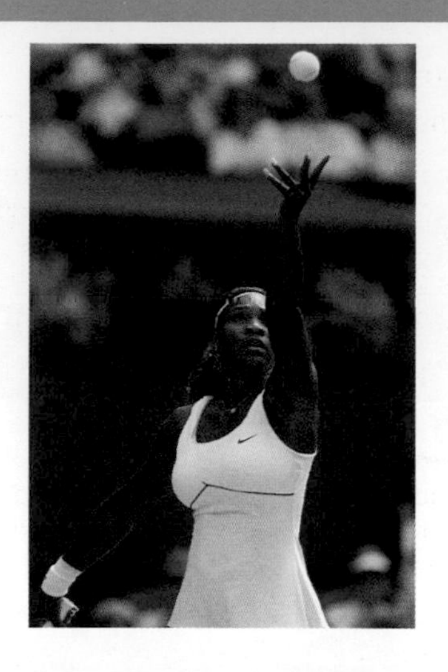

FIGURE 1: Good coordination takes practice.

Nerves and hormones

Nerves and nerve cells

How good is your hand–eye **coordination**? Try putting the middle finger of your right hand onto the tennis ball, in the Figure 1 photo. Your eyes see the ball and send information to the brain about where it is.

The information travels from the eyes to the brain along **nerves**, as fast-moving electrical impulses (signals). Nerves contain special cells called nerve cells. The brain then sends more impulses to the muscles in your arm and hand, telling them how and where to move.

Nerves transmit information to and from the brain and spinal cord. The brain and spinal cord make up the **central nervous system**.

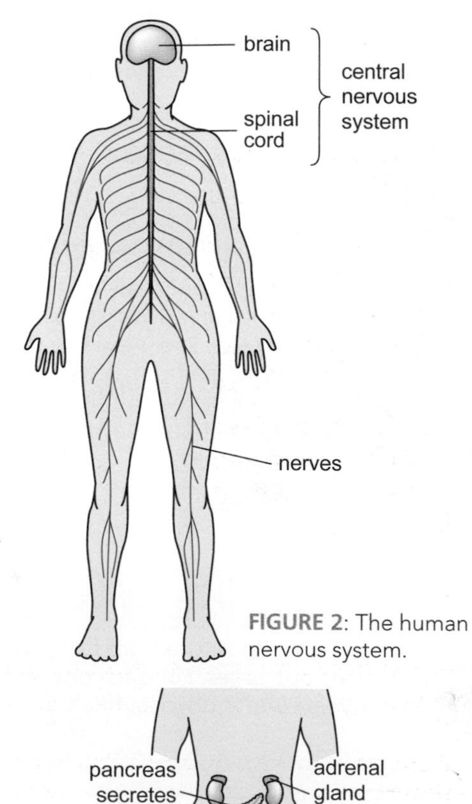

FIGURE 2: The human nervous system.

Hormones

The human body contains many organs. They must work together and need to communicate with one another.

As well as nerves, **hormones** carry information between organs. Hormones are chemicals. They are made by certain **glands**. Glands are organs that make and release useful substances.

The release process is called **secretion**. Glands secrete hormones into the blood. The bloodstream carries the hormones around the body.

An example of a hormone is **adrenaline**. In the excitement of a match, adrenaline is secreted in the tennis player's body.

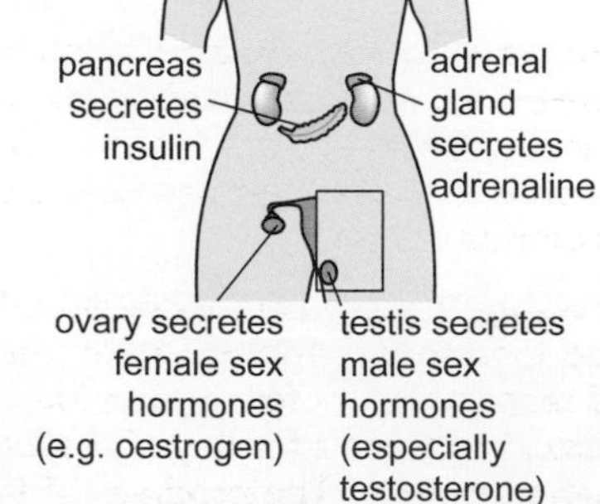

FIGURE 3: Four glands that secrete hormones.

QUESTIONS

1 List some of the organs that the tennis player is using as she hits the ball.

2 Adrenaline makes the heart beat faster. How is adrenaline carried to the heart?

… nervous system for students AND hormones GCSE

Targets

Nerves and behaviour

Think about touching the ball, on the Figure 1 photo, with your finger.

As your arm and fingers move, the brain constantly monitors their positions in relation to the ball. It adjusts their movements so that your middle finger lands directly on the ball. While you are making this simple movement, huge amounts of information are buzzing along nerves between your eyes, brain and muscles. Constant tiny adjustments are being made.

Hormones and target organs

In the nervous system, nerves carry information between one organ and another in the form of electrical impulses. Hormones move around the body in a different way. Hormones, made in glands, dissolve in the **blood plasma**, and are carried all around inside blood vessels.

Most hormones affect just a few different organs. These are called their **target organs**. The hormone adrenaline has more target organs than most other hormones. Adrenaline affects the heart, breathing muscles, eyes and digestive system.

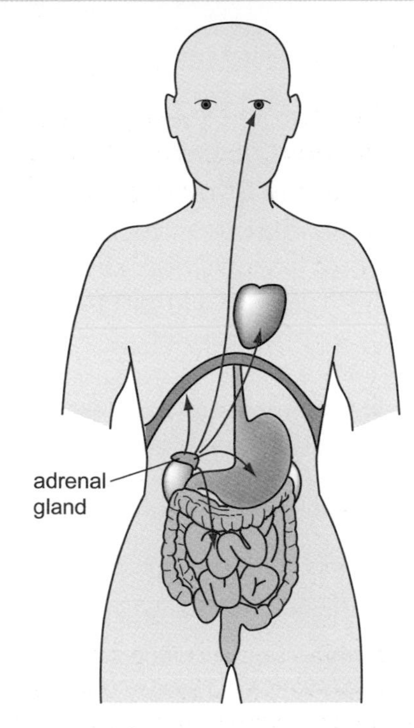

FIGURE 4: The target organs for adrenaline. Try to identify them by name.

QUESTIONS

3 What information was your brain receiving about your surroundings, as you thought about touching the ball, on the photo?

4 Suggest why nerves, and not hormones, transfer information, when you carry out an action like this.

5 Adrenaline is sometimes called the 'fight or flight' hormone. Think about being frightened or excited. How is adrenaline affecting its target organs?

Did you know?

When a fly touches a Venus fly trap leaf, electrical impulses – just like the ones in our nerves, only a lot slower – make the leaf close over the fly and trap it.

Response duration

A simple piece of behaviour, such as touching a photo, takes only a short time. Impulses travelling along nerves are an ideal way of achieving this. They are fast and short-lived. Where a longer-term response is needed, hormones are often a more appropriate method of communication.

For example, if you are frightened by a bull chasing you, it's probably a good idea to stay frightened at least until you have run across the field and leapt over the fence. However, you would not want to go on feeling frightened for days. To avoid this, the liver gradually breaks down hormones. The products of this breakdown return to the blood and are excreted in urine.

QUESTIONS

6 Draw a table with two columns, headed NERVES and HORMONES. Complete the table to compare these two methods of communication and coordination in the body.

Q ... hormones target organs GCSE

Receptors

You will find out:

> receptors have different positions in the body
> receptors detect stimuli and effectors respond to them

A painless life

The young boy with this hand was born with a rare condition called CIPA. He has no sense of pain. He bites his tongue so badly that the tip has gone. He burnt his hand like this, by holding on to a very hot radiator. He did not realise that he was hurting himself.

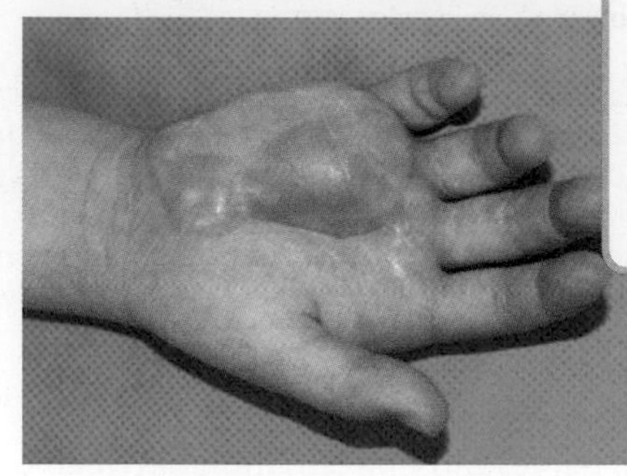

FIGURE 1: The hand of a boy who cannot feel pain.

Receptors and effectors

The boy did not feel his fingers burning because his brain did not get the right information from the **receptors** in his fingers. Receptors are special cells that detect stimuli.

A **stimulus** is a change in the **environment**.

In this case, the stimulus was the very high temperature of the radiator. If you touch a hot radiator, receptors in your fingers send electrical impulses along cells (neurones) in nerves to your brain. This did not happen in the boy's body.

Your brain then sends impulses speeding along other neurones to the muscles in your arm. The muscles contract. They pull your hand away from the radiator.

All your muscles are **effectors**. An effector is an organ that does something in response to a stimulus. As well as muscles being effectors, glands are also effectors. For example, salivary glands respond to a stimulus by secreting saliva.

> Receptors detect stimuli from the environment, and send information to the central nervous system.

> Effectors (muscles and glands) do something in response to the stimulus.

We have many different kinds of receptors. Figure 2 shows the positions of receptors in the head.

Examples of stimuli are: light, sound, touch, movement, temperature change, pressure, pain, position and chemicals.

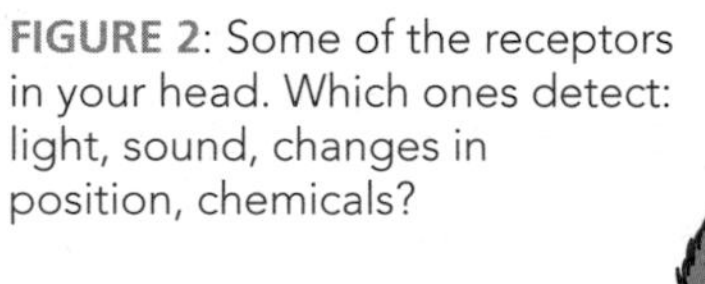

FIGURE 2: Some of the receptors in your head. Which ones detect: light, sound, changes in position, chemicals?

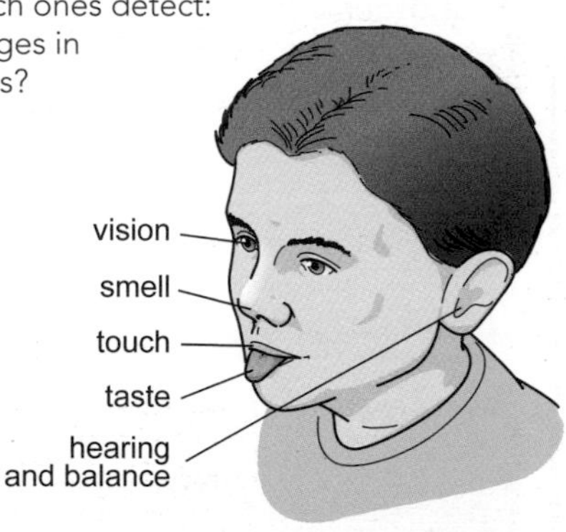

QUESTIONS

1 Explain what a receptor does.

2 Explain what an effector does.

3 It is the largest organ in your body – so large that most people don't think of it as being an organ at all. It is the most visible organ you have. Without it, most of your cells would dry out very quickly. It contains receptors that detect temperature, touch, pressure and pain. What is this organ?

Neurones

Information is carried in the nervous system as electrical impulses. The cells that transmit these impulses are called nerve cells. Another name for them is **neurones**.

The neurones that transmit impulses from receptors to the central nervous system are called **sensory neurones**. The neurones that transmit impulses from the central nervous system to effectors are called **motor neurones**.

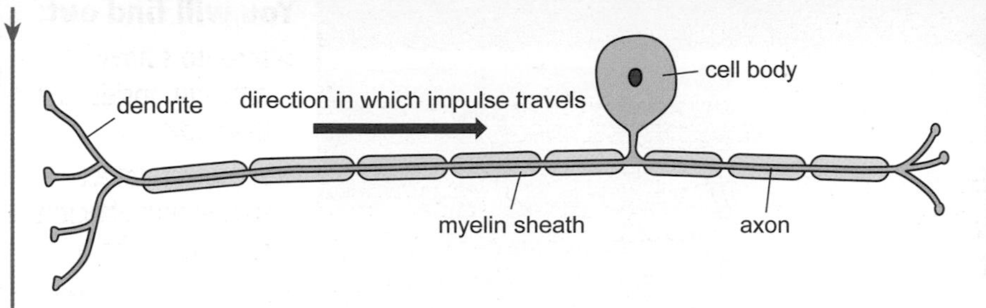

FIGURE 3: A sensory neurone.

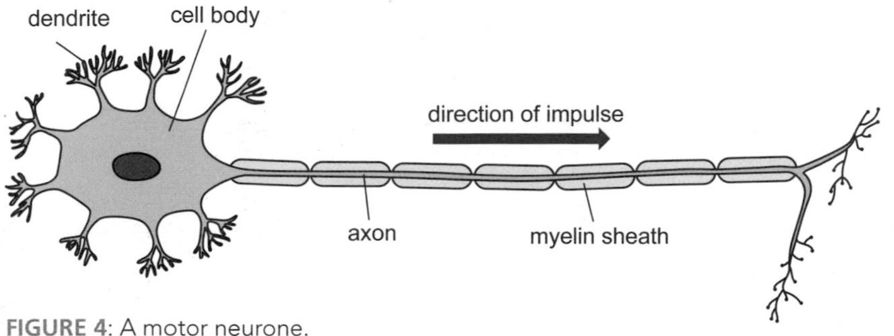

FIGURE 4: A motor neurone.

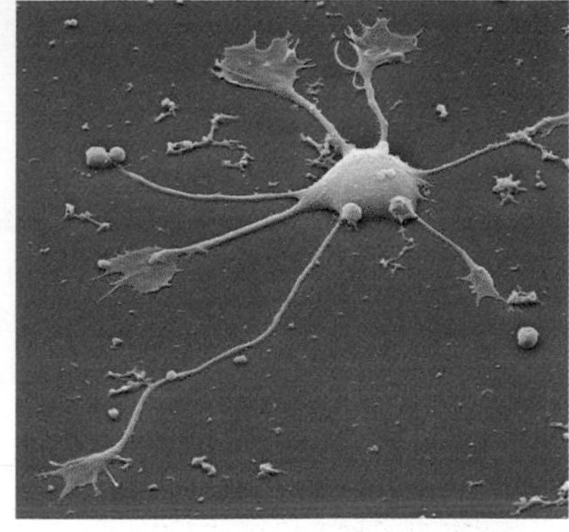

FIGURE 5: A micrograph of a neurone. Which part contains the nucleus of the cell?

Remember

Neurones themselves do not move around. They stay in one place while electrical impulses sweep along them.

QUESTIONS

4 Suggest the features that a sensory neurone shares with other animal cells (think about the structure of the cells).

5 Suggest the features that a motor neurone shares with a sensory neurone. How are they different?

Did you know?

Each of your fingers has about 3000 touch receptors. This is more than the whole of your back and chest put together.

Rod cells

At the back of your eyes, in the retina, are specialised cells called **rod cells**. They are very sensitive to light. If only one photon of light falls onto a rod cell, this is enough to make it generate an electrical impulse. It is sent along the optic nerve to the brain. The brain uses the pattern of impulses, arriving from rod cells in different parts of the retina, to construct a 'picture' of the world you are looking at.

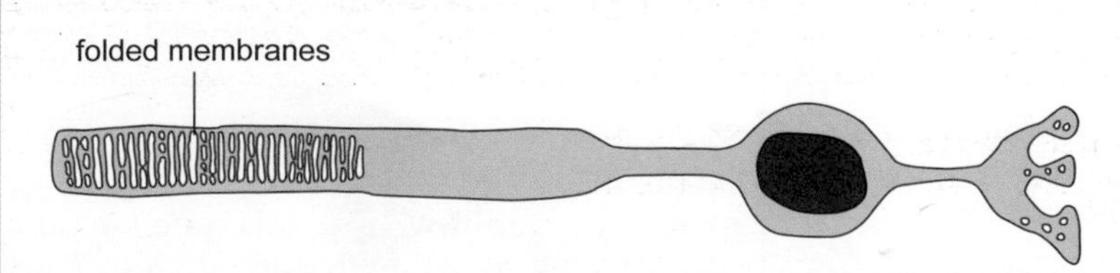

FIGURE 6: A rod cell. Copy this diagram and label the nucleus, cytoplasm and cell membrane.

QUESTIONS

6 The folded membranes in the rod cell are covered with a pigment (coloured substance) called rhodopsin. This is what makes it sensitive to light. Suggest why the rod cell has these membranes, rather than just keeping the rhodopsin in its cytoplasm.

Q ... types of neurons GCSE

Reflex actions

You will find out:

> a reflex action is fast and automatic

> reflex actions involve sensory, relay and motor neurones in a reflex arc

> what synapses are and how they work

Reflexes

If doctors think you may have damaged your spinal cord, they might test your knee-jerk reflex. They will tap your knee with a small hammer. If you are OK, then your lower leg should kick upwards. You don't think about it – it just happens.

FIGURE 1: Reflex actions are automatic and fast.

Stimulus and response

The tap on the knee is a **stimulus**. It is detected by **receptors** in the thigh muscle that is connected to your knee. The receptors send nerve impulses to your spinal cord.

The spinal cord then sends nerve impulses to your leg muscles. The leg muscles respond by contracting. This pulls your lower leg upwards.

This is a **reflex action**. A reflex action is a fast, automatic response to a stimulus.

Most reflexes protect you. For example, if something moves fast towards your face, you blink. This protects your eyes.

In any reflex action, the sequence of events is as follows:

> A receptor detects a stimulus.

> The receptor sends an electrical impulse along a sensory neurone to the central nervous system (CNS).

> The CNS sends an electrical impulse along a motor neurone to an effector. The effector could be a muscle or a gland.

> The effector does something as a response to the stimulus.

It is not correct to say that reflex actions do not involve the brain. The nerve impulse could go through either the brain or the spinal cord – but it will not involve the conscious parts of the brain.

QUESTIONS

1 In the blinking reflex, what is the receptor? Choose from: skin, eye, muscle, brain, finger.

2 In a 'hand pulling away from hot object' reflex, what is (a) the receptor and (b) the effector?

3 Think of another reflex action. It must be an automatic response that you do not even think about. Why is this reflex action useful? What is (a) the receptor and (b) the effector?

Impulse pathways

A reflex arc

A **reflex arc** is the pathway taken by a nerve impulse as it passes from a receptor, through the central nervous system, and finally to an effector. Figure 2 shows the positions, in the body, of the cells and organs that make up a reflex arc.

Your spinal cord is inside your spine. The object at the right of Figure 2 is what your spinal cord would look like, if you viewed a slice across it, end on.

Put your finger on the receptor in Figure 2. Run your finger along the sensory neurone, into the spinal cord, along the relay neurone, and then along the motor neurone until you arrive at the effector. It takes a nerve impulse only a fraction of a second to go along this route. That is why reflex actions are so quick.

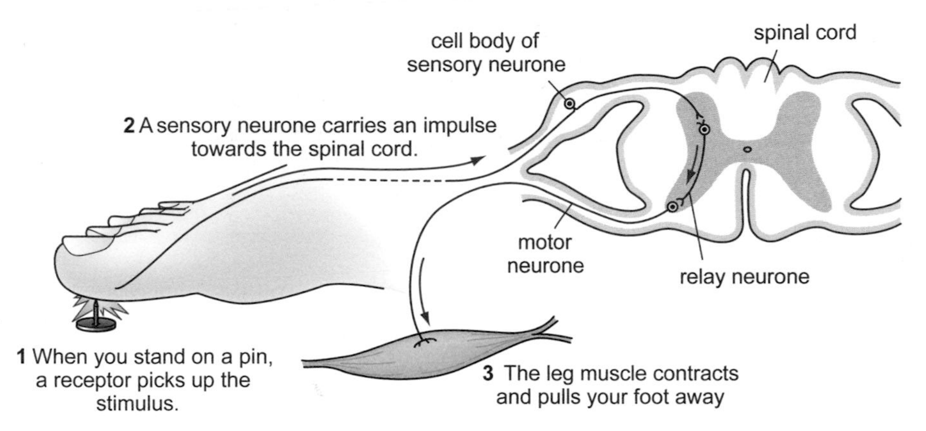

FIGURE 2: A reflex arc.

Synapses

If you look carefully at Figure 3, you will see that there is a tiny gap between the end of one neurone and the start of the next. This gap is a **synapse**.

Electrical impulses cannot jump across these gaps. Instead, when an impulse gets to the end of a neurone, it causes a chemical to be secreted. The chemical diffuses across the gap, but slower than an electrical impulse travelling the same distance.

When the chemical arrives at the beginning of the next neurone, it starts an electrical impulse along that neurone.

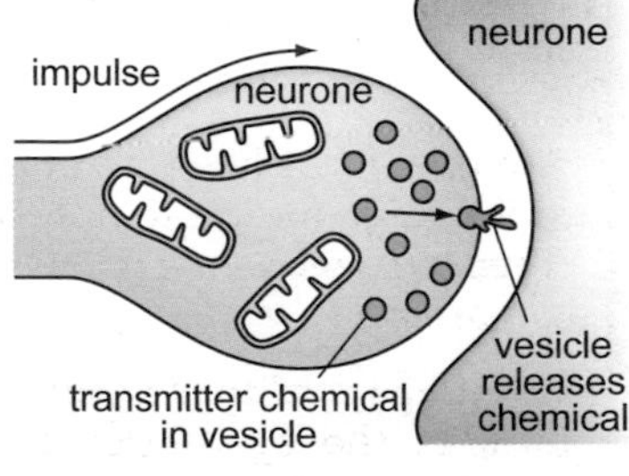

FIGURE 3: At a synapse.

QUESTIONS

4 In the reflex arc, there are how many (a) neurones (b) synapses?

5 Explain the difference between a reflex action and a reflex arc.

6 Explain why synapses slow down the passage of a nerve impulse.

Did you know?

You were born with 1000 trillion synapses, and you have already lost at least half of them. Luckily, your brain does make new synapses as you learn new things.

Conscious control

Synapses enable us to respond to a stimulus in more than one way. For example, the relay neurone, in the spinal cord, will have synapses to other neurones that can carry nerve impulses down from the brain. This allows us to take conscious control of our response to a stimulus.

QUESTIONS

7 Suggest why conscious control might be useful when picking up a hot drink. Give examples of other situations where you might want to override an automatic reflex.

Controlling the body

You will find out:

> conditions inside the body are controlled
> water and ion content, temperature and sugar concentration are controlled

Keeping the balance

Exercise hard on a hot day and, in one hour, you could lose up to 2500 cm^3 of water in your breath and sweat. Ions are also lost in sweat, so it's best to drink water with ions in it. Isotonic drinks have the same balance of water and ions as the body.

FIGURE 1: A marathon runner needs to replace water and ions lost by sweating.

Controlling the conditions in cells

Chemical reactions take place in the cells in your body. They must happen at the right time and at the right speed. For this, the conditions around each cell must be perfect – and constant. These conditions include:

> water content
> ion (salt) content
> temperature
> concentration of sugar in the blood.

Controlling water and ions

You gain water from food and drink. You lose water in your breath, sweat and urine.

Your blood has many different substances dissolved in it. Some are ions, such as sodium and chloride, both found in salt.

Too much salt and not enough water in the blood can lead to high blood pressure. People who eat too much salt can increase their risk of having a heart attack.

The kidneys help to keep the balance of water and ions just right. They do this by varying the amount of water and salt excreted from your body in urine.

FIGURE 2: How the body loses water.

Did you know?

You can survive for weeks without food, but only days without water.

QUESTIONS

1 List three ways in which the body loses water.

2 Why does a person running a marathon, on a hot day, lose much more water from their body than usual?

Temperature control

The normal human body temperature is about 37 °C. This is the temperature at which our **enzymes** work best. If body temperature drops too much, then metabolic reactions happen too slowly. If body temperature rises too much, then the enzymes are damaged so badly that they can no longer act as catalysts. The body's metabolism is disrupted.

The body gains heat from outside – for example, from infrared radiation from the Sun, or from touching warm objects – and from heat inside us, produced when cells use glucose in respiration.

The body loses heat by radiation from the skin, and from the **evaporation** of **sweat**.

Sweating

Sweating keeps us cool. Sweat is made by glands in the skin. The glands take water and ions out of the blood, to make sweat. The sweat travels up through a sweat duct, and lies on the surface of the skin.

Sweat is a mixture of water, ions and a small amount of urea. So, when you sweat, you lose all of these from the body.

The water in sweat evaporates. Sweat cools you down because, as its water evaporates, energy transfers from the skin to the water and into the air. This cools your body. It is not because sweat is cold.

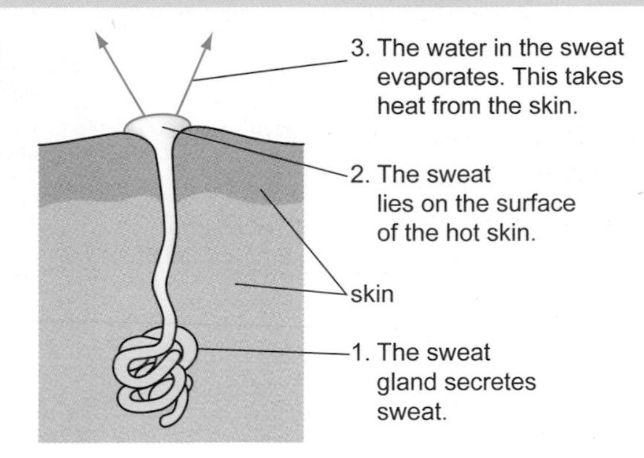

FIGURE 3: How sweating helps you to lose heat.

QUESTIONS

3 Explain why you need to drink more water on a hot day than on a cold day.

4 Most people produce more urine on a cold day than on a hot day. Explain why.

5 Respiration is the breakdown of glucose to release energy. Suggest why it is important to maintain a constant level of glucose in the blood.

Survival in the desert

Your car has broken down in a desert. You do not know where you are. You have been driving all day and have not seen anyone else. You do not have much water. You cannot get a signal on your mobile phone. All you can do is wait and hope that someone comes along before you die of dehydration or heatstroke.

An SAS survival manual gives this advice:

> avoid exertion

> keep cool and stay in the shade

> don't lie on the hot ground

> don't eat, because digestion uses up fluids

> talk as little as possible and breathe through your nose rather than your mouth.

QUESTIONS

6 Explain how following the SAS advice could help you to survive. Use what you know about heat transfer, as well as the information on this page.

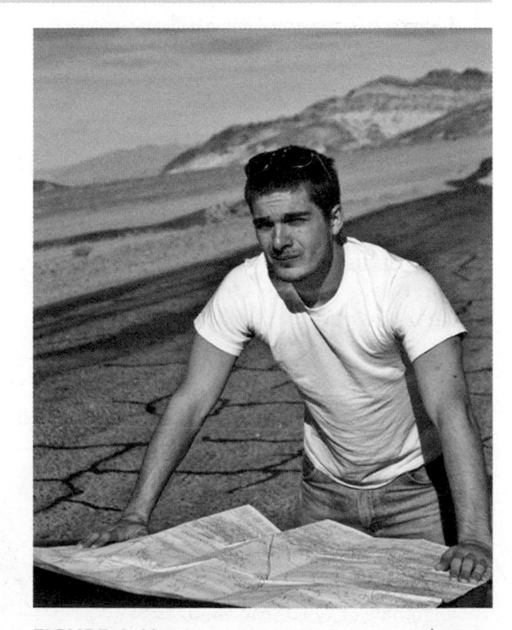

FIGURE 4: How can someone increase their chance of surviving in a desert?

Q ... desert survival for kids

Reproductive hormones

You will find out:

> hormones from the pituitary gland and ovaries control the menstrual cycle

> FSH causes egg maturity and oestrogen secretion while LH causes ovulation

The master gland

Figure 1 is a computer-enhanced MRI scan of a person's head. The little green C-shaped blob is the pituitary gland. This gland controls many of the other glands in the body. It also has overall control of sex cell production by the testes and ovaries.

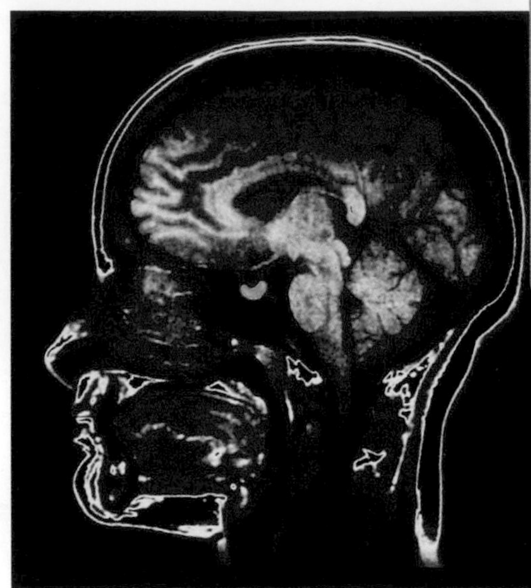

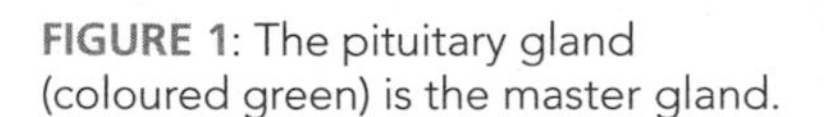

FIGURE 1: The pituitary gland (coloured green) is the master gland.

The menstrual cycle

In the **menstrual cycle**, one egg is released from a woman's **ovaries** about every 28 days. Before the egg is released, the lining of the womb (also called the uterus) thickens. So, if the egg is fertilised, the womb is ready to receive the tiny embryo.

If the egg is not fertilised, the lining of the womb breaks down. It is lost through the vagina. This is called **menstruation**.

The menstrual cycle is controlled by hormones. These include:

> hormones called **FSH** (follicle stimulating hormone) and **LH** (luteinising hormone), secreted by the pituitary gland

> a hormone called **oestrogen**, secreted by the ovaries.

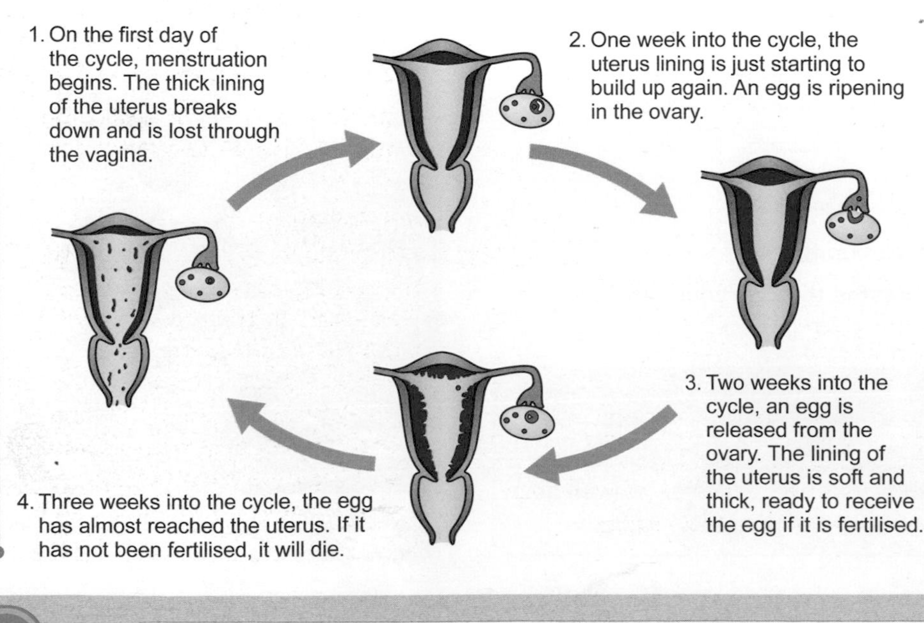

FIGURE 2: The menstrual cycle.

QUESTIONS

1 A woman starts to menstruate on the second day of March. Predict the date when an egg will next be released from one of her ovaries.

2 Where is FSH made?

Q ... pituitary gland AND ovaries for kids

Hormones and the menstrual cycle

At the start of the menstrual cycle, the pituitary gland secretes FSH. This causes an egg to mature in one of the woman's ovaries.

The FSH also stimulates the ovary to secrete oestrogen. The oestrogen makes the inner lining of the uterus grow thicker. The lining grows extra blood vessels. This prepares it for the arrival of a fertilised egg. If one did arrive, it would be able to sink into the uterus lining and start to grow into a fetus.

As more and more oestrogen is secreted, the high levels eventually affect the pituitary gland. They make it stop secreting FSH.

As the level of FSH drops, the ovary stops secreting oestrogen. This cuts off the inhibition of FSH secretion, so the cycle starts all over again.

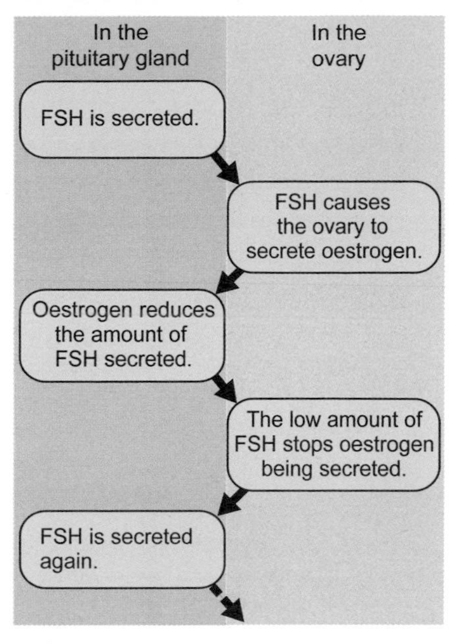

FIGURE 3: FSH and oestrogen secretion.

QUESTIONS

3 What is the stimulus that causes the ovary to secrete oestrogen?

4 Name one target organ for oestrogen.

Coordinating the menstrual cycle

During the menstrual cycle, changes occur in the concentrations of FSH and oestrogen in the blood. It is the change in concentration of oestrogen that causes the changes in thickness of the uterus lining. The rise in oestrogen concentration causes the uterus lining to thicken. When oestrogen concentration falls below a certain point, the uterus lining breaks down.

The release of an egg from the ovary usually happens at about day 14 of the cycle. It is called **ovulation**. Ovulation is caused by a surge of another hormone, called LH, that is secreted by the pituitary gland.

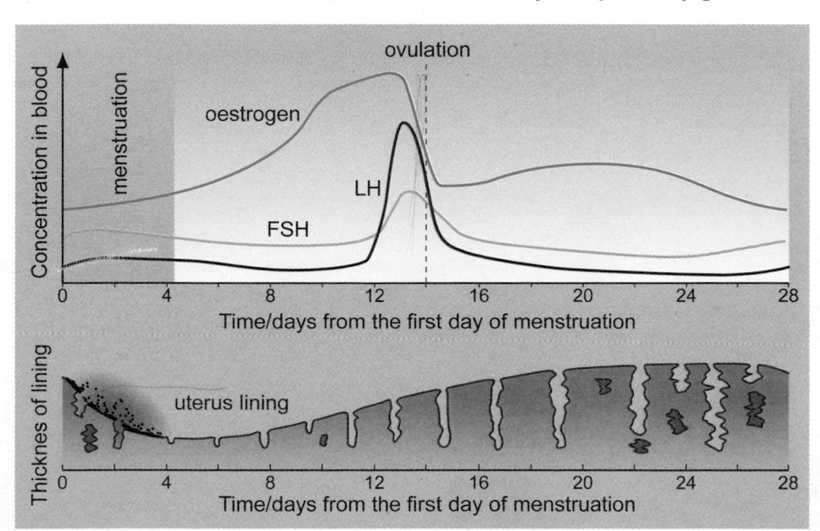

FIGURE 4: Changes in hormone levels and the uterus lining during the menstrual cycle.

QUESTIONS

5 Use the data in Figure 4 to answer these questions.

(a) Describe the pattern of oestrogen secretion throughout the 28 days of this cycle.

(b) Explain why the concentration of oestrogen rises between days 1 and 13.

(c) What causes the egg to be released from the ovary on day 14?

(d) Although the oestrogen level starts to drop on day 13, the lining of the uterus does not start to break down until day 28. Explain why it is important for there to be a delay between the beginning of the fall in oestrogen levels and the breakdown of the uterus lining.

Q ... menstrual cycle GCSE

Controlling fertility

You will find out:

> FSH and LH can help an infertile couple to have a baby
> oral contraceptives contain hormones

Multiple births

On average, one in eighty women who conceive naturally has a multiple pregnancy. It's usually twins, sometimes triplets, and on rare occasions more. In contrast, one in four women who have fertility treatment has a multiple pregnancy.

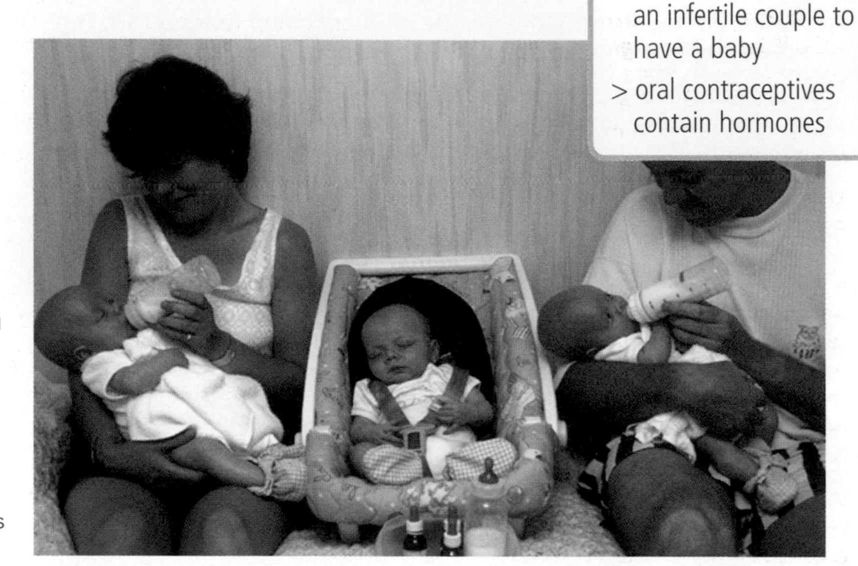

FIGURE 1: IVF enabled this couple to have a family.

Fertility

Fertility treatment

Many couples trying for a baby are not successful. They may ask for help at their local fertility clinic.

Sometimes, the problem is that the woman is not producing eggs. Her ovaries probably do have eggs in them, but the eggs are not maturing. To help them mature, she may be given a course of treatment containing the FSH and LH hormones. The hormones are sometimes called **fertility drugs**.

The hormone FSH stimulates the woman's eggs to mature in her ovaries. LH stimulates an egg's release. When an egg is released from her ovary into an oviduct, she can conceive in the normal way.

It is difficult to get the dose of the fertility drugs just right. Too low and the woman still will not get pregnant. Too high, and she could make several mature eggs at the same time, becoming pregnant with twins, triplets or even quadruplets.

Another possibility is to use **IVF**. Some of the woman's eggs are put into a dish with some of her partner's sperm. One of the fertilised eggs is then placed in her uterus (womb), where it grows into a fetus, as normal.

Did you know?

In 2007, more than 90 000 babies were born in Europe as a result of IVF.

Oral contraceptives

Other women may have the opposite problem. They want to be able to have sexual intercourse without risking getting pregnant.

A woman can be prescribed **oral contraceptives** – the contraceptive pill. The pills contain hormones such as oestrogen. These stop FSH being produced, so that her eggs do not mature.

Most oral contraceptives contain oestrogen plus another hormone called **progesterone**. They can produce side effects, such as putting on weight. Nowadays, the pills contain much less oestrogen than they used to, so there are fewer side effects. Some women take pills that contain only progesterone, as these produce the fewest side effects.

QUESTIONS

1 What does fertility mean?

2 Name a hormone that can be used as a fertility drug.

3 Explain how a fertility drug helps a woman to get pregnant.

4 Explain how the contraceptive pill works.

IVF

IVF stands for *in vitro* fertilisation. *In vitro* is Latin for *in glass*. The name comes from the fact that fertilisation happens inside a piece of laboratory equipment – which may or may not be made of glass.

The woman is given hormones, such as FSH, to make her ovaries produce several eggs. The eggs are removed by a simple operation, done under anaesthetic. Some eggs are put into a dish with a fluid at just the right temperature and nutrients to keep them healthy. Then, some of her partner's sperm are added and given time to fertilise the eggs.

When the eggs have been fertilised, they divide to make tiny balls of cells – **embryos**. One of the embryos is chosen and placed in the woman's uterus. With luck, it will sink into the uterus lining and develop as a fetus.

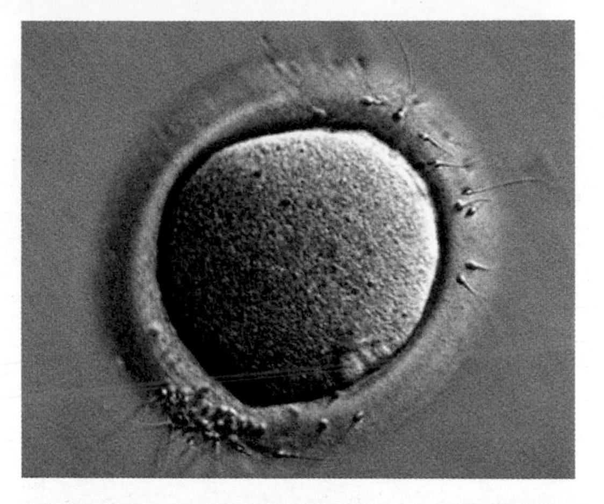

FIGURE 2: This egg is in a Petri dish, and several sperms are trying to fertilise it.

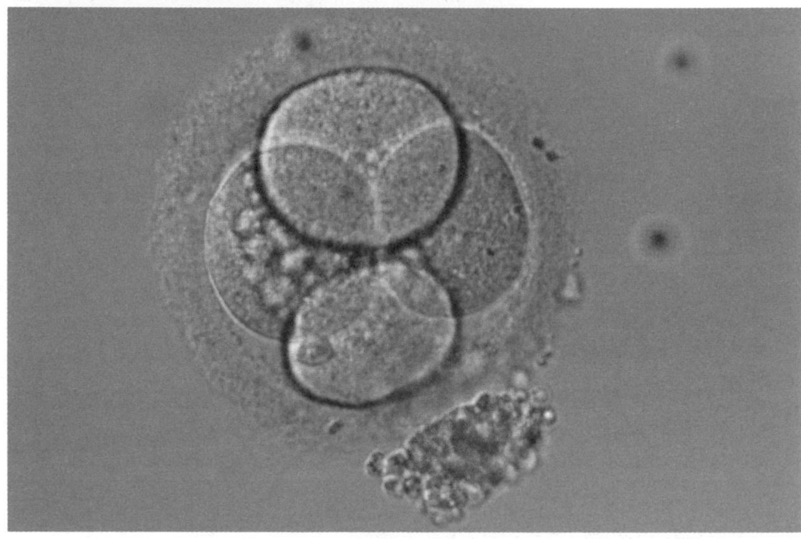

FIGURE 3: This embryo has been produced by IVF, and is now ready to be implanted into the mother's uterus.

QUESTIONS

5 Babies produced by IVF are sometimes called 'test tube babies'. Explain why that is not really true.

6 On which day (or days) of the menstrual cycle should the embryo be put into the uterus?

Multiple births

Most women would prefer to have one baby at a time, not twins or triplets. It is easier to care for one baby than several.

Twins or triplets are more likely to have problems developing in the uterus than a single fetus. They are more likely to be underweight at birth.

In the UK, almost 98% of women who conceive naturally give birth to a single child. Only 2% of births are twins, 0.05% are triplets and only 0.001% are quadruplets or more. For women who have fertility treatment with FSH, the figures are very different. About 20% of births are twins, 8% are triplets and 3% are more.

QUESTIONS

7 Explain why fertility treatment increases the chance of multiple births.

8 Discuss the benefits and problems associated with the use of hormones to increase fertility.

Control in plants

> **You will find out:**
> - plants respond to light and gravity by growing towards or away from them
> - auxin helps plants to respond to light and gravity
> - plant growers make use of plant hormones

Sensitive plants

When you touch the leaves of a particular kind of mimosa, the leaves fold up. It's sometimes called the sensitive plant. It is probably a way of protecting the leaves from things that might damage or eat them. This is very unusual for plants – they usually respond to stimuli much more slowly, by growing.

FIGURE 1: How is the response of a mimosa leaf to touch similar to a reflex action in an animal?

Responding to light and gravity

Tropisms

Plants are sensitive to light, moisture and gravity. They respond by growing in particular directions. These growth responses are called **tropisms**.

Shoots grow towards light and away from gravity. This helps to get plenty of light for photosynthesis.

Roots grow towards gravity and, usually, grow away from light. They also grow towards water – downwards, into the soil. Water is then transported from the roots to the rest of the plant. It also means the root can anchor the plant firmly into the ground.

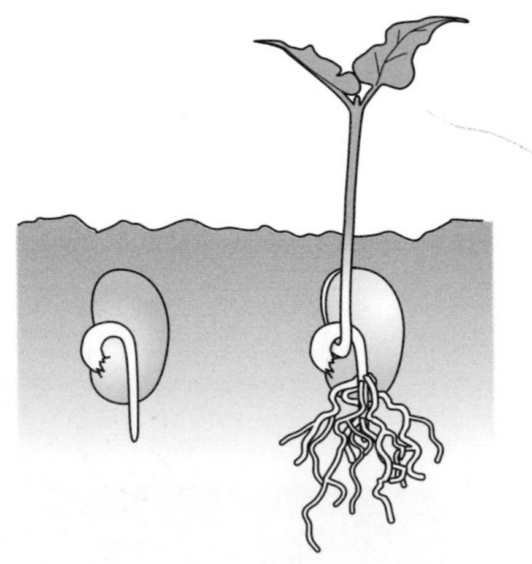

FIGURE 2: Even if you plant a bean seed upside down, its shoot will grow upwards and its root will grow downwards. How does it work out which way to grow?

> **Did you know?**
>
> Mangrove trees grow in seawater. Some of their roots grow upwards to get air – like snorkels.

Auxin

Animals, including humans, produce chemicals called hormones. Hormones move through the body to make organs respond. Plants have hormones, too. One of these plant hormones is called **auxin**.

Auxin makes the cells in shoots get longer. When light shines onto a shoot, the auxin builds up on the shady side. This makes the cells on that side get longer. So, the shoot bends towards the light.

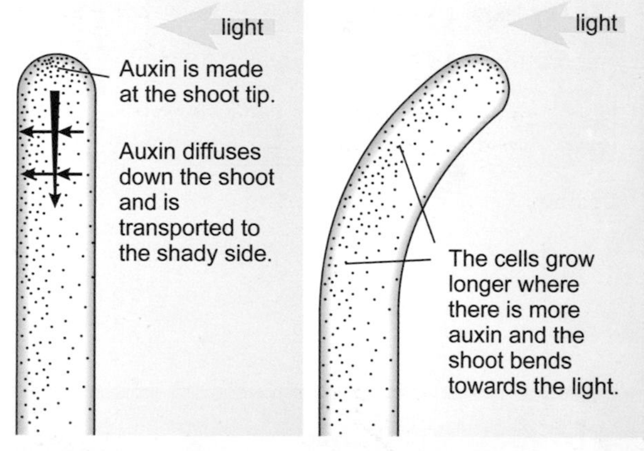

FIGURE 3: How auxin makes a shoot grow towards the light.

QUESTIONS

1 What does the term 'tropism' mean?

2 Give one example of a tropism shown by plant shoots.

3 How does the tropism, in your answer to question 2, help the plant to survive?

Q ... tropisms plant hormones GCSE

Controlling plant growth

Phototropism and gravitropism

A growth response to light is called **phototropism**. Shoots show positive phototropism, meaning that they grow towards the light. Roots that show negative phototropism grow away from light.

A growth response to gravity is called **gravitropism** or geotropism.

Auxin is involved in gravitropism. Auxin tends to accumulate on the lower side of a root. The effect of auxin on root cells differs from its effect on shoot cells. In roots, auxin reduces the rate of growth. So, the lower side of the shoot grows more slowly than the upper surface. This causes the root to bend downwards.

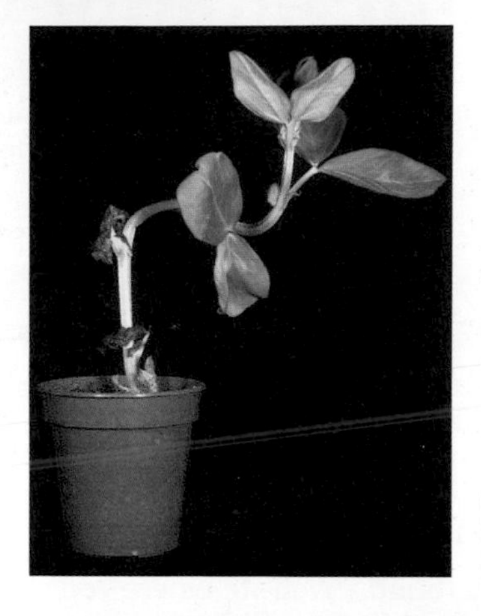

FIGURE 4: This broad bean plant has been exposed to light coming from different directions, at different stages of its growth. Suggest an explanation, including what the plant's auxin has been doing, for its growth pattern.

Using plant hormones

Growers often use plant hormones to make plants grow and develop in ways that they want.

Gardeners often grow new plants from **cuttings**. They usually dip the base of the cutting into a powder or gel called rooting hormone. This contains a plant hormone that makes the cutting grow roots.

Plant hormones are also used as weedkillers. For example, many people spray selective weedkillers on lawns, to kill weeds such as dandelions and daisies but not the grass. These weedkillers contain hormones. The hormones make the weeds grow very fast and then die. The hormones do not affect the grass because grass has a different metabolism to the weeds.

QUESTIONS

4 How does negative phototropism in roots help a plant to survive?

5 Name a part of a plant that shows negative gravitropism.

6 Auxin causes roots to show positive gravitropism. Draw a diagram, similar to Figure 3, to explain this.

Where are the receptors in a plant?

Plants do not have eyes. So, which part contains receptors that sense light? Figure 5 shows an experiment to investigate this question.

Method

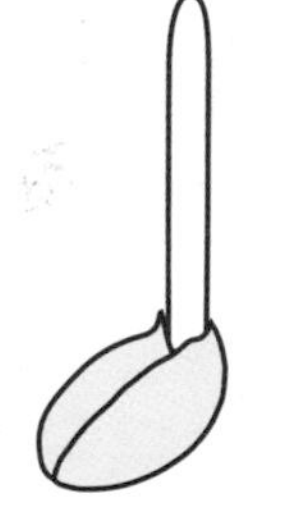

A normal

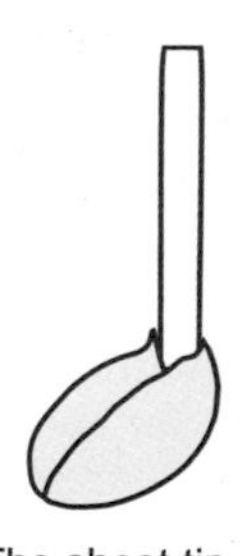

B The shoot tip has been cut off.

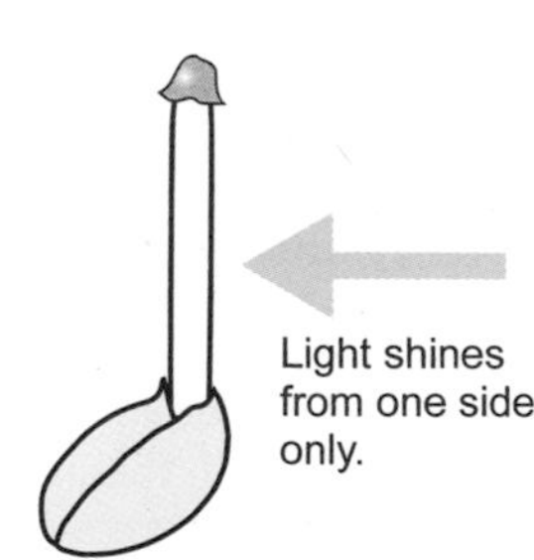

C The shoot tip has been covered by foil.

Results

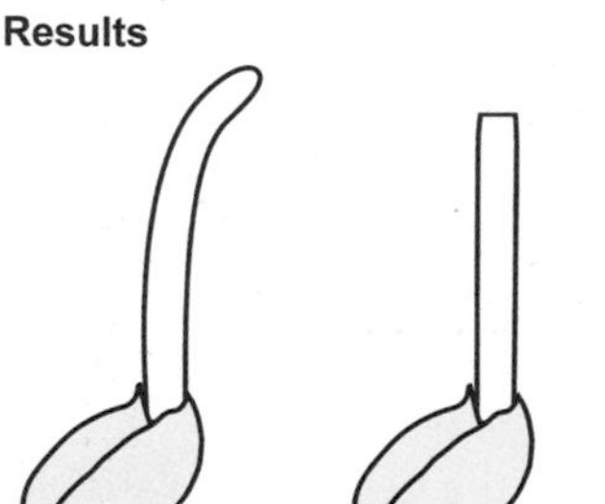

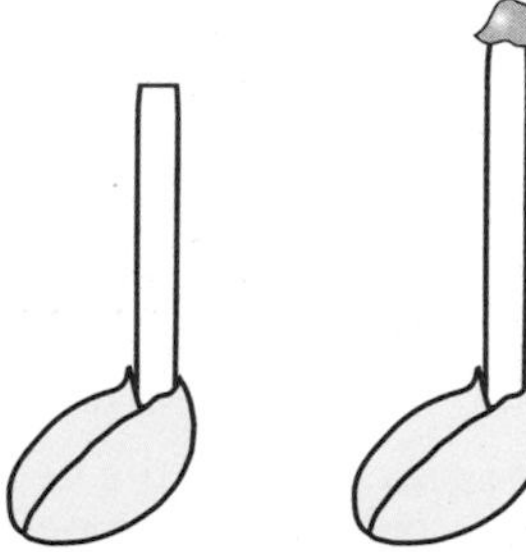

FIGURE 5: Investigating which part of a plant shoot senses light.

QUESTIONS

7 From the results of the experiment in Figure 5, suggest which part of the shoot is (a) needed for growth (b) sensitive to light.

Preparing for assessment: Planning an investigation

To achieve a good grade in science, you not only have to know and understand scientific ideas, but you need to be able to apply them to other situations and investigations. These tasks will support you in developing these skills.

Investigating the effect of caffeine on reaction time

Jay often drank a cup of coffee if she wanted to be more alert or to stay awake. She wanted to investigate the hypothesis that drinking caffeine makes people react faster. Jay researched in a textbook and on the internet. She decided on two methods for measuring reaction time.

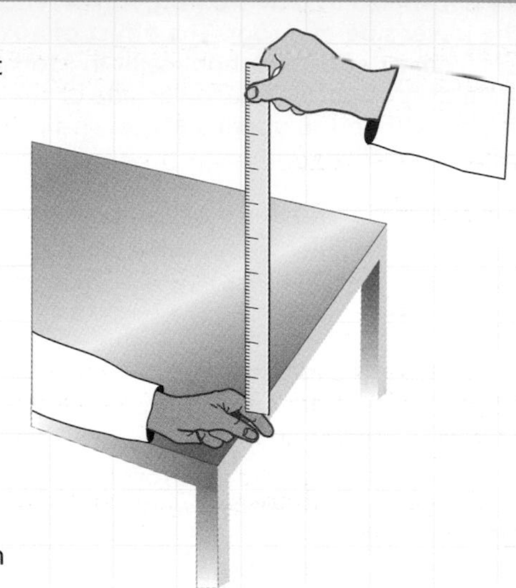

Method 1 Jay found a website where you clicked a button to start, then waited for a traffic light to turn green. As soon as it did, you clicked the button again. The website measured how long it took between the appearance of the green light and you clicking the button. It gave the time to the nearest millisecond (ms).

Method 2 A friend stood on a bench and held a ruler vertically, with the 0 mark exactly level with the bench top. Jay held her [illegible] resting on the bench.

[illegible] ned the ruler and Jay caught it, keeping her [illegible] She read off the point on the ruler at which [illegible] Jay then used this chart to convert the [illegible] ruler had dropped into a reaction time.

[illegible]er dropped (cm)	6	8	10	12	14	16	18	20	22	24	26	28
[illegible]me (ms)	110	130	140	160	170	180	190	200	210	220	230	240

Planning

1. Which method – the website or the ruler-dropping method – would be the better means of obtaining data? Explain your choice.

> The only difference between the two methods is the way of measuring the reaction time. Which one has the higher resolution? Is it possible to tell which one gives the more accurate results?

2. What is the independent variable in Jay's investigation?

> The independent variable is the one that you change.

3. What is the dependent variable in Jay's investigation?

> The dependent variable is the one that is measured for each change in the independent variable.

4. Suggest two variables that Jay should keep the same in her investigation. How could she do this?

> When experiments involve people, it is often really difficult to control some of the variables that you would like to control. There are some that Jay definitely should be able to control.

5. Outline any risks that Jay should be aware of in her investigation. What should she do to control these risks?

> Be sensible when suggesting hazards and risks – try to suggest real ones, not things that are very unlikely to cause any harm.

Processing and evaluating secondary data

On Thursday, Jay tried the traffic light website. She had five tries. Then she drank a cup of strong coffee and tried again. These were her results, in milliseconds (ms).

before coffee 224, 221, 215, 209, 208

after coffee 207, 201, 197, 195, 196

On Friday, Jay tried the ruler-dropping method. She read the distance the ruler dropped to the nearest half centimetre. These were her results.

before coffee 24.5, 24.0, 23.0, 23.5, 19.0

after coffee 20.5, 20.0, 19.0, 18.5, 18.0

6. Construct a results table and complete it, to show Jay's results from Thursday.

7. Use the data in the chart on page 42 to draw a graph showing the relationship between the distance the ruler dropped (*x*-axis) and reaction time (*y*-axis).

8. Use your graph to convert each of Jay's Friday results into reaction times.

9. Now construct a results table like the one in question 6, showing Jay's Friday results.

10. Once Jay had looked carefully at her results, she decided that it was not appropriate to calculate a mean for any of them. Do you agree with her? Why?

When planning your investigation, you need to think about the way in which you will process the data. Looking at data from previous investigations will help you plan effectively.

Make sure that the headings of each row and column are complete, and contain units where needed.

Draw a smooth curve through the points.

It is a good idea to use a ruler to draw careful vertical and horizontal construction lines when reading values off a graph.

If you look carefully, you will see a pattern in the results. Think about what happens when you practise doing something …

Connections

How Science Works

> Planning an investigation

> Assess and manage risks when carrying out practical work

> Select and process primary and secondary data

> Analyse and interpret primary and secondary data

Science ideas

B1.2 Nerves and hormones (pages 28–33)

B1.3 The use and abuse of drugs (pages 44–49)

Drugs

Caffeine, cola and coffee

Caffeine is a drug that many people take each day. It's found in coffee, tea, cola and other drinks. Caffeine stimulates the heart and the nervous system, helping you to stay alert. However, it's wise to avoid excessive quantities.

You will find out:

> drugs change the body's chemical processes

> all drugs have side effects

> some drugs can cause dependency and addiction

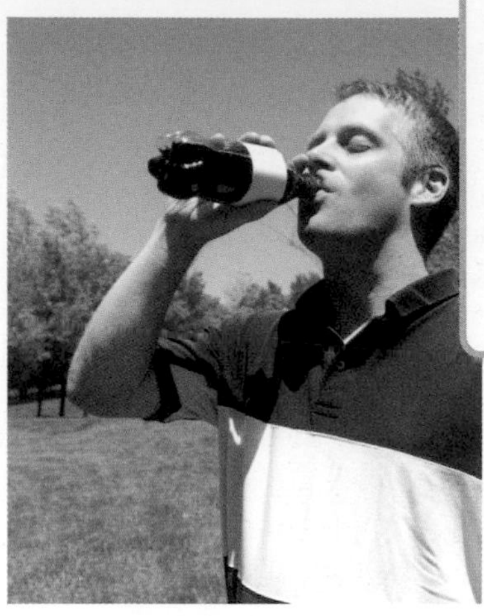

FIGURE 1: Caffeine is a drug that can help you to stay alert. This bottle of cola contains about half the caffeine in a mug of instant coffee.

Using drugs

Almost everyone takes drugs sometimes. A drug is something that changes the chemical processes in the body. Many drugs are helpful. For example, antibiotics help us to recover from infections. On the other hand, some drugs are dangerous, especially if used wrongly.

Why use drugs?

Drugs have been used for thousands of years. Many come from plants and other natural substances. An example is aspirin, which is found in willow bark. People used to chew the bark to relieve pain.

Medical drugs are very helpful. They can help to make people feel better when they are ill, and cure diseases.

All drugs have side effects. Medicines come with instructions on how much to take and a list of possible side effects.

Dangers of drugs

Some people take drugs, not because they need them, but because they make them feel different. This is called **recreational drug use**.

Many recreational drugs can be harmful. These include:

> alcohol

> cannabis

> tobacco

> cocaine and heroin.

Some drugs are illegal because of the harm they can cause. Cannabis is illegal because it can cause mental illness. However, misuse of legal drugs can also cause harm.

Some people become **addicted** to a drug. They cannot manage without it. A drug addiction can have dangerous long-term effects. Over time, the lungs, brain and liver can be seriously damaged. The liver is damaged because it has the job of destroying harmful chemicals in the body.

FIGURE 2: Medicinal plants for sale in Bolivia.

QUESTIONS

1 What is a drug? Name two useful kinds.

2 Explain the difference between medical and recreational use of drugs.

3 Suggest how the misuse of legal drugs can cause harm.

4 Explain why the liver can be damaged, if a person takes a lot of drugs.

Addiction and dependency

All drugs change the way the body works. Sometimes, these changes make someone feel they cannot manage without the drug. If someone is **dependent** on a drug, they constantly crave it. They may have started taking the drug to help them through a bad patch or may associate it with something they enjoyed. Whatever the reason, they cannot manage without it later.

If someone is **addicted** to a drug, they feel really ill if they stop taking it. They may suffer from very unpleasant withdrawal symptoms.

There is not any hard and fast dividing line between addiction and dependency. It's really difficult to escape from either. People often need a lot of help to get away from their drug habit, to live a better life.

Cocaine and heroin are very addictive drugs. A high proportion of people who take them will become addicted to them. Nicotine, the drug in tobacco, is also addictive. Some people also become addicted to alcohol.

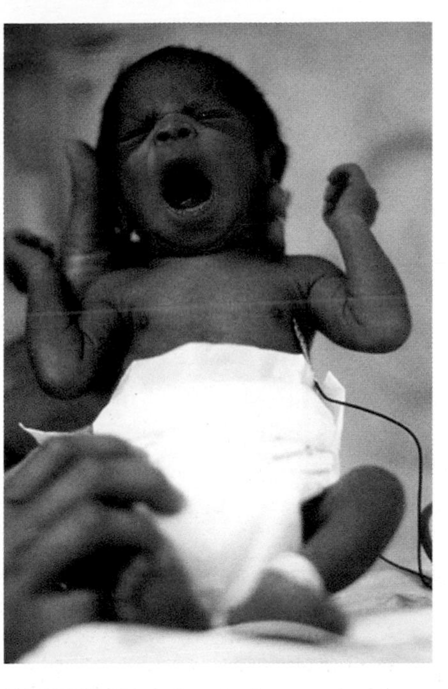

FIGURE 3: This baby was born with a drug addiction, and a low birth weight, because its mother took crack cocaine during pregnancy. Explain how this happens. (Use your knowledge of how a baby grows in the uterus.)

Did you know?

Almost one in thirteen people in Britain are thought to be dependent on alcohol.

QUESTIONS

5 Explain why people find it difficult to give up smoking cigarettes.

6 Discuss the difference between addiction and dependency.

Deaths from drug use

Each year, thousands of people in Britain die from misusing drugs. Some deaths are from poisoning. Some happen because drugs can affect the brain, making people behave in a dangerous way. Most deaths occur in people between the ages of 20 and 39.

Figure 4 shows the number of deaths from misuse of drugs, other than alcohol, in England and Wales, between 1993 and 2008.

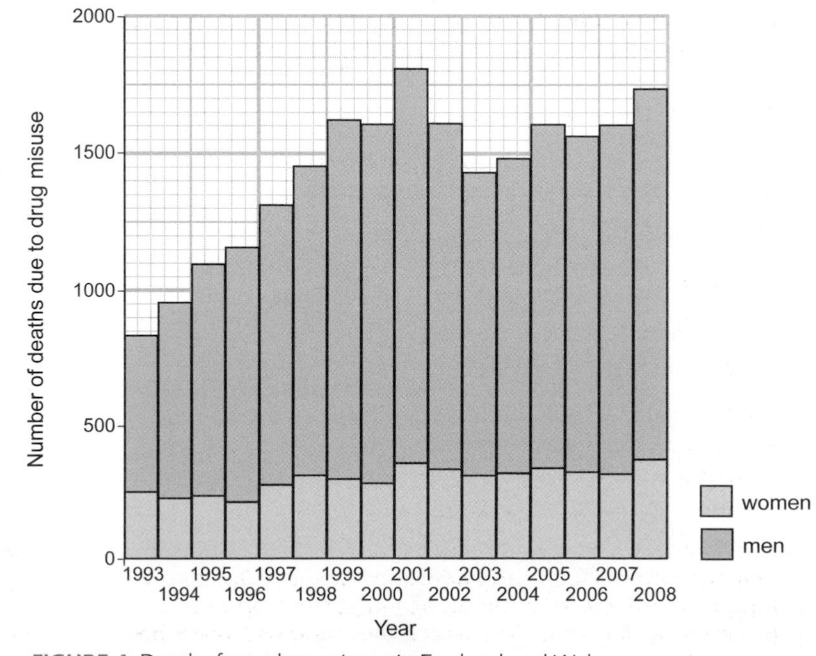

FIGURE 4: Deaths from drug misuse in England and Wales.

QUESTIONS

7 Describe trends and patterns in the data presented in Figure 4.

8 Select two trends or patterns and suggest reasons for them.

Developing new drugs

You will find out:

> how new drugs are tested before they can be used

> thalidomide harmed fetuses, even though it had been tested

Thalidomide

In the 1960s, many pregnant women were prescribed the drug thalidomide, to treat morning sickness. No one thought that the drug was dangerous. However, women who took thalidomide in early pregnancy often gave birth to babies with short arms or no arms. Thalidomide was banned worldwide, but not before more than 10 000 children had been affected.

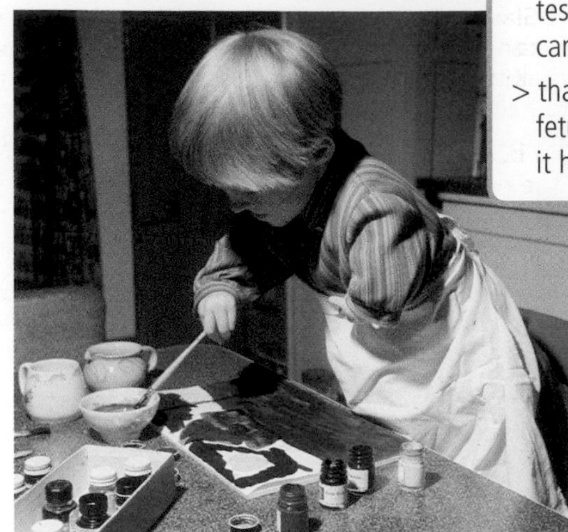

FIGURE 1: A child affected by thalidomide.

Testing drugs

Before doctors are allowed to prescribe any new drug, it has to be thoroughly tested.

There are three trial stages:

> *Is it safe?* The drug is tested in a laboratory, to find out if it is toxic (poisonous). It may be tested on human cells or **tissues** growing in a culture solution. It may be tested on live animals.

> *Is it safe for humans?* Human volunteers are given different doses, to find out what is the highest dose that can be taken safely. Any side effects are recorded.

> *Does it work?* In clinical trials, the drug is tested on its target illness. It is given to people who have the illness, to see if it makes them better.

If the drug is more effective than currently available drugs, then it may be sold commercially.

Trialling can take years and may not result in a new drug. Even if successful, it will be at least five years before the drug is available to patients.

Thalidomide had been thoroughly trialled as a sleeping pill, but no one had thought to test it on pregnant women. Now, thalidomide is banned for use as a sleeping pill, but it is being used to treat serious diseases such as leprosy.

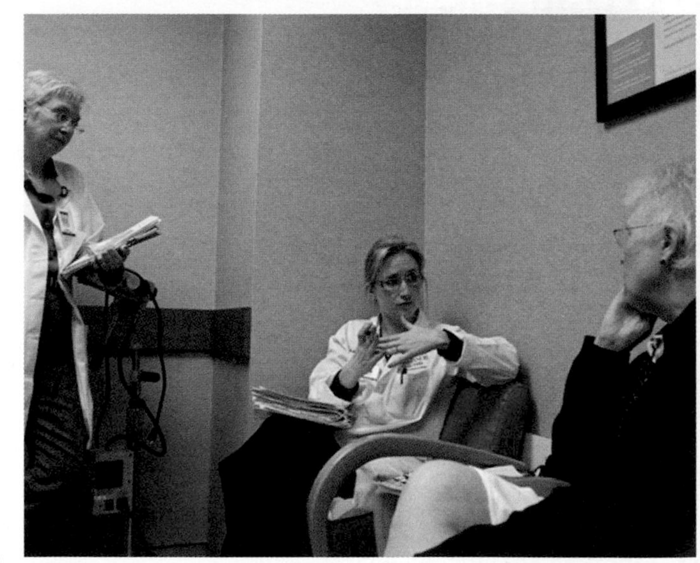

FIGURE 2: Doctors must make sure that this patient fully understands she is part of an experiment, before she agrees to take part in a drug trial.

QUESTIONS

1 Explain what thalidomide was used for, before it was banned in the 1960s.

2 What is thalidomide used for today?

3 Explain how scientists discover whether or not a new drug works.

... stages of testing new drugs

Testing an antiviral drug

In the early 1990s, GlaxoSmithKline developed a flu drug called zanamivir (now sold as *Relenza®*.) It passed the early stages of the trials with flying colours. GlaxoSmithKline then tested it thoroughly, on several groups of volunteers.

They used a double blind trial. Volunteers were divided into two groups. One group was given the drug, and the other group a **placebo** (a pill, or other treatment, that does not contain the drug). In a double blind trial, neither the volunteers nor the doctors know which they are using.

Table 1 shows some results from the trial on young Finnish soldiers. When a soldier went down with flu, they were given either zanamivir or a placebo.

	Given zanamivir	Given a placebo
number of subjects	293	295
mean age in years	19	19
number of days until their temperature went down to normal	2.00	2.33
number of days until they lost all their symptoms and felt better	3.00	3.83
number of days until they felt just as well as before they had flu	4.5	6.3
average score the volunteers gave to their experience of the major symptoms of flu	23.4	25.3

TABLE 1: Effect of zanamivir and a placebo on soldiers suffering from flu.

Did you know?

It costs a pharmaceutical company about £600 million to develop each new drug.

QUESTIONS

4 Other trials gave similar results to those in Table 1. Do you think zanamivir should be distributed? How would you support your decision?

Evaluating statins

Statins are drugs that help people to reduce their blood cholesterol level. People with high blood cholesterol, who take statins, greatly reduce their risk of getting heart or circulatory disease or dying from a heart attack.

When statins were first introduced, in the 1990s, it was suggested that most people over the age of 50 should take them, even if they did not have high blood cholesterol levels. Trials had shown almost no side effects.

However, since then, side effects have been discovered. For example, around 5% of people taking statins get painful muscles. More importantly, people taking statins are slightly more likely to develop Type 2 diabetes.

QUESTIONS

5 What are the arguments for and against everyone over 50 taking statins, whatever their cholesterol level?

6 Suggest why the side effects, now known to be caused by statins, were not detected in the drug's trials.

... side effect of statins

Legal and illegal drugs

You will find out:

- > both legal and illegal drugs can cause harm
- > cannabis increases the risk of schizophrenia
- > alcohol causes large numbers of deaths

Which drugs are illegal?

Several drugs have been made illegal, because they cause serious harm to the person who is taking them. These illegal drugs include cannabis, heroin and ecstasy. They all affect the brain and can do serious, permanent damage. They may also have adverse effects on the heart and circulatory system.

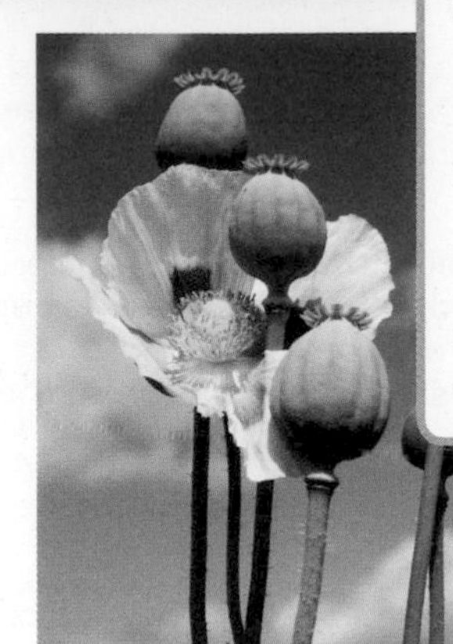

FIGURE 1: Heroin is made from opium poppies. Heroin is an important medical drug for the treatment of severe pain. It is illegal to use it recreationally.

Cannabis

Cannabis is made from the dried leaves of the cannabis plant. Some people smoke cannabis as others smoke tobacco. Just like tobacco, cannabis can cause bronchitis and lung cancer.

Cannabis makes the user feel relaxed and happy. Some people think that cannabis is safe to use. People with long-term illnesses, such as multiple sclerosis, claim that cannabis helps them feel better. Cannabis-based drugs are now licensed in the UK. However, whether or not they are effective remains controversial.

Doctors now think that young people who smoke cannabis are more likely to develop a serious mental illness called schizophrenia. It looks as though cannabis is much less safe than many people thought in the past.

Almost everyone who takes even more harmful drugs, such as cocaine or heroin, has previously smoked cannabis. This suggests that using cannabis may lead a person to use other drugs. However, there could be other explanations for this link.

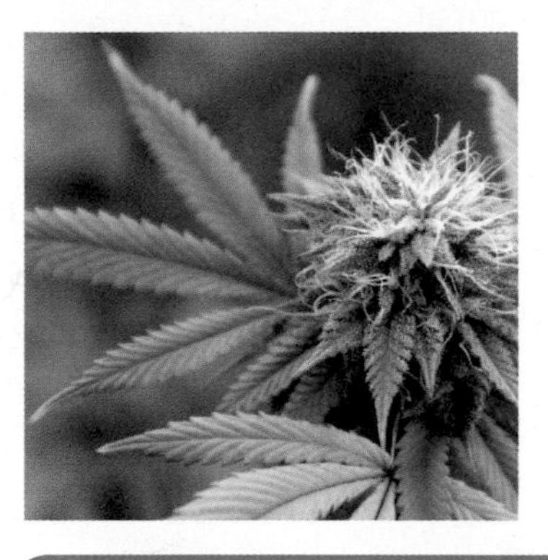

FIGURE 2: Cannabis comes from a resin made by cannabis plants.

QUESTIONS

1 List three ways in which cannabis can be harmful.

2 Suggest other explanations for the link between smoking cannabis and using cocaine or heroin.

Dangers from recreational drugs

Alcohol and nicotine are legal drugs. In actual fact, they cause far more illness and deaths each year than all the illegal drugs put together.

People do not worry so much about them because they have been around for so long, and because so many people use them.

Year	2000	2001	2002	2003	2004	2005	2006	2007	2008
Men	4483	4938	5069	5443	5431	5566	5768	5732	5999
Women	2401	2561	2632	2721	2790	2820	2990	2992	3032

TABLE 1: The number of deaths from drinking alcohol, in each year from 2000 to 2008. These data are for England and Wales.

Did you know?

In 1954, 80% of adults in Britain smoked. Today, it is fewer than 20%.

Is cannabis safe?

This is what two young people have to say about cannabis:

KEISHA aged 18

"I'm either very up or very down. I thought cannabis would make me relax. But it made me paranoid. I didn't like to admit it because everyone else seemed fine. A friend told me that if there's a history of mental health problems in your family, smoking dope might trigger it in you. My dad has suffered from depression for years. I don't smoke now and I'm much better."

Her message

"If you feel paranoid or edgy, don't use it. Check out your family history, too."

MAHESH aged 16

"I GOT A MOPED FOR MY 16TH BIRTHDAY. A FEW WEEKS LATER, I SMOKED A LOAD OF PUFF. I WAS SHOWING OFF THE MOPED TO MY MATES. I THOUGHT I COULD HANDLE IT. I LANDED UP ON THE PAVEMENT AND CRASHED INTO A WALL. I SPENT THE NIGHT IN A&E WITH A BROKEN WRIST. IT SHOOK ME UP. AND I LOST MY LICENCE."

His message

"DRUG-DRIVING IS AS BAD AS DRUNK-DRIVING. IT'S ILLEGAL AND YOU'LL LOSE YOUR LICENCE."

QUESTIONS

3 Find out what 'paranoid' means. Explain why Keisha decided not to smoke cannabis.

4 Using the figures in Table 1, draw a graph like the one on page 45.

5 Compare the data in your graph with those in the graph on page 45.

6 Suggest why, despite these data, alcohol is not an illegal drug in most countries.

Drugs in sport

Many drugs could potentially help athletes to perform better in their sport. For example:

> Steroids can stimulate the body to grow larger, stronger muscles – useful for shot putters or weight lifters.

> Beta blockers can help someone to stay calm and steady – useful in snooker or archery.

> Stimulants can increase heart rate – useful in any sport where muscles need to work hard.

However, most people think it is wrong for professional athletes to take performance enhancing drugs. Taking these drugs, when your body does not need them, often causes long-lasting damage. There have been many cases of professional sportspeople who have died as a result of taking them.

The World Anti Doping Agency, WADA, carries out scientific research and provides information about drug use in sport.

FIGURE 3: Some professional cyclists have taken EPO, a drug that increases the number of red blood cells. How might this help their performance?

QUESTIONS

7 Use your own ideas, and information from reliable sources such as WADA, to discuss whether or not athletes should be allowed to use performance enhancing drugs.

Q ... performance enhancing drug articles

Checklist B1.1–1.3

To achieve your forecast grade in the exam you will need to revise.

Use this checklist to see what you can do now. Refer back to the relevant topic in this book if you are not sure. Look across the three columns to see how you can progress. **Bold** text means Higher tier only.

Remember that you will need to be able to use these ideas in various ways, such as:

> interpreting pictures, diagrams and graphs
> applying ideas to new situations
> explaining ethical implications
> suggesting some benefits and risks to society
> drawing conclusions from evidence that you are given.

Look at pages 276–297 for more information about exams and how you will be assessed.

To aim for a grade E	To aim for a grade C	To aim for a grade A
Understand what is meant by a balanced diet and how lack of a balanced diet can affect health.	Understand that different people have different energy needs (indicated by their metabolic rate).	
Recall that exercise and inherited factors can affect metabolic rate and cholesterol level.	Describe how lifestyle can affect health.	Explain how inherited factors in combination with diet and exercise can affect health.
Recall what is meant by a pathogen, and that bacteria and viruses can be pathogens.	Explain how bacteria and viruses cause illness.	
Understand that white blood cells defend the body against pathogens.	Explain the different roles of phagocytes and lymphocytes in defence against pathogens.	Explain how the immune system reacts to pathogens and how a person becomes immune to an infectious disease.
Describe how Semmelweiss reduced the number of deaths in hospitals.	Explain the key evidence that led Semmelweiss to his conclusions.	Relate Semmelweiss's work to modern-day methods of reducing the spread of infection in hospitals.
Recall that antibiotics are drugs that kill bacteria, and that some bacteria have developed resistance to antibiotics.	Explain the limitations of antibiotics and the implications of their overuse.	**Explain how antibiotic-resistant strains of bacteria develop and the implications for general antibiotic use.**
Describe how vaccination, for example the MMR vaccine, makes people immune to diseases.	Explain how vaccination produces immunity against a disease.	Evaluate the advantages and disadvantages of vaccination.
Describe how to safely grow cultures of microorganisms in the laboratory.	Explain the steps that must be taken to ensure the cultures are uncontaminated and cannot harm humans.	Explain why in hospital laboratories it is common to culture pathogens.

To aim for a grade E	To aim for a grade C	To aim for a grade A
Name the main receptors in the human body and the stimuli that they detect.	Describe the pathway taken by a nerve impulse during a reflex action, including how the impulse crosses a junction between two neurones.	Describe in detail how an impulse crosses a synapse.
Recall that water content, ion content, temperature and blood sugar levels inside the body are controlled.	Explain why it is important to keep temperature and blood sugar level constant.	Explain the link between the control of water, ions, temperature and blood sugar and survival in harsh environments, such as desert.
Recall the meaning of the term hormone, and know that FSH, LH and oestrogen are involved in regulating a woman's menstrual cycle.	Describe the roles of FSH (follicle stimulating hormone), LH (luteinising hormone) and oestrogen in regulating the menstrual cycle. Identify where in the body FSH, LH and oestrogen are produced.	
Recall that hormones can be used to control fertility, for example the oral contraceptive.	Explain how oestrogen and progesterone prevent mature eggs being produced.	Explain the balance of hormones oestrogen and progesterone in the contraceptive pill.
Recall that *in vitro* fertilisation (IVF) involves eggs being fertilised in a dish in a laboratory, after which the embryo is placed in the mother's uterus.	Explain why FSH and LH are used in *in vitro* fertilisation (IVF) treatment.	Explain in detail the process of *in vitro* fertilisation (IVF) and the role that FSH and LH play.
Recall how plants respond to moisture, light and gravity, and recall that plant hormones are used as weed killers and as rooting hormones.	Understand plant growth responses (tropisms) as illustrated by phototropism and gravitropism.	Explain in detail the role of auxin in phototropism and gravitropism.
Recall that new drugs are tested for toxicity, efficacy and dose.	Explain how drugs are tested in the laboratory and in clinical trials.	Explain the value of the 'double blind' trial.
Recall that statins are used to lower the risk of heart disease and circulatory disease.	Explain how statins can reduce the risk of heart disease and circulatory disease.	
Recall that nicotine and alcohol are legal drugs. Recall some examples of illegal drugs and how their misuse causes serious harm to the body, including effects on the brain, heart and the circulatory system.	Explain why the misuse of legal drugs causes greater overall harm than the use of illegal drugs. Describe the difference between dependency and addiction, and explain how the use of many drugs can lead to dependency and/or addiction.	Explain why, if alcohol and nicotine were discovered today, they would both probably be made illegal. Discuss the possible progression from the use of recreational drugs to the use of hard drugs.
Recall that some athletes may take drugs, such as steroids, to enhance performance, and that the use of these drugs is prohibited.	Discuss the implications of the use of performance-enhancing drugs by athletes.	

Exam-style questions: Foundation B1.1–1.3

1. Measles is an infectious disease caused by a virus. Today, most children are vaccinated against measles when they are very young.

The graph shows the number of measles cases per year in the USA between 1950 and 2000. The graph also shows when vaccination against measles was first introduced to the USA.

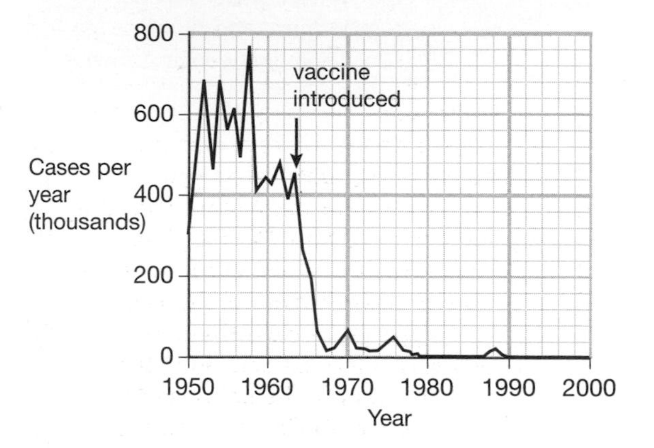

(a) The use of the measles vaccine reduces the number of measles cases.

AO3 **(i)** Describe **one** piece of evidence from the graph that supports this statement. [1]

AO2 **(ii)** Explain why this graph alone does not prove that the use of the measles vaccine reduces the number of measles cases. [2]

AO2 **(b)** If most children are vaccinated against measles, this reduces the chance of any unvaccinated children getting measles. Explain why. [2]

AO2 **(c)** Explain why measles cannot be treated with antibiotics. [2]

2. This article appeared in a newspaper in a developing country.

"Doctors are concerned that the number of people in this country who have Type 2 diabetes is increasing very rapidly. They do not know why this is, but they think it is related to the increase in the number of people who are obese.

The doctors have suggested that the increase in obesity is caused by the increased consumption of fast food, such as burgers and fries. In the past, most people ate a lot of rice and grains, but now they eat more meat and fatty foods.

The doctors are worried that the increase in obesity may also lead to an increase in the number of people with heart disease. They are advising people to reduce the amount of fast food in their diet, and to increase the amount of exercise that they do."

AO1 **(a)** Type 2 diabetes is a condition in which the blood sugar level is not properly controlled.

Explain why it is important to control blood sugar level. [1]

AO2 **(b)** Suggest why eating fast foods, rather than rice and grains, could increase the risk of a person becoming obese. [2]

AO2 **(c)** Explain how taking more exercise could help a person to reduce their risk of getting heart disease. [3]

3. Some couples who are unable to have children naturally can have IVF treatment. The woman is given hormones to make her ovaries produce eggs. The eggs are collected and put into a dish. Sperm are added to the dish. Any fertilised eggs can grow into a tiny ball of cells, which can be put into the mother's uterus.

AO1 **(a)** Name the **two** hormones that are given to the woman in IVF treatment. [1]

AO1 **(b)** What is the correct term for the tiny ball of cells that develops in the dish? [1]

The graph shows the percentage of women of different ages who become pregnant following IVF treatment.

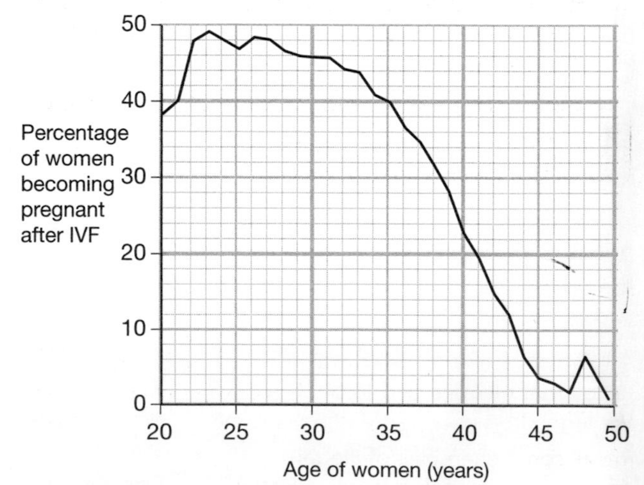

AO2 **(c)** At which age is IVF most likely to be successful? [1]

AO3 **(d)** Use the information in the graph to suggest why most couples need at least two IVF treatments before they have a child. [1]

AO1 recall the science AO2 apply your knowledge AO3 evaluate and analyse the evidence

WORKED EXAMPLE – Foundation tier

A group of researchers investigated the relationship between body mass and metabolic rate.

- The researchers measured the body mass of five people.
- They measured their body mass excluding fat.
- They measured their metabolic rate while they were asleep.

The results for the five people are shown in the table.

Person	Total body mass (kg)	Fat-free body mass (kg)	Metabolic rate while asleep, (kJ per hour)
A	61.2	53.2	350
B	58.7	46.4	295
C	65.2	43.9	271
D	51.3	40.0	256
E	54.6	36.5	228

(a) What is meant by the term *metabolic rate*?

Tick (✓) **one** box.

the amount of energy a person uses ☑ each hour

the amount of exercise a person ☐ does each day

the amount of food a person eats ☐ each day [1]

(b) Which person had the highest metabolic rate? [1]

350

(c) The researchers concluded that metabolic rate while sleeping is affected by fat-free body mass, rather than by total body mass.

Explain why the data in the table support their conclusion [2]

When the person's fat-free body mass gets higher, their metabolic rate gets higher too.

How to raise your grade!

Take note of these comments – they will help you to raise your grade.

If the word 'rate' appears, then it is important that there should be something about *time* in your answer. If you cannot remember the meaning of a term like this, it is always worth looking at how it is used in the question, as this can often give you a clue. Here, the heading in the last column of the table shows you the units used to measure metabolic rate, which is a big clue! This is a correct answer, and the candidate gets one mark.

To answer this question, you need to look carefully at the table. The question asks which *person* has the *highest* metabolic rate. The candidate has chosen the correct row in the table, but they have given the person's metabolic rate, rather than stating the person. They should have added 'person A'. This will not get the mark.

This answer is right – but it does not go far enough. Notice that there are two marks for this question, so you need to go a little bit further with the answer. The candidate should also have said something about the total body mass – that it does not follow the same pattern as the metabolic rate, or that the person with the highest **total body mass** (person C) does not have the highest metabolic rate.

Exam-style questions: Higher B1.1–1.3

1. In a sprint race, a gun is fired to give the signal to the runners to start.

A runner with a fast reaction time has an advantage. The reaction time is made up of:

- the time it takes for the sound of the gun to reach their ears
- the time it takes for a nerve impulse to pass from their ear to their brain
- the time it takes for a nerve impulse to pass from their brain to their leg muscles.

The table shows the reaction times of the runners in lane 1 and lane 6 in eight heats of a 100 m race.

	Reaction time (ms)							
	heat 1	heat 2	heat 3	heat 4	heat 5	heat 6	heat 7	heat 8
lane 1	0.145	0.132	0.164	0.162	0.139	0.157	0.139	0.147
lane 6	0.223	0.227	0.195	0.188	0.178	0.163	0.128	0.167

AO3 **(a) (i)** There is one result in the table that does not fit the pattern of the others. State the heat in which this occurred. [1]

AO2 **(b) (i)** What is the receptor involved in the reaction of the runner to the starting gun?

(ii) What is the effector involved in this response? [1]

AO1 **(c) (i)** Name the type of neurone that carries the impulse from the ear to the brain.

(ii) Name the type of neurone that carries the impulse from the brain to the leg muscles. [1]

AO3 **(d)** The reaction time of each runner is measured using an electronic sensor attached to their starting blocks, which responds to the push of their feet against the blocks.

Explain why the reaction times could not be measured by a person with a stop watch. [2]

AO2 **(e)** In general, tall runners are found to have slightly longer reaction times than shorter runners. Suggest why. [1]

2. Many populations of the bacterium *Staphylococcus pneumoniae* have become resistant to penicillin. The graph shows the mean percentage of people who use antibiotics each day in six countries, and the percentage of *S. pneumoniae* populations that are resistant to penicillin.

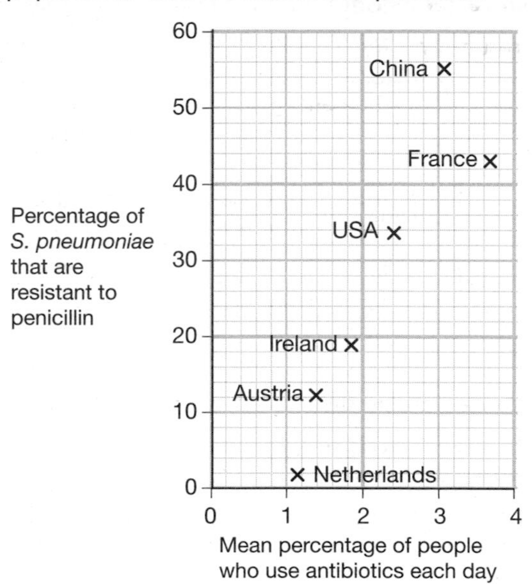

AO3 **(a)** Describe the relationship between the use of antibiotics and the occurrence of penicillin-resistant bacteria. [2]

AO2 **(b)** Suggest an explanation for the relationship you have described. [3]

AO3 **(c)** Explain why these data do not prove that there is a causal relationship between the use of the antibiotics and the occurrence of antibiotic resistance. [2]

AO2 **3.** *In this question you will be assessed on using good English, organising information clearly and using specialist terms where appropriate.*

Plants produce hormones, such as auxin, that enable them to respond to light and gravity.

Explain the role of auxin in the natural response of a plant shoot to light and to gravity. [6]

AO1 recall the science AO2 apply your knowledge AO3 evaluate and analyse the evidence

WORKED EXAMPLE – Higher tier

Most children are vaccinated with a triple vaccine to protect against measles, mumps and rubella.

Researchers carried out a trial to compare three different triple vaccines, A, B and C.

Almost 700 six-year-old children were tested to see if they had antibodies against these three diseases in their blood. Each child was then vaccinated with one of the three triple vaccines. The children were tested for antibodies again, 30 days after vaccination.

The results are shown in the table.

Disease	Percentage of children with antibodies against the disease			
	before vaccination	after vaccination with A	after vaccination with B	after vaccination with C
measles	79.2	100.0	99.5	100.0
mumps	69.4	99.5	94.5	92.0
rubella	55.4	92.6	91.3	88.6

(a) Explain what is meant by the term *antibody*. [2]

It's a substance produced by white blood cells that kills pathogens, like bacteria.

(b) What percentage of children did **not** have protection against measles before vaccination? [1]

20.8

(c) Many of the children had antibodies to measles, mumps or rubella in their blood before they were vaccinated. Suggest why. [1]

They could have already had the disease and recovered from it.

(d) The researchers did not test the children again for antibodies until 30 days after they were vaccinated. Suggest why. [1]

To give time for them to make the antibodies.

(e) Which vaccine, A, B or C, would be the most effective at preventing these three diseases? Explain your answer. [2]

A, because it produced the highest number of antibodies.

How to raise your grade!

Take note of these comments – they will help you to raise your grade.

The candidate has made two clear and correct statements, explaining exactly what the term 'antibody' means. He or she could also have said that each kind of antibody is *specific* for a particular kind of pathogen.

The children without protection against measles are the children without antibodies against it. The actual answer is correct – the candidate has worked this out by taking 79.2 from 100. However, working should always be shown.

That is a sensible idea, which gets the mark. There is often more than one possible answer to 'suggest' questions – can you think of any other possible answers to this one?

That is correct. You should remember that white blood cells take a while to make antibodies after they have met an antigen.

A is correct, but not the reason. We do not know *how many* antibodies were produced – we only know about the *percentage of children* in which antibodies were found. It is really important to think what exactly the data you are given might mean. Always look very carefully at the headings in tables or graphs.

Biology B1.4–1.8

What you should know

Variation

All living things show variation. This may be inherited, environmental or the result of selective breeding.

Genetic information is carried in the form of genes on chromosomes in the nucleus of cells, and can cause differences between individuals.

What is the name of the chemical from which genes are made?

Environment

Living things are adapted to survive in their environment.

Food chains show how energy from the Sun passes from one organism to another.

The environment affects the distribution and type of living organisms found in a habitat.

Human activity can damage the environment.

How does sulfur dioxide pollution affect plants?

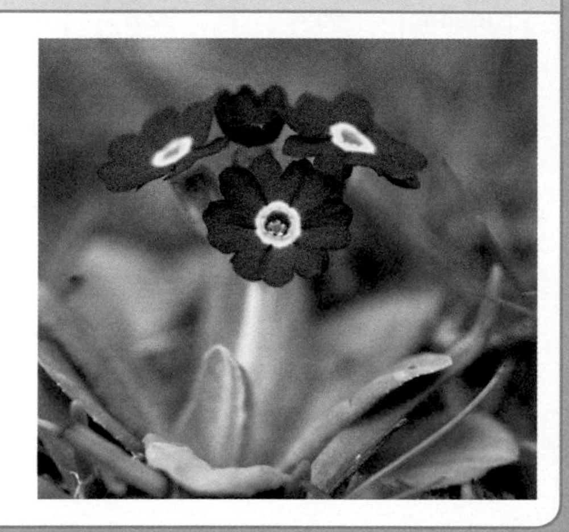

Reproduction

Animals and plants produce young by sexual reproduction.

Sexual reproduction involves two parents.

The new life begins when a male sex cell and a female sex cell fuse together.

What is the correct term for the fusion of the two sex cells?

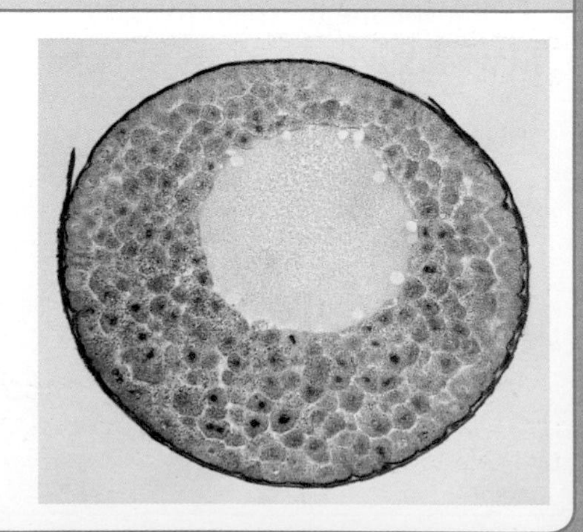

You will find out

Interdependence and adaptation

> Organisms interact with one another in their environment. Where two organisms need the same resource, which is in short supply, they must compete for it. They need to have adaptations that enable them to do this successfully, in order to survive.

> The environment often changes, and animals and plants may not always be able to survive these changes.

Energy and biomass in food chains

> Energy from the Sun flows along food chains in biomass.

> Energy is wasted during the transfers, so there is less biomass towards the end of the chain than at the start.

Waste materials from plants and animals

> Microorganisms break down (decay) waste materials from organisms. This releases substances that plants need to grow.

> Carbon is cycled between living organisms and the environment.

Genetic variation and its control

> Individuals belonging to the same species differ because of differences in their genes and their environment.

> Sexual reproduction causes genetic variation among offspring, whereas asexual reproduction produces genetically identical offspring.

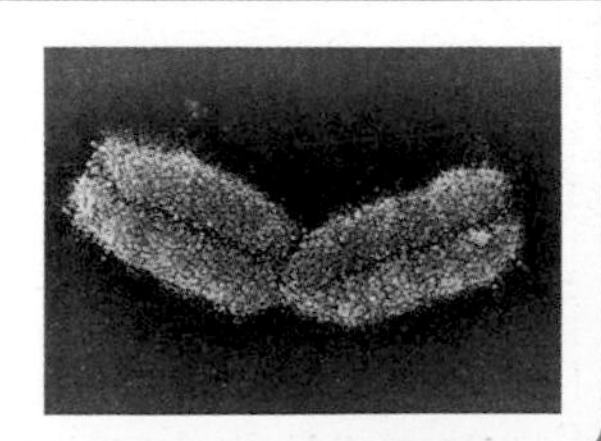

Evolution

> Darwin was the first person to put forward the theory of evolution by natural selection.

> Organisms that are best adapted to their environment are more likely to survive and reproduce, passing on their genes to their offspring.

Competition

You will find out:

> organisms compete with each other for resources

> organisms compete with each other for mates

> some organisms can live in extreme environments

Silent fight

Coral reefs are stunningly beautiful and look very peaceful. Yet fierce battles are going on. Each different coral is competing for space on the reef. At their edges, they secrete chemicals that try to stop their neighbours growing into their own patch.

FIGURE 1: This is a battleground for survival.

Resources

All living organisms need resources from their environment in order to stay alive.

For example, plants must have:

> **light** – to photosynthesise and make food

> **water** – to keep cells alive, to transport substances around and for photosynthesis

> **space** – for room to put down roots and spread out leaves to capture light

> **nutrients** – such as nitrates from the soil.

Competition

Often, resources that organisms need are in short supply. There are not enough for every plant or animal to get what it needs.

When this happens, organisms have to **compete** for resources. This does not often mean that they actually fight over them. They just have to find ways of being better at getting them than others are.

For example, plants compete for light – the ones that grow tallest win the competition.

The individuals best at competing are the most likely to survive. Those not good at getting resources are the most likely to die.

FIGURE 2: Why do farmers want to kill weeds growing in their crops?

Did you know?

One tiger may have a territory of up to 75 square kilometres.

QUESTIONS

1 Suggest which of the resources, listed above, are also needed by animals.

2 Think of at least two more resources that most animals must get from their environment.

3 Suggest the resources that plants are most likely to compete for in (a) a desert (b) a tropical rainforest.

Competing to reproduce

Many animals reproduce sexually. If there are not enough females to go around, then the males will compete for a mate.

Animals may also compete for a **territory** – a space in which they can find food and a place to breed.

Sometimes, the males do actually fight over mates and territory. However, in many species of animals, the males use some kind of display to attract the females.

FIGURE 3: Stags fight over the right to mate with the females in the herd.

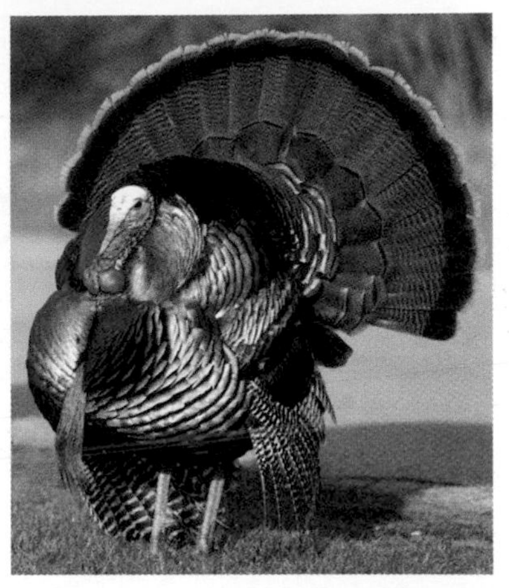

FIGURE 4: Male turkeys display to attract females.

QUESTIONS

4 What features will give stags an edge in the competition for mates?

5 Suggest features that will give male turkeys the best chance of getting a mate.

6 Suggest advantages of the way that turkeys compete for mates, compared with the way that stags compete for mates.

Avoiding competition

Competition wastes energy. Having to share resources with other organisms reduces an organism's chances of surviving and having large numbers of offspring.

Many organisms have become able to live in places where few others can survive. Although these places make survival tough, there is no need to share resources. This can increase chances of success.

For example, flamingos are able to feed in lakes that are so alkaline that almost nothing else can survive there – except the shrimps and small aquatic insects that they eat.

QUESTIONS

7 Suggest two resources for which a flamingo has to compete with other flamingos.

8 Suggest two resources for which flamingos do not have to compete with other species.

FIGURE 5: Flamingos feed and nest where few other organisms can survive.

Adaptations for survival

You will find out:

> organisms have adaptations that help them survive in their habitat

> how animals in the Arctic are adapted to conserve heat

> how plants and animals in deserts are adapted to conserve water

Colours aid survival

Animals are coloured for different reasons. Some are camouflaged so well that they blend in perfectly with their surroundings. Some male animals are vividly coloured in an attempt to attract a mate. Others are very brightly coloured to warn off predators.

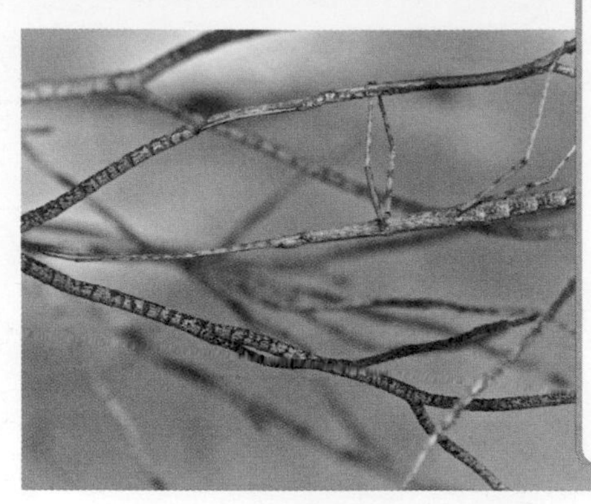

FIGURE 1: Can you spot the stick insect?

Adapting to compete

Plants and animals must have features that allow them to survive in their habitat. These features are **adaptations**.

Well-adapted organisms can compete successfully for the things they need. If they are not adapted for life in a particular habitat they will either die or move elsewhere.

Living in dry places

Plants that live in dry places usually have:

> long, wide-spreading roots. The roots grow deep into the soil, to reach water.

> small leaves. The small surface area reduces the amount of water evaporating into the dry air.

> tissues (groups of cells) that can store water.

Animals that live in dry places:

> must be able to manage without drinking much water. Gerbils do not need to drink at all. They get enough water by eating seeds and other plant material.

Desert animals:

> often have large ears. A large surface area helps the animal lose body heat and stay cool.

Living in cold places

Animals that live in very cold places, such as the Arctic, often:

> have thick fur and thick layers of fat. This insulation helps the animal reduce heat loss from its body to the icy air.

> are coloured white, for camouflage against snow and ice.

Special adaptations

Many plants and animals have thorns, poisons and warning colours to deter predators. Throughout the animal world, red and black or yellow and black mean 'I am poisonous – don't eat me'. Many animals cheat: they have warning colouring even though they are not poisonous.

FIGURE 2: How is this cactus adapted for survival in a dry desert?

FIGURE 3: An Arctic fox.

QUESTIONS

1 Explain what 'adaptation' means.

2 How are gerbils adapted for living in a desert?

3 Give one way in which an Arctic fox is adapted to live in its habitat.

4 List some animals that have warning coloration and are poisonous. Suggest some that have warning coloration but are not poisonous.

Camels

Figure 4 shows how camels are adapted for life in deserts.

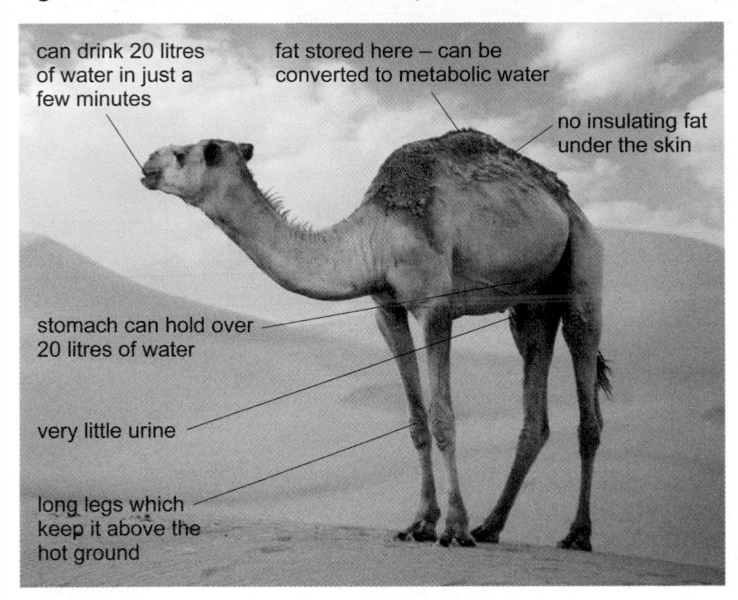

FIGURE 4: How a camel is adapted for desert life.

Did you know?

The animal with the longest fur is the musk ox. It lives in very cold northern areas of Europe. Its hairs can be 70 cm long.

QUESTIONS

5 Draw an outline of a polar bear. Add annotations, like those on Figure 4, to explain how the polar bear is adapted to live in the Arctic.

Extreme environments

In some habitats, conditions are so extreme that it is difficult to imagine how anything at all can be adapted to live there. Yet, so far, living things have been found in every environment we have been able to explore on Earth. Organisms that can live in very difficult environments are called **extremophiles**.

In the really extreme places, it is often only microorganisms that can survive. They have been found around deep sea volcanic vents at 400 °C and in the Arctic at –20 °C. In the deepest part of the Pacific Ocean, some microorganisms survive at pressures 1000 times greater than at the surface. Others survive in water where the salt concentration is very high.

For most organisms, these conditions would be lethal. Enzymes and other protein molecules quickly lose their shape if they get too hot. These microorganisms must have very stable protein molecules that are not affected by such high temperatures.

QUESTIONS

6 Enzymes from microorganisms that live in hot springs have many uses in industry. One example is an enzyme added to washing powder. Suggest why these enzymes are especially useful.

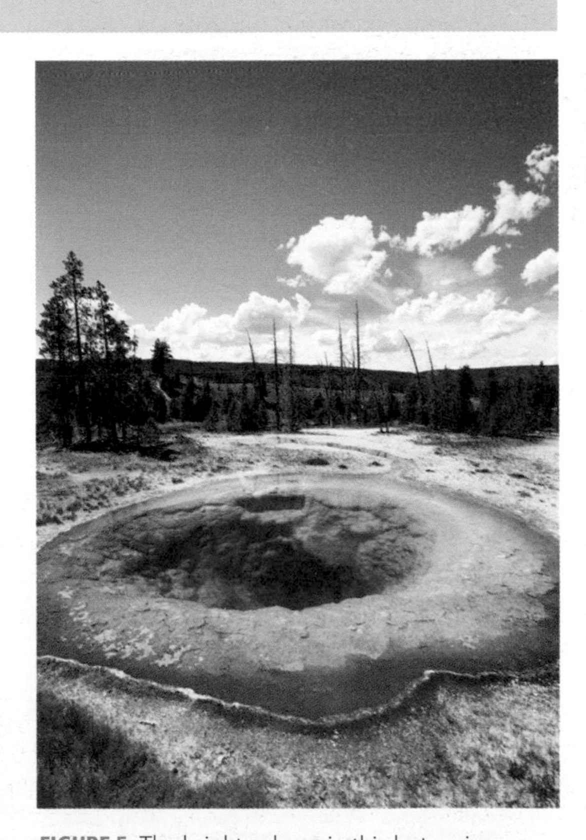

FIGURE 5: The bright colours in this hot spring are growths of microorganisms. The temperature of the water may be as high as 90 °C.

Environmental change

Coping with change

This Arctic hare is beautifully camouflaged – or is it? It is adapted to blend in with the snow, but this year there is no snow. Changes in an animal's environment can make survival very difficult.

FIGURE 1: Adaptations cannot always keep up with a changing environment.

You will find out:

- > an organism's environment can change
- > changes can be caused by living or non-living factors
- > environmental changes can affect the distribution of organisms

Getting warmer

The environment is always changing. This was happening even before people lived on Earth. Now, we have extra changes to contend with, such as global warming.

All living organisms must be adapted for survival in a particular environment. If that environment changes, they may no longer be able to survive.

For example, in the UK, average temperatures are slowly increasing. For some species, that is good news. Species that are adapted to live in warm places are spreading northwards. They are spreading further than they have ever been able to live before.

For other species, such as the Scottish primrose, the news is not so good. It grows only on the coast of the far north of Scotland.

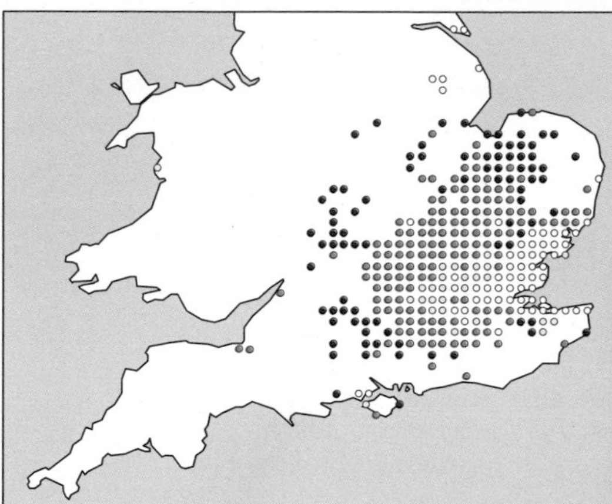

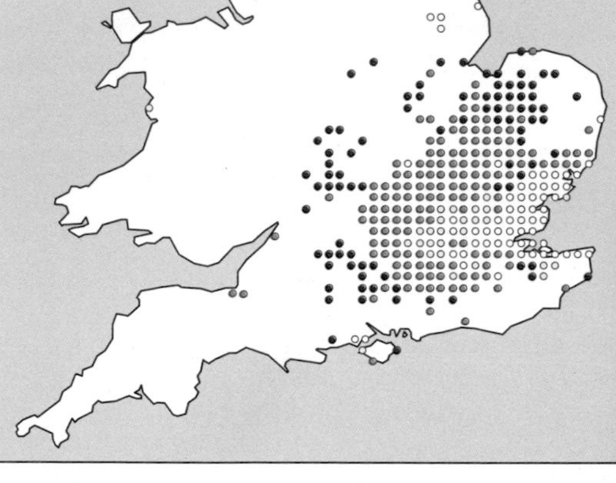

FIGURE 2: Roesel's bush cricket is spreading across England.

FIGURE 3: What will happen to the Scottish primrose when the climate gets warmer?

QUESTIONS

1 Explain why changes in the environment can make it difficult for plants or animals to survive.

2 Describe how an environmental change has caused a change in the distribution of a species.

Causes of change

Some changes are caused by **non-living factors** in the environment. For example, global warming is affecting temperature and rainfall. In the UK, the mean annual temperature is rising. It is difficult to pick out changes in the pattern of rainfall in the UK, as this is always very variable. However, in other places, such as central Australia, there seems to be a lot less rain than there used to be.

Other changes are caused by living factors. For example, at the end of the 19th century, all squirrels in the UK were red squirrels. Then the grey squirrel was introduced from North America. They seem to be better adapted than red squirrels, for survival in the UK, and outcompete them for food. Also, grey squirrels seem to be immune to a virus that they carry. The virus is deadly to red squirrels. In England, red squirrels are now only found in a few places. They are in decline in the rest of the UK, as well.

Did you know?

Some researchers predict that the UK's average temperature will go up by 4 °C by 2080.

FIGURE 4: A red squirrel in Dorset.

QUESTIONS

3 Give one example of the effect on the distribution of a species of (a) a changing non-living factor (b) a changing living factor.

The disappearing bees

Many people keep bees for honey production. Honeybees are also very important in pollinating flowers that will develop into food crops, such as beans and rapeseed.

In recent years, there has been a decline in the numbers of honeybees. Sometimes, whole colonies seem to have been wiped out overnight. Various suggestions have been put forward to explain this dramatic fall in the numbers of honeybees. They include:

> an increase in parasites, such as the varroa mite and a virus called IAPV

> increased use of insecticides

> climate change

> reduction in the number of different plant species (plant diversity). Bees that feed on pollen and nectar from only a few species of plants seem to have less effective immune systems.

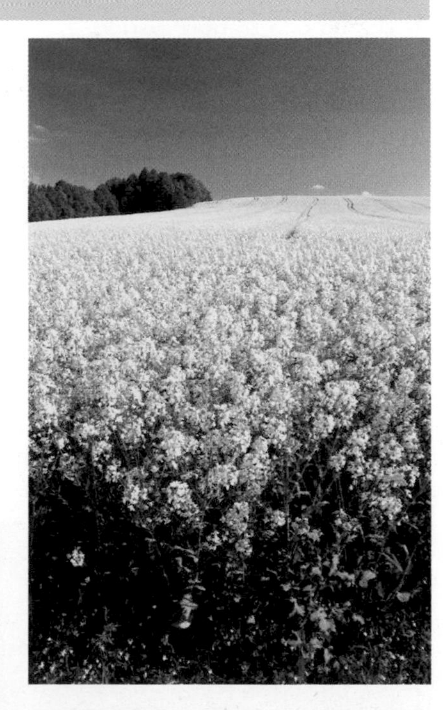

FIGURE 5: How might the increased amount of rapeseed that is grown in the UK contribute to the decline in numbers of honeybees?

QUESTIONS

4 So far, the idea that feeding on just one sort of pollen, such as rapeseed pollen, might be contributing to the reduction in bee numbers is only a hypothesis. Suggest how this hypothesis could be tested.

5 Bee numbers in the UK may continue to fall. Suggest how this could affect the numbers and distribution of other species.

Pollution indicators

You will find out:

- > meters are used to measure changes in environmental temperature, oxygen levels and rainfall
- > living organisms can be indicators of pollution

Anyone for a swim?

For hundreds of years, the Thames was used as a dump for all sorts of waste. This cartoon is from 1855. It shows Michael Faraday, the famous scientist, talking with 'Father Thames'. London's sewers were first built in 1859. They are being constantly repaired. At least the Thames is now much cleaner, and even seahorses have been found living there.

FIGURE 1: The smell of sewage and dead animals in the Thames must have been overpowering.

Measuring changes in the environment

In the UK, the composition of the air and of the water in rivers and streams, air temperature and rainfall, are constantly being measured. This makes sure that any changes can be tracked.

> **Oxygen meters** are for measuring the concentration of dissolved oxygen in the water. Unpolluted water contains a lot of dissolved oxygen.

> Thermometers are for measuring temperature. Maximum–minimum thermometers record the highest and lowest temperature reached during one 24-hour period.

> Rain gauges are for measuring rainfall. The depth of water collected tells you how much rain has fallen.

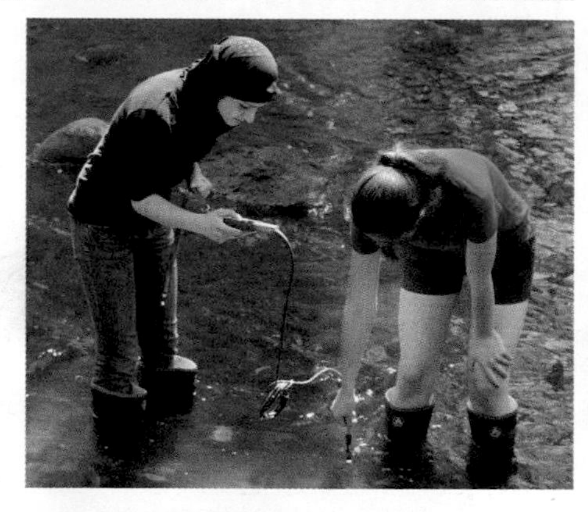

FIGURE 2: Using an oxygen meter.

Living pollution indicators

When a habitat is polluted, it changes. Organisms that normally live there will either die or move away. Sometimes, other organisms move in. They have adaptations which let them live in polluted conditions.

Scientists can use the distribution of living organisms to find out about **pollution**. For example:

> If there is a lot of **sulfur dioxide** in the air, many species of **lichens** will not be able to grow.

> If there is not very much **oxygen** in a river, there will be no oxygen-loving **mayfly larvae** in the water. Instead, there will be just **rat-tailed maggots** and **bloodworms**.

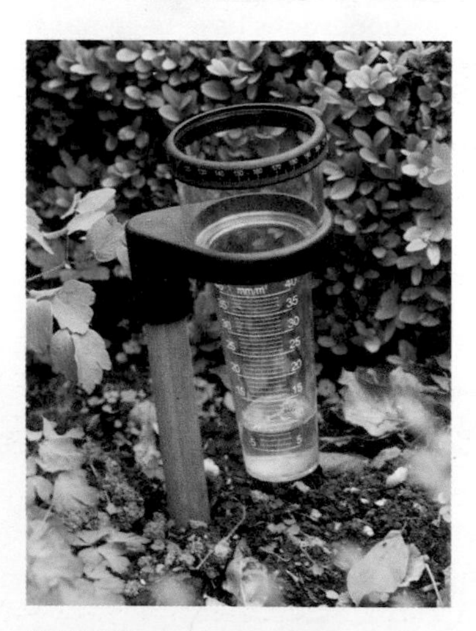

FIGURE 3: A rain gauge.

QUESTIONS

1 Explain how measuring the oxygen content of river water can show if the river is polluted.

2 Describe a way of measuring the oxygen content of a river using (a) a scientific instrument (b) a living pollution indicator.

3 For your answers to question 2, explain which method you think is better.

Sulfur dioxide, cities and lichens

A scientist surveyed the distribution of lichens. She counted the number of different species that grew on tree trunks at various distances from the centre of a polluted city. She also measured how much sulfur dioxide (SO_2) there was in the air. Her results are shown in Table 1.

TABLE 1: Lichen distribution and SO_2 levels around a city.

Distance to city centre (km)	0	1	2	3	4	5	6	7	8	9	10
Number of different species of lichen	0	1	2	6	8	10	30	44	51	56	56
SO_2 level (arbitrary units)	180	160	145	119	93	71	49	35	11	5	0

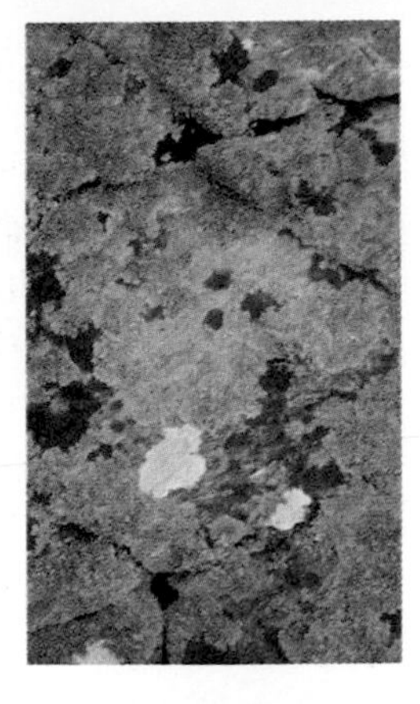

FIGURE 4: How can lichens indicate pollution?

QUESTIONS

4 Draw a line graph of the results in Table 1.

5 Describe the relationship between the number of lichen species and the sulfur dioxide concentration.

6 Suggest an explanation for this relationship.

7 Suggest how these results could enable people without access to scientific equipment to monitor pollution in a city.

Sewage pollution and invertebrates

Polluted water often contains very little dissolved oxygen. Some species of invertebrate are able to live in this water, but others are not. Figure 5 shows some of these species.

Did you know?

The Thames was declared 'biologically dead' in 1957. Now 125 different species of fish live in it.

QUESTIONS

8 A scientist found tubifex worms, mosquito larvae and flatworms in a river. What can he conclude about the river water?

9 Chironomid larvae and rat-tailed maggots can survive perfectly well in oxygenated water. Suggest why they are not normally found in well-oxygenated, unpolluted streams.

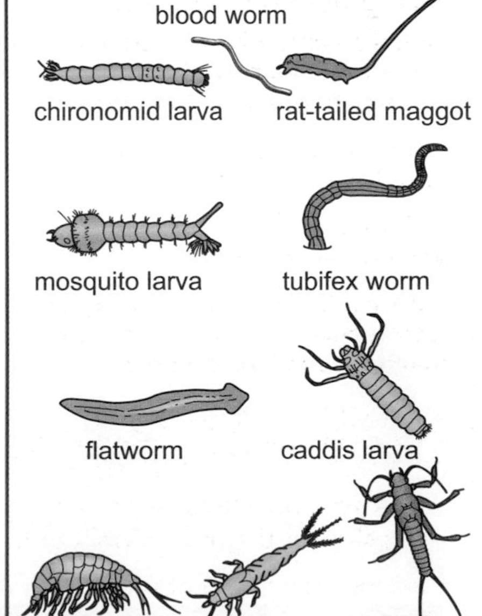

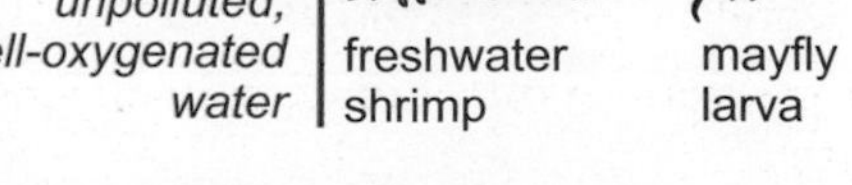

FIGURE 5: Invertebrates that can be used as pollution indicators.

Food chains and energy flow

You will find out:

- > energy is transferred from one organism to another along food chains
- > plants do not use all the energy in the sunlight that falls onto them

Death in the sunshine

This lion killed and is now eating an antelope. All that is left are a few bones, some scraps of meat and its horns. The jackal is waiting its turn. Where do these animals get their energy? It all came from sunlight originally. The grass is like a huge power station, transferring energy from a store that animals cannot use into one that they can.

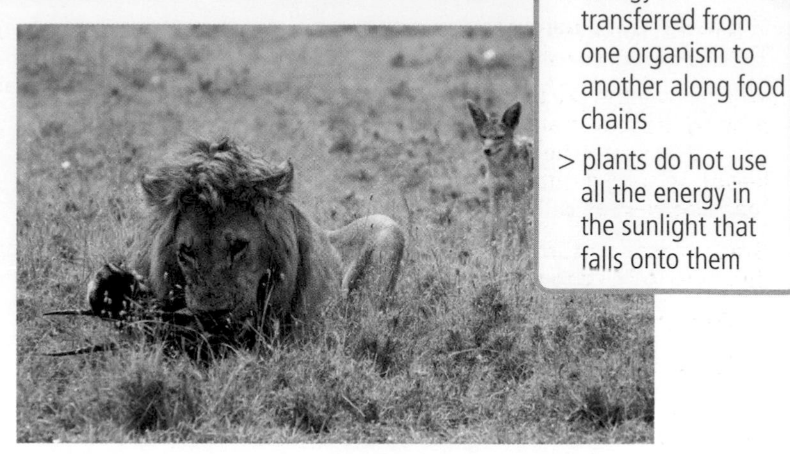

FIGURE 1: All of this activity is originally powered by sunlight.

Energy flow

You can show how energy passes from one organism to another, in a **food chain**, like this:

grass → antelope → lion

The arrows show energy transfer from one organism to the next. The energy comes from chemicals in the food that the animals eat. Carbohydrates, fats and proteins all contain energy.

The grass in this food chain is a **producer**. It uses energy from sunlight to produce carbohydrates and other nutrients. All food chains begin with producers. The antelope and the lion are **consumers**. They consume food that has originally been made by plants.

The antelope is a **herbivore**. Herbivores eat plants.

The lion is a **carnivore**. Carnivores eat other animals.

The lion is also a predator. Predators kill and eat other animals for food. The animals that they kill are their prey.

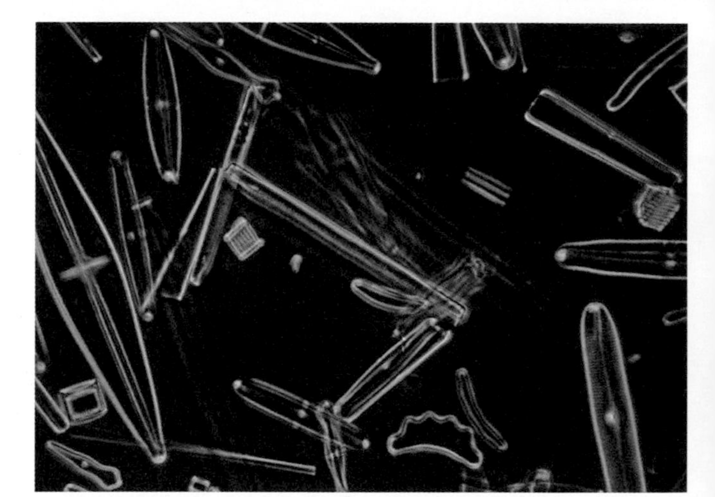

FIGURE 2: These microscopic single-celled algae photosynthesise, just like plants.

Did you know?

One square metre of grass captures about 15 000 kJ of energy from the Sun in one year.

QUESTIONS

1 What is the initial source of energy for all the living organisms in a food chain?

2 Explain what the arrows, in a food chain, indicate.

3 In a freshwater pond, tadpoles eat algae and diving beetles eat tadpoles. What is the producer in this food chain?

4 Draw a food chain to show how the diving beetles get energy.

Energy wastage

Figure 3 shows sunlight filtering through the leaves of a tree. The photographer was standing on the ground looking upwards.

You can see the sky between the leaves, so some of the light is missing the leaves altogether, going straight down to the ground beneath the tree. You can see *through* some of the leaves, so some light must be going right through them and reaching your eyes.

Green plants capture only a small amount of energy from the light that falls onto them. This is because some light:

> misses the leaves altogether

> hits the leaf and reflects back from the leaf surface

> hits the leaf, but goes all the way through without hitting any chlorophyll

> hits the chlorophyll, but is not absorbed because it is of the wrong wavelength (colour).

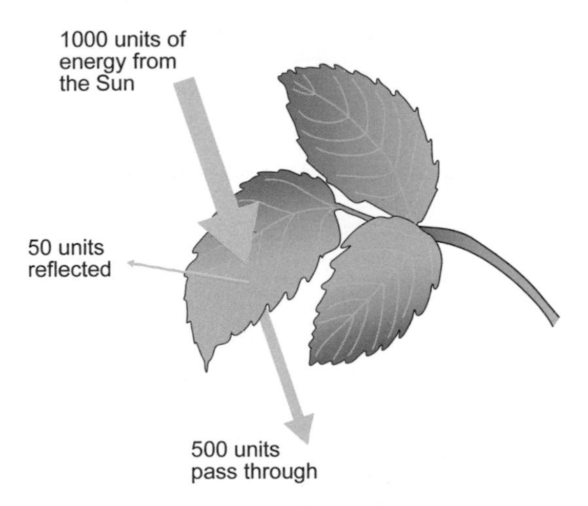

FIGURE 4: This is what happens to energy from the Sun that falls onto a leaf.

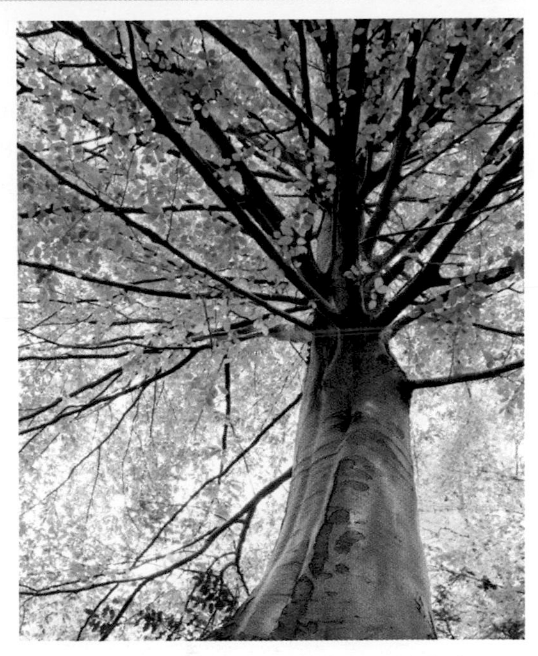

FIGURE 3: Do the leaves of a tree catch all of the sunlight?

QUESTIONS

5 Look at Figure 4. Out of 1000 units of energy, from the sunlight, how much can the plant use for photosynthesis?

6 Chlorophyll is green. Suggest the colours of light it reflects.

7 Suggest what colours of light can be absorbed by chlorophyll.

Efficiency

Whenever energy is transferred, some of it is wasted. To calculate the efficiency of energy transfer:

$$\text{efficiency} = \frac{\text{useful energy transferred}}{\text{original amount of energy}} \times 100\%$$

Imagine that 200 units of energy hit a leaf, and the leaf uses 40 units in photosynthesis:

$$\text{efficiency} = \frac{40}{200} \times 100 = 20\%$$

The plant uses some of the energy that it captures to make carbohydrates and other substances. The energy is stored in these substances, inside the cells of the plant. It is therefore available to animals that eat the plant.

QUESTIONS

8 Calculate the efficiency of energy transfer in the leaf in Figure 4.

9 Plants use carbon dioxide in photosynthesis. Imagine that a plant is growing in conditions where the carbon dioxide concentration is very low. Explain how this would affect the efficiency of energy transfer in photosynthesis.

Biomass

You will find out:

> the mass of living material is called biomass
> each successive level in a food chain receives less energy and therefore has less biomass
> how energy is lost to the environment as it passes along a food chain

Gentle giant

This is a whale shark, the largest living species of fish. Whale sharks can be up to 20 metres long. The largest one that has been caught had a mass of 40 tonnes.

Despite their size, whale sharks feed on plankton, which are tiny organisms that float in the water. The shark filters them out of the water passing over its gills.

Did you know?

6000 dm^3 of seawater are filtered through a whale shark's gills each hour.

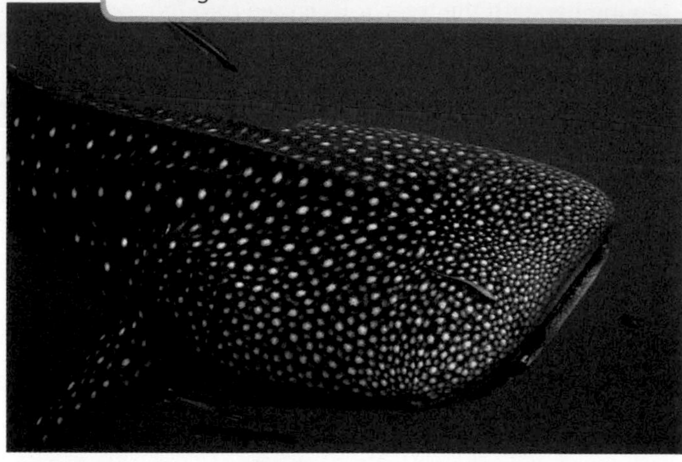

FIGURE 1: Why does a whale shark have such an enormous mouth?

Pyramids of biomass

Think back to this food chain:

grass → antelope → lion

> How many antelopes are needed to support one lion?
  – Several hundred in the lion's lifetime.

> How much grass is needed to support one antelope?
  – Much more than the antelope's own body mass.

If you could measure the mass of all the antelopes needed to supply one lion, and the mass of all the grass needed to support these antelopes, you might find something like Figure 2.

The mass of living material is called **biomass**. Figure 2 shows a **pyramid of biomass**. The size of each block represents the biomass at each step in a food chain. At each step in a food chain, there is less biomass than in the step before.

Why is this? Firstly, the antelopes do not eat all the grass. For example, the roots of the grass are under the ground.

Secondly, not all the antelopes are killed and eaten by lions. Even if the lion does kill an antelope, it does not eat absolutely all of it. Not all the antelope biomass is passed on to lions.

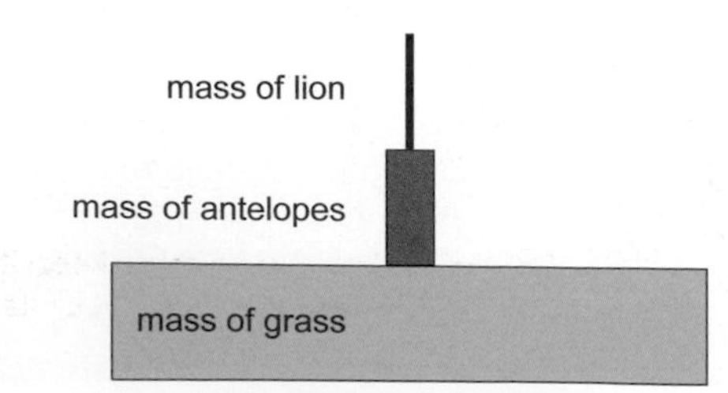

FIGURE 2: A pyramid of biomass.

Remember

A food chain is a flow diagram showing how energy is passed from one organism to another.

QUESTIONS

1 Antelopes do not eat plant roots. Suggest one other reason why not all the grass biomass is passed on to antelopes.

2 Which parts of an antelope are not eaten by lions?

3 Draw a pyramid of biomass that has a whale shark at the top.

Energy losses

Whenever energy is transferred, some is wasted (not used for useful work). This happens in food chains. At each step of the chain, energy is wasted.

This is why biomass gets less along a food chain. At each step, there is less energy available for the organisms to use. Less energy means less biomass.

The food chain loses energy because:

> Some materials and energy are lost in the waste materials produced by each organism, such as carbon dioxide, urine and faeces.

> Respiration in each organism's cells releases energy from nutrients, so that the organism can use it for movement and other purposes. Much of this energy is eventually lost, heating the surroundings.

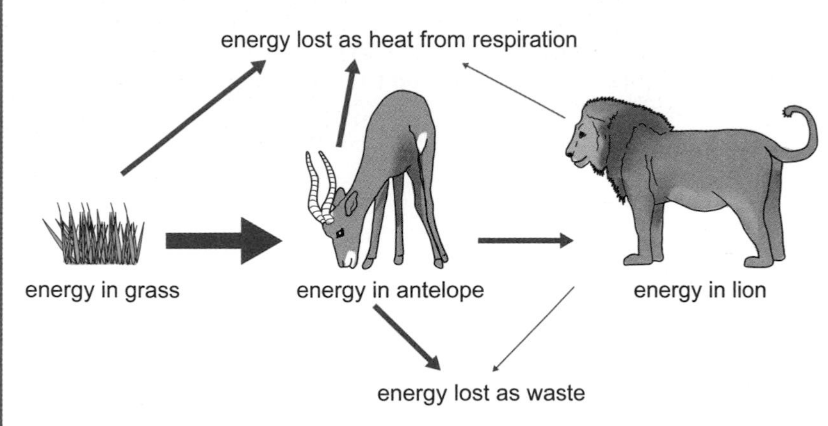

FIGURE 3: How energy is wasted as it passes along a food chain.

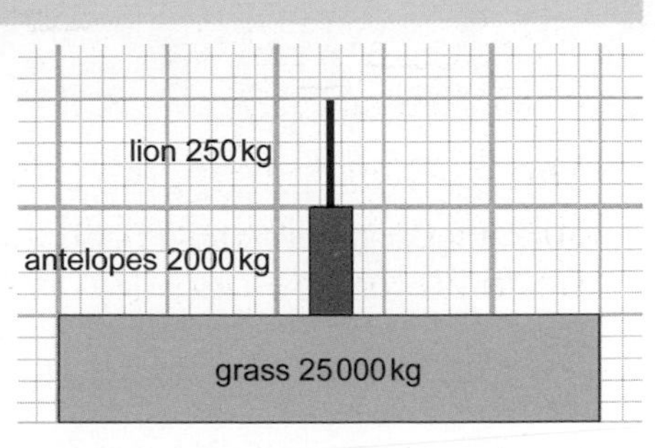

FIGURE 4: A pyramid of biomass drawn to scale.

QUESTIONS

4 Explain how energy is wasted along a food chain.

5 Use your answer to question 4 to explain why biomass gets less at each successive step in a food chain.

6 The biomass of algae in a pond is 1000 g. The biomass of tadpoles is 100 g. The biomass of diving beetles is 10 g. On graph paper, draw a pyramid of biomass for this food chain. Think of the pyramid as a kind of sideways-on bar chart.

More about energy loss

Mammals and birds use glucose to provide energy to keep their body temperatures high, even when the temperature of their surroundings is low.

This means that energy losses from birds and mammals are even greater than energy losses from other organisms in a food chain. Other animals, such as snakes, frogs and fish, just stay the same temperature as their environment.

QUESTIONS

7 Explain why snakes need to eat only about once a month, whereas most mammals and birds need to eat much more regularly.

8 Write down a food chain containing at least one mammal and another containing at least one reptile, fish or amphibian. Using what you know about energy waste, sketch pyramids of biomass for these two food chains to show how their shapes would differ.

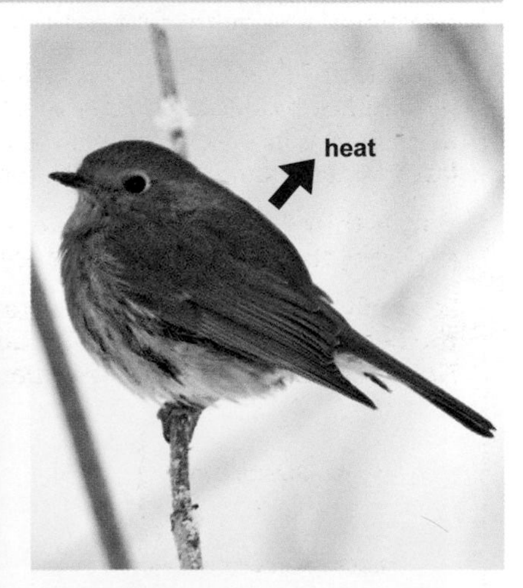

FIGURE 5: How does a robin keep its body temperature above that of its surroundings?

Preparing for assessment: Applying your knowledge

To achieve a good grade in science, you not only have to know and understand scientific ideas, but you need to be able to apply them to other situations and investigations. These tasks will support you in developing these skills.

Mammals on the move

Ashley was watching a wildlife programme on TV. It was about the Serengeti National Park in Africa and featured animals called wildebeest. He was fascinated by the odd appearance of these animals, with their large heads, curved horns, big shoulders and spindly rear legs; then the programme explained that their journey is the biggest mammal migration in the world.

One reason for the migration is the need for fresh grazing areas – short grass is their preferred diet. They cannot go without water for more than a day or so. They have several predators, including hyena and lions. An adult will live for up to 20 years.

The females calve in May, giving birth to a single calf; they don't seek shelter but give birth surrounded by the herd. Most of the females in the herd give birth within two or three weeks of each other. The calf can stand and run within minutes of being born. Within a few days, it can run fast enough to keep up with the herd.

The herd starts to migrate north soon after that and travels at a relentless pace through day and night; many are lost, injured or even killed. In November the return journey south commences.

Ashley was intrigued by the programme and could not understand why the wildebeest lived the way that they did. The young seemed to have a really rough time; they had to be up and on the move in a very short time and sometimes perished before reaching adulthood.

Surely they would stand a better chance of survival if they were not all born at about the same time or, if the herd stayed in the same area for several months, until the calves were older and stronger?

It made a really good programme, but he was so glad that humans didn't raise their young like that.

Task 1

(a) What is the preferred diet of wildebeest and why must they live close to water?

(b) Why do wildebeest have to move on repeatedly?

Task 2

If the wildebeest lived in smaller herds all the year round (instead of just during the breeding season), thought Ashley, perhaps they would be able to settle in one area. Why do you think they do not do that?

Task 3

(a) Why do you think the females give birth in the middle of the grassy plain, surrounded by the herd instead of finding a sheltered place?

(b) Most of the females give birth at around the same time. Why does this give an advantage, in terms of survival?

Task 4

Ashley found the mass migration a stunning sight, with a great number of animals relentlessly pressing on. They did not stop, even if some members of the herd were injured or left behind. Surely they would stand a better chance of survival if they cared more for each other? What do you think?

Task 5

Use some of the ideas you have gathered to explain how the need for survival has influenced the adaptation of the wildebeest.

Maximise your grade

Grade	Answer includes showing that you…
	understand one reason why the wildebeest keep moving on.
	can suggest one reason why large group size aids survival.
E	understand several reasons why the wildebeest keep moving on.
	can suggest several reasons why large group size aids survival.
	understand why being out in the open aids survival.
C	understand why many females giving birth at the same time aids survival.
	can explain, with reference to one feature of the wildebeest's behaviour, how survival has influenced them.
A	can explain, with reference to both examples of wildebeest behaviour and features, how survival has influenced them.

Decay

You will find out:
> materials decay when they are broken down by microorganisms
> decay happens faster when temperature, moisture content and oxygen levels are at optimum levels

Bog body

This body is about 2000 years old. It was found in a peat bog in Cheshire. It is a man who died during Roman times. He had been hit on the head with great force, before having his throat slit. His stomach contained the remains of his last meal – some bread. The body is so well preserved that his clothing, skin, hair and internal organs are all still there.

FIGURE 1: Lindow man died 2000 years ago.

Decay

What happens when something decays? Why didn't the body in the bog decay?

Decay happens when **microorganisms** feed on a dead body, or waste material from animals and plants, or food. The microorganisms include **bacteria** and **fungi**. The microorganisms produce enzymes that digest the food material. The material gradually breaks down and dissolves.

This is a very important process. If things did not decay, we would be knee-deep in dead bodies and waste. Even more importantly, the decay process releases substances into the soil which plants need to grow.

Most microorganisms that help with decay:

> need oxygen

> use oxygen for aerobic respiration

> cannot cause decay without oxygen.

Peat bogs are very wet. There is so much water in them that microorganisms cannot get enough oxygen for respiration. This is why bodies can be preserved in a peat bog. Once the body is buried in the peat, it will decay only very slowly.

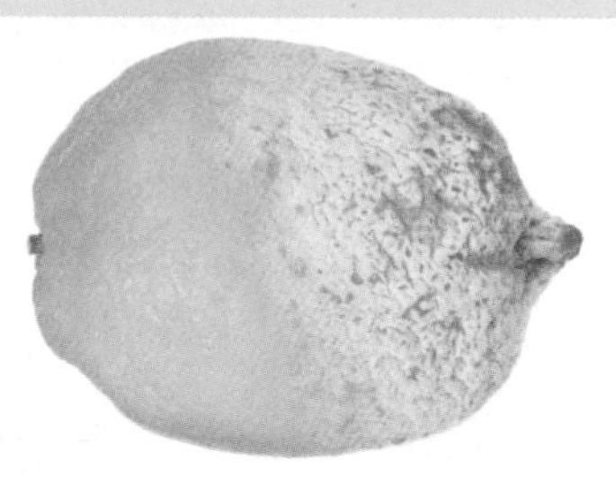

FIGURE 2: The mould is using enzymes to break down this lemon.

QUESTIONS

1 What happens when something decays?

2 Think of three substances that can decay, and three that cannot decay.

3 Explain why the body in the bog did not decay.

Speeding or slowing decay

Decay is caused by microorganisms. Anything that affects the microorganisms can affect the rate of decay.

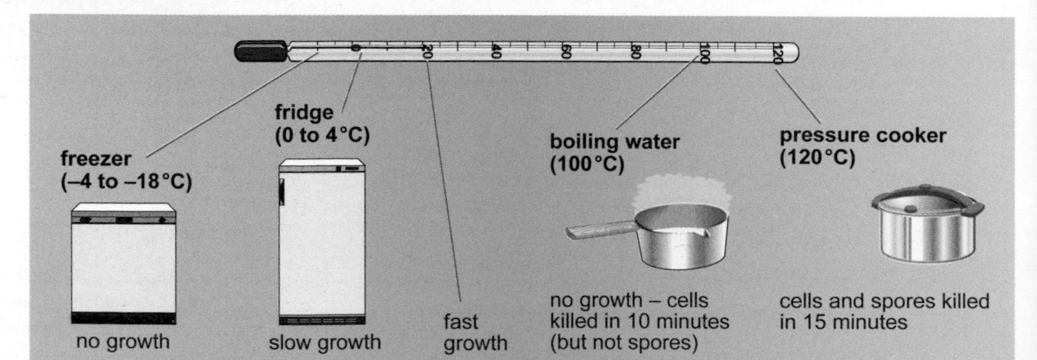

FIGURE 3: How temperature affects the activity of microorganisms.

Q ... biological decay GCSE

Temperature

Most microorganisms function most rapidly at warm temperatures.

This happens because **enzymes** catalyse their metabolic reactions, just like ours. Decay will happen fastest at the **optimum temperature** for their enzymes. Different microorganisms have different optimum temperatures. For most, the optimum is between 25 °C and 45 °C. Heating up the microorganisms a lot kills them. However, some microorganisms produce spores. Spores can survive at very high temperatures.

Moisture

Microorganisms need moisture in order to reproduce and feed. Some types can survive when it is very dry, but they will not be able to reproduce until some water is available.

Oxygen

Many microorganisms need oxygen for respiration. They are more active when oxygen is available.

FIGURE 4: This fruit has been dried. Why can it be kept for a long time, without decaying?

FIGURE 5: This salmon is packed in a vacuum. Why does it take longer to go bad?

Did you know?

The first part of a dead human body to decay is usually the intestines, because of all the enzymes and bacteria they contain.

QUESTIONS

4 A fridge is usually kept at about 4 °C and a freezer at –18 °C. Explain why food can be kept for longer in a freezer than in a fridge.

5 Explain what would happen to some frozen food if it is thawed and left in a warm kitchen for a while.

6 Explain why frozen food should not be put back into the freezer, once it has thawed. Use what you know about microorganisms, temperature and decay.

Preventing decay

If food is not to decay, it can be treated to slow down or stop the activity of microorganisms. Figure 6 shows some examples.

FIGURE 6: Preserved foods.

QUESTIONS

7 Jam is made of fruit that has been cooked with a lot of sugar. This draws water out of the fruit by a process called osmosis. Explain how the high concentration of sugar could stop microorganisms growing in the jam.

8 Onions can be cooked and then put into vinegar. Enzymes are damaged in a very low pH. Explain how the vinegar stops microorganisms growing.

9 Canned fish is heated to a very high temperature and then sealed. Explain how this preserves the fish.

Recycling

Cowpats

Cow dung contains undigested material from the food that the cows ate. We are not knee-deep in cowpats because microorganisms and other small organisms, such as dung beetles, break down and digest the material in the cowpats.

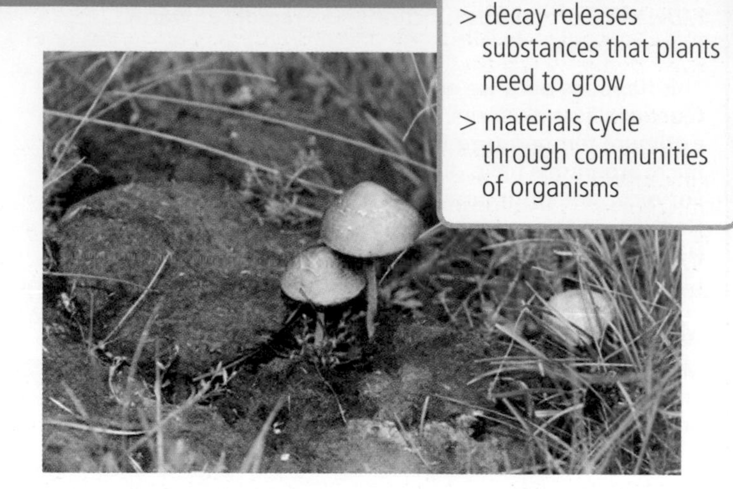

FIGURE 1: The fungi have underground cells that are producing enzymes to break down the cowpat. How will this help the grass?

You will find out:

> decay releases substances that plants need to grow

> materials cycle through communities of organisms

Decay and recycling

What if there were no microorganisms or other organisms to cause decay? Every dead body or piece of body waste (such as a cowpat) would stay just as it was. None of the substances locked up in the body or the waste would be released. Decay releases substances, such as **nitrates**, that plants need to grow. With no decay, plants would not get so many nutrients. They would not grow well.

Recycling

Living organisms take materials from the environment. When they die or produce waste, the material returns to the environment. Some animals, such as earthworms, eat dead bodies and waste material. Microorganisms also help to make the material decay, releasing the substance back into the environment again.

All the organisms that live in one place – a **community of organisms** – are constantly reusing materials that have been part of other organisms. If the community is stable, then there is a balance between the processes that remove and return materials, to and from the environment.

QUESTIONS

1 How can decaying waste help plant growth?

2 How might a substance in a grass plant become part of a different grass plant, going through a cow on the way?

Did you know?

One cow produces about 55 kg of dung every day.

Recycling and food chains

When looking at food chains and food webs, you may not have included microorganisms. This food web shows how they fit into a simple food chain.

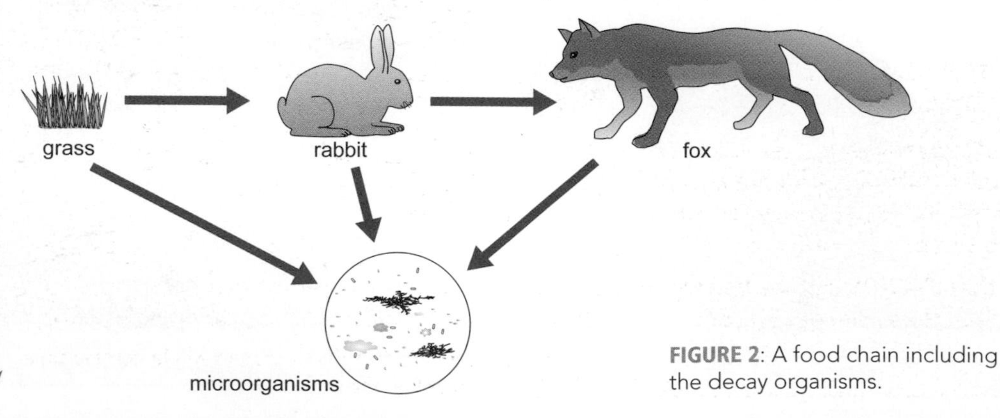

FIGURE 2: A food chain including the decay organisms.

... recycling in nature GCSE

You can see that these decay microorganisms feed on every organism in the chain. They will break down most of the waste material that the plants and animals produce, and then finally their bodies when they die.

Compost heaps

Gardeners often make **compost** heaps of fallen leaves, weeds, unwanted plants – as well as waste food. Inside the heap, microorganisms digest the materials in the plants. After a while, it is broken down into a crumbly brown substance. This compost can be put onto the garden, to provide materials that help the plants to grow.

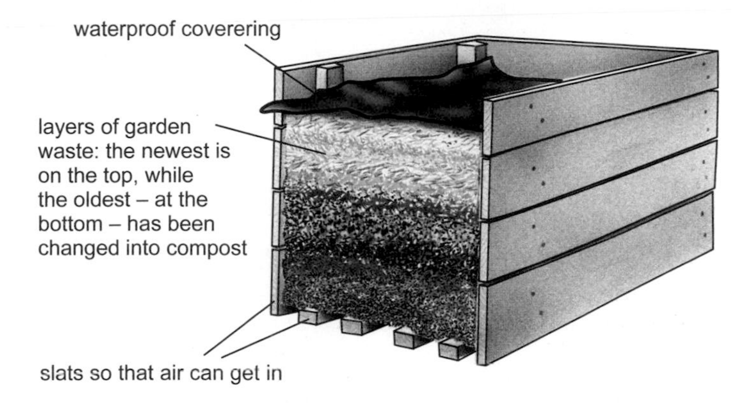

FIGURE 3: Suggest how the gaps in the sides help the microorganisms to make the compost.

QUESTIONS

3 Suggest why food webs do not usually show decay microorganisms.

4 Think back to what you know about pyramids of biomass. Explain how the biomass of the decay microorganisms might compare with the biomass of the animals in the food chain.

5 Explain why it is good for a compost heap to be warm and moist inside.

Dead whales

If a whale dies at sea, its body will probably sink to the bottom.

Here, there is no light, so nothing photosynthesises. Yet, whole communities of organisms use the whale carcass as food.

Many different animals, including crabs, worms and fish, eat the whale's body. Microorganisms gradually decay the whale's tissues. The bones are the last to be decayed. The whole process can take decades.

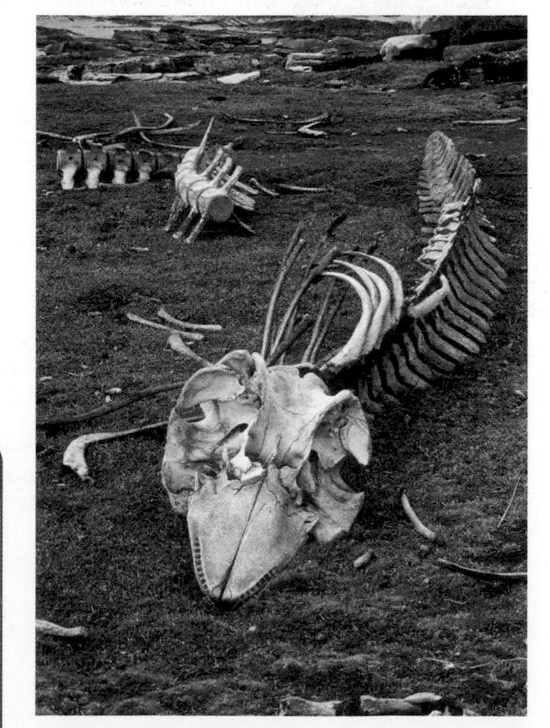

FIGURE 4: A decomposed whale carcass on a beach in the Falkland Islands.

QUESTIONS

6 For the organisms involved with a dead whale food chain: (a) draw the food chain (b) sketch a pyramid of biomass. You will need to think about what the producers must have been.

7 Bodies of whales that die on beaches usually decay much more quickly than those at sea. Suggest why this is so.

The carbon cycle

You will find out:
- > plants take in carbon from the air in photosynthesis
- > all organisms return carbon to the air in respiration

Sing to your plants

Some people sing to their plants. They say it makes them grow better. That is not a completely crazy idea. The carbon dioxide in your breath is used by plants for photosynthesis. So, breathing over a plant could maybe make it grow a little bit faster.

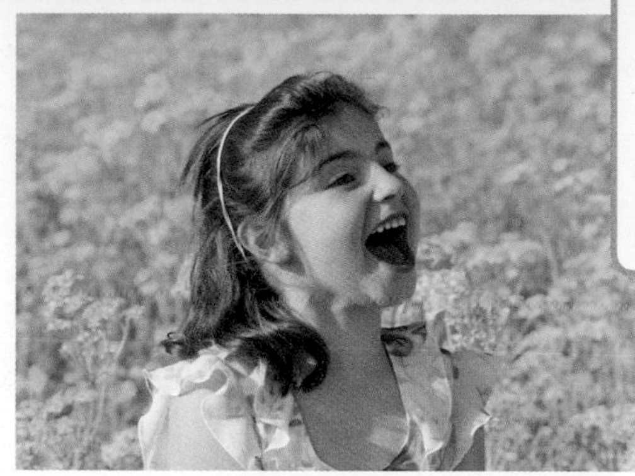

FIGURE 1: How could you test the hypothesis that singing to plants makes them grow faster?

The cycle

Photosynthesis

Many of the different substances that make up your body contain carbon. They include **carbohydrates, fats** and **proteins**. The carbon in your body was originally part of a carbon dioxide molecule in the air.

Somewhere in the world, a plant absorbed **carbon dioxide** from the air. It used the carbon dioxide to make carbohydrates, by **photosynthesis**.

carbon dioxide + water → glucose + oxygen

Glucose is a carbohydrate. The plant also used some glucose to make fats and proteins. These all contain some of the carbon atoms that the plant had taken from the air.

Respiration

You ate food from that plant (or from an animal that ate the plant). That food contained carbohydrates, fats or proteins.

Your food goes into your cells. Here, some of it is broken down by **respiration**. Carbon in the glucose molecules becomes part of a carbon dioxide molecule.

glucose + oxygen → carbon dioxide + water

When you breathe out, you return this carbon dioxide to the air.

Plants respire, all the time, too. They do this to get usable energy from the food that they make by photosynthesis. So, although plants take carbon dioxide out of the air, they put some back again.

This constant cycling of carbon is called the **carbon cycle**.

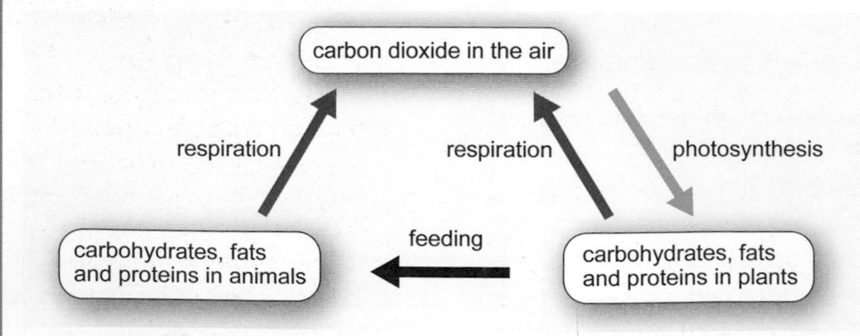

FIGURE 2: Plants and animals in the carbon cycle.

Did you know?

About 18% of your body mass is made of carbon atoms.

QUESTIONS

1 Explain why living organisms need substances made from carbon atoms.

2 Where do these get carbon atoms from: (a) plants (b) your body?

3 Name the process that happens in your cells and in plant cells which returns carbon dioxide to the air.

Q ... carbon cycle

The complete carbon cycle

Figure 3 shows how animals, plants and decomposers all interact with each other in the carbon cycle.

Some of the dead organisms, however, do not decay. They become buried and compressed, deep underground. This changes them into fossil fuels. Combustion of wood or fossil fuels is yet another way that carbon dioxide is returned to the air.

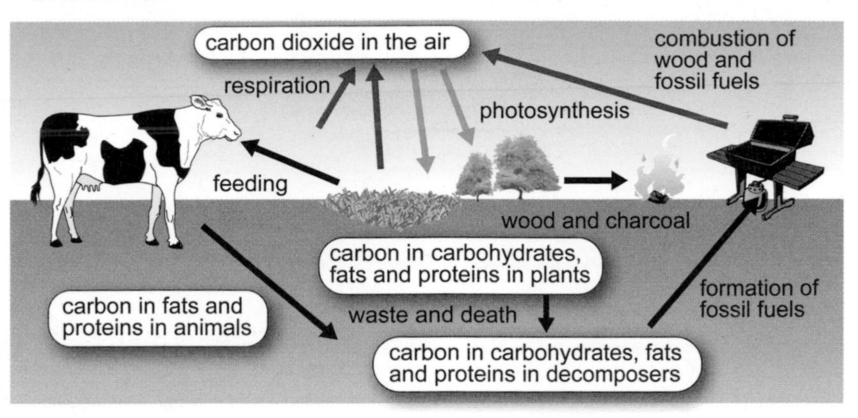

FIGURE 3: Describe how decomposers help in the carbon cycle.

Researchers in Canada have looked carefully at what happens to carbon in a forest. Figure 4 shows what they found.

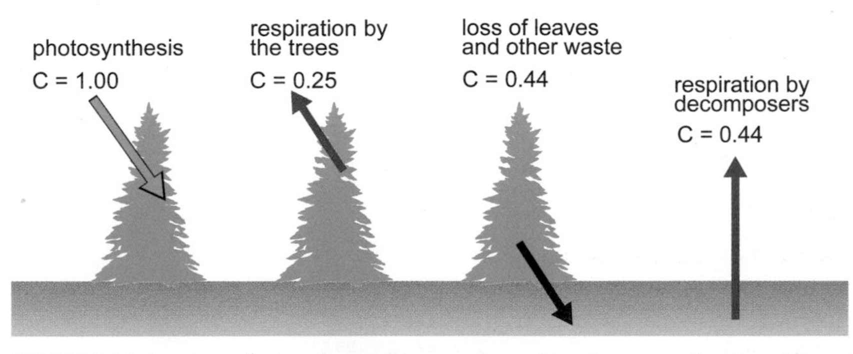

FIGURE 4: Carbon transfer in a forest. The numbers show the mass of carbon taken in or given out each year, in kilograms per square metre.

Remember

Every kind of living thing respires, even plants and microorganisms.

QUESTIONS

4 Using the data in Figure 4:

(a) How much carbon do the trees use in photosynthesis?

(b) How much carbon do the trees give out in respiration?

(c) How much carbon do the trees lose in dead leaves and other waste?

(d) What happens to this carbon that the trees lose?

(e) How much carbon do the trees in the forest store each year? (Use your answers to questions a, b and c to work this out.)

Energy in the carbon cycle

It is not only carbon that moves from one organism to another in the carbon cycle. Energy is being transferred as well.

When a plant photosynthesises, energy from sunlight is transferred to energy stores as chemicals in carbohydrates. Some of this energy is transferred to other organisms – such as animals or decomposers – when they feed on the plant. Some of the energy is wasted, heating the soil and air.

In your cells, some of the energy from the plants and animals that you have eaten is released by respiration, so that your cells can use it. Most of this energy eventually goes to heat the air.

So, unlike the carbon atoms, the energy gets 'lost' as it goes round the cycle. But it does not vanish. It is all eventually transferred, heating up the environment.

QUESTIONS

5 Explain, in your own words, why it is correct to say that carbon *cycles* between living organisms and their environment, but energy flows *through* them. If you prefer, you could do this in a diagram rather than words.

Genes and chromosomes

You will find out:

- > genes are carried on chromosomes inside the nuclei of cells
- > different genes control different characteristics of an organism
- > differences between organisms are caused by their genes and their environment

The ultimate information store

Chromosomes store the information living things need to survive. Humans have 46 chromosomes in each cell. There is a species of fern that has 1260 chromosomes, and a species of ant is able to function on just one chromosome.

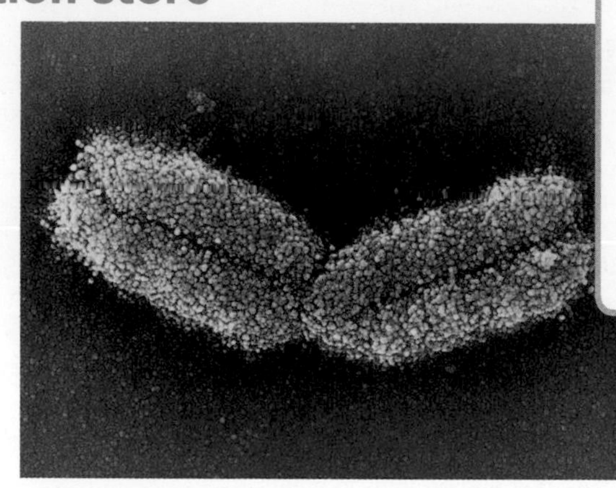

FIGURE 1: Human chromosome, magnified about 8500 times.

Just like your Mum or Dad?

Some children look just like their Mum or Dad. They might have the same colour hair, or the same shape of nose as one of their parents.

Living organisms look like their parents because they inherit information from them. This information is passed on from parents to offspring as **genes**. Genes are linked together in long chains called **chromosomes**.

One set of chromosomes comes from the mother's egg cell. The second set comes from the father's sperm cell. Eggs and sperms are special cells called **gametes**.

An egg and a sperm join together during fertilisation. The new cell that is formed contains genes from the father and the mother. This cell divides repeatedly to form a new person. Each of their cells contains all of these genes.

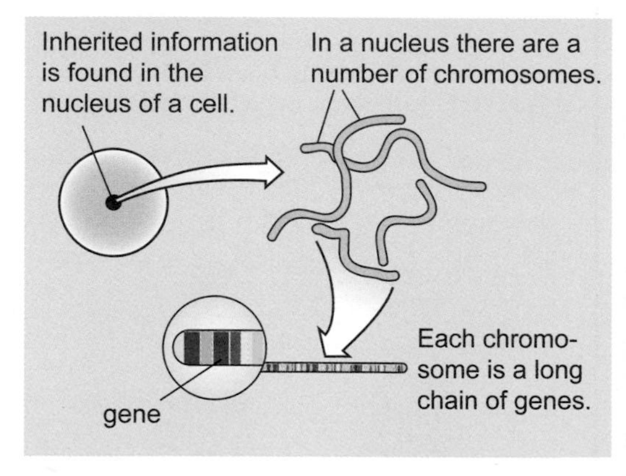

FIGURE 2: Each chromosome is made up of many different genes.

Why are we different?

The genes we inherit control the development of many characteristics. We all have different combinations of genes – it makes us all different from one another.

Some features are not inherited. Tattoos, hairstyles and scars from cuts are examples of features that are not inherited.

QUESTIONS

1 Which part of a cell contains chromosomes and genes?

2 Explain what gametes are.

3 Explain why young plants and animals look like their parents.

4 State two features of your best friend that they did not inherit from their parents.

Did you know?

Chromosomes are made of tightly curled DNA molecules. If you could stretch out the DNA in one of your cells, it would be about three metres long.

Chromosome numbers

The nucleus of a cell normally has two complete sets of chromosomes in it. Human cells contain 46 chromosomes, two sets of 23. Each of these chromosomes is made up of hundreds of different genes.

Gametes contain half of the number of chromosomes that normal cells have – that is, only one set.

Fertilised eggs therefore have two sets of chromosomes – one set from the mother and one set from the father.

Different forms of genes

We have thousands of different genes, each controlling the development of different characteristics. For example, some genes control hair colour. Other genes control eye colour.

Most of these genes come in two or more forms. For example, a gene that controls hair colour might have one form that produces brown hair and a different form that produces red hair.

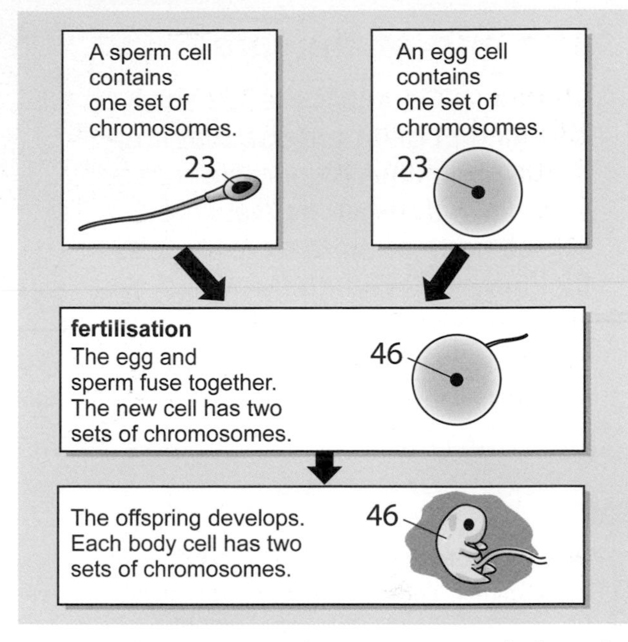

FIGURE 3: Chromosome numbers in gametes and other cells.

FIGURE 4: The cells in these two dogs all contain two sets of dog chromosomes. Explain why the dogs look so very different.

QUESTIONS

5 A body cell of a dog has 78 chromosomes. How many *sets* of chromosomes does it have?

6 Suggest the number of chromosomes in a sperm cell from a dog.

Causes of variation

A student investigated the variation in the surface area of leaves from different parts of a tree.

He found that, on average, the surface area of the leaves on the shady side of the tree was larger than the surface area on the sunny side.

QUESTIONS

7 Suggest what caused the differences in the characteristics of the leaves on the shady and sunny sides of the tree. It may be genes alone, environment alone, or both genes and environment. Explain your answer.

FIGURE 5: Variation in leaf size.

Reproduction

You will find out:

> sexual reproduction involves gametes and produces variation

> asexual reproduction does not involve gametes and produces genetically identical offspring

The end of the line

Lonesome George is the last survivor of a kind of giant tortoise that lived in the Galapagos Islands. He is about 90 years old. When he dies, there will be none of these tortoises left. Biologists have found George a female companion from a closely related subspecies. It is hoped that they may have offspring.

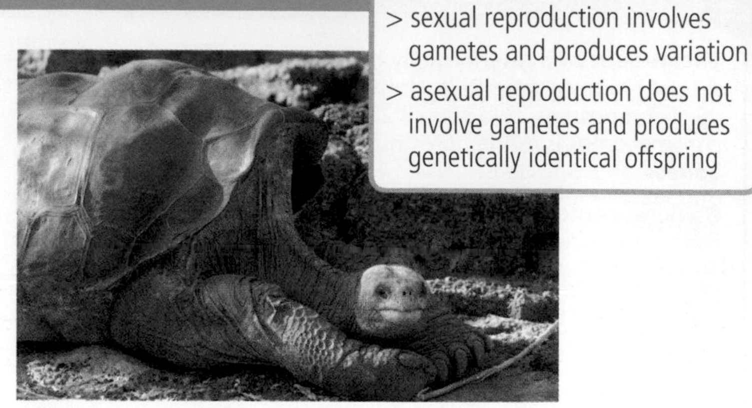

FIGURE 1: George is the last member of his subspecies.

Sexual or asexual

Sexual reproduction

Most animals and flowering plants reproduce sexually. The organisms make sex cells called **gametes**.

> In animals – whether they are beetles, humans or frogs – there are two sorts of gamete. The male gametes are **sperm cells**, and the female gametes are **egg cells**.

> In flowering plants, the male and female gametes are made inside flowers.

When a male and female animal mate:

> A male gamete and a female gamete can join together – this is **fertilisation**.

> Fertilisation produces a cell that is the start of a new life.

> The new organism contains a mixture of genes from its two parents.

Asexual reproduction

There is another way of reproducing, which does not require two different parents.

Some animals, and many plants, just grow young from their bodies or by splitting into many pieces.

> There is no mixing of genes.

> The new organisms have exactly the same genes as their parent.

FIGURE 2: In frogs, fertilisation happens in the water outside the female's body. Where does fertilisation happen in a mammal?

FIGURE 3: Why will the new plants that grow from this spider plant be very like their parent?

QUESTIONS

1 Which of these examples are sexual reproduction, and which are asexual reproduction?

(a) A fish releasing egg cells into the water, and a male releasing sperm cells into the water to fertilise them.

(b) A flower on a tomato plant being fertilised, and growing into a tomato fruit with seeds inside it.

(c) A single strawberry plant producing long stems called runners, which each grow into new strawberry plants.

2 For each of the examples in question 1, suggest whether the offspring will have the same genes as their parent or a different mixture of genes.

Q ... sexual AND asexual reproduction

How it works

How sexual reproduction works

In sexual reproduction, gametes and fertilisation are always involved.

Usually, the two gametes are different. The male gamete can move. The female gamete stays where it is and waits for a male gamete to find it.

The new cell that is produced by fertilisation is a **zygote**. It divides repeatedly to produce a little ball of cells. This develops into an embryo and finally into an adult animal.

The zygote:

> contains genes from both parents

> has a different mix of genes from its parents

> has a different mix of genes from all its brothers and sisters.

Sexual reproduction produces **variety** in the offspring.

Sexual reproduction does not always need two parents. Some plants have flowers that produce both male and female gametes, so they can fertilise themselves.

How asexual reproduction works

Asexual reproduction does not need sex cells. An individual just splits in two (as in bacteria), or a part divides off. This is the offspring. There are no gametes, no fertilisation and no zygotes.

There is no variation. The new organisms all have exactly the same genes as their parent, and as each other. They are **genetically identical**. These genetically identical individuals are **clones**.

Did you know?

A female codfish produces between four and six million eggs each year. Only one or two of these will survive to adulthood.

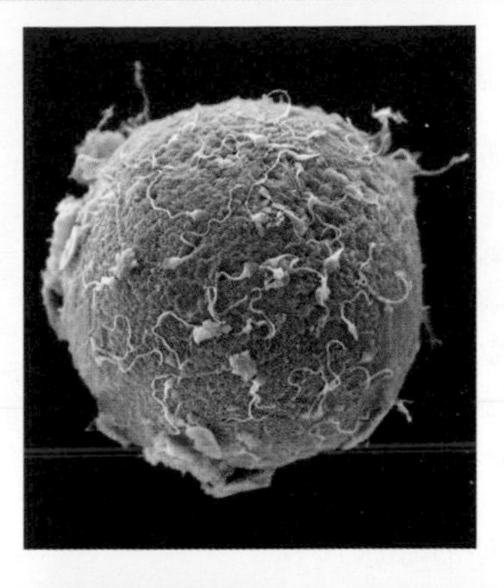

FIGURE 4: These sperm cells are trying to be the first to get into the egg and fertilise it. What happens next?

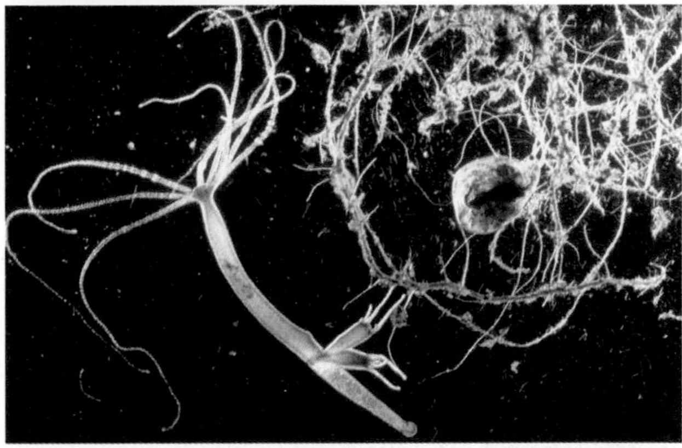

FIGURE 5: This is a tiny animal called *Hydra*, which lives in ponds. How will the genes of the new *Hydra* compare with those of its parent?

QUESTIONS

3 Make a list of differences between sexual and asexual reproduction.

4 Explain what clones are and how they are produced.

Different kinds of fertilisation

In birds and mammals, the male sperm are deposited inside the female's body. The egg is fertilised inside her body. This is called **internal fertilisation**.

In other animals, such as fish, the male and female shed sperms and eggs into water. This is called **external fertilisation**. The fertilised eggs develop outside the female's body.

QUESTIONS

5 Suggest, and explain, which type of fertilisation gives the best chance of survival to the young.

6 Fish produce thousands of eggs at one time, whereas mammals only produce a few. Suggest why.

Cloning plants and animals

Human clones

Is it possible to clone humans? At the moment, no one has cloned a human, and in any case this is banned in most countries. Yet, natural human clones do exist. Identical twins are produced when a single zygote divides completely into two, and then each one develops into an embryo.

FIGURE 1: Identical twins contain exactly the same genes – so why are there always differences between them?

You will find out:

- > clones can be produced using cuttings, tissue culture and splitting embryos
- > adult cell cloning has been done successfully in just a few kinds of animals

Did you know?

The chance of a mother having identical twins is about 1 in 250.

Producing plants

Growing plants from cuttings

Taking cuttings is a way of making new plants from one original plant. The new plants will be genetically identical to the original, parent plant. Figure 2 shows the stages involved in making cuttings.

Tissue culture

Tissue culture is another way of producing lots of genetically identical plants. A **tissue** is a group of similar cells.

It starts with a very small piece of tissue from a root, stem or leaf of the parent plant.

The tissue is then grown on a jelly containing all the nutrients it needs. Everything has to be kept sterile, so this is usually done in a laboratory. Eventually, each tiny group of cells grows into a complete adult plant.

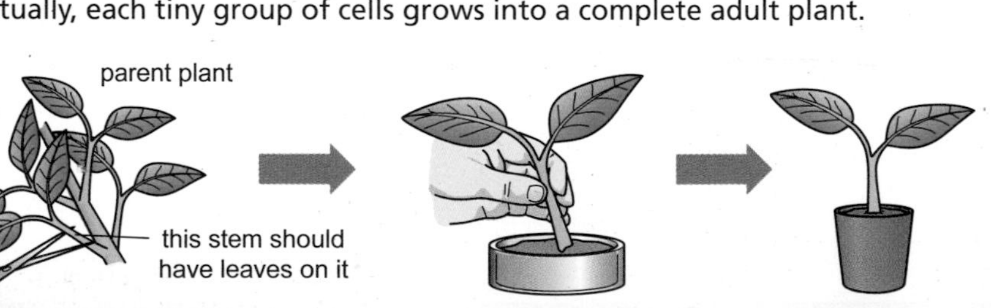

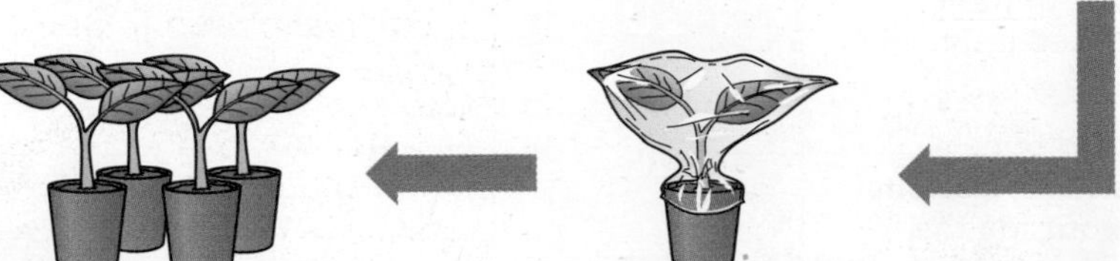

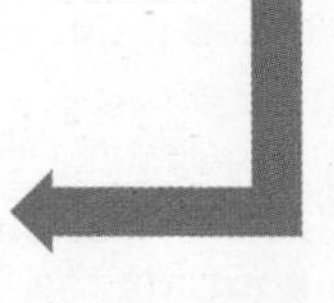

FIGURE 2: How to take cuttings.

FIGURE 3: These tiny sundew plants have been grown by tissue culture. Why are they all genetically identical to one another?

QUESTIONS

1 Explain why taking cuttings and growing plants by tissue culture are both examples of asexual reproduction.

2 Are new plants, produced from cuttings or tissue culture, clones? Explain your answer.

Q ... plant tissue culture

Embryo transplant

It is fairly easy to produce clones of plants. It is much more difficult to clone animals.

One technique is called **embryo transplants**. This is sometimes done with farm animals, such as cows.

1 Take egg cells from a cow, put them into a Petri dish and add sperm cells to them, to fertilise them.
2 Let the zygotes divide a few times to produce tiny embryos.
3 Choose one embryo and split it into two (or more) before its cells become specialised.
4 Let both embryos grow, and transplant each one into a host mother.

Each embryo that has come from the same zygote has the same genes.

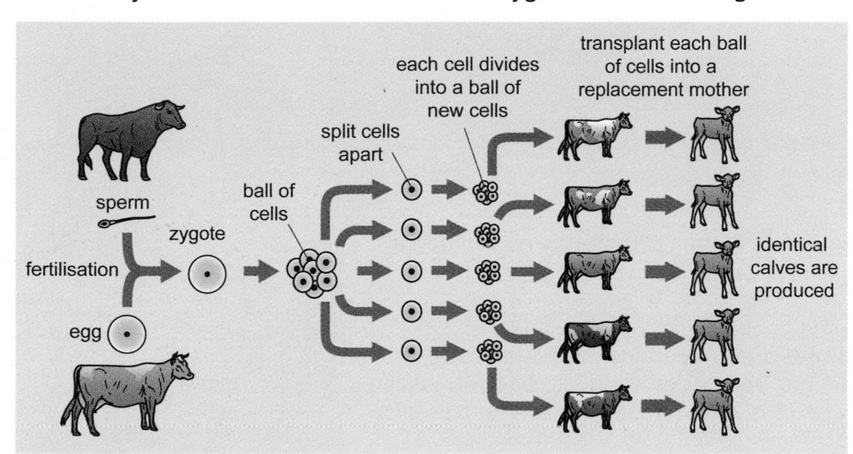

FIGURE 4: How to transplant embryos.

QUESTIONS

3 Explain why all the embryos that develop from one zygote are genetically identical.

4 A farmer has a herd of cows, in which individual animals produce different quantities of milk. He wants to have a herd in which all the cows are good milk producers. Suggest how the embryo transplant technique could help him to achieve this goal.

Adult cell cloning

Embryo transplants start off with a fertilised egg – a zygote – and produce clones from that. The zygote has a mixture of genes from the egg cell and sperm cell that produced it.

What if you want to produce a lot of cells that are genetically identical to just one parent?

To do that, a cell has to behave as though it is a fertilised egg. Figure 5 shows how it is done.

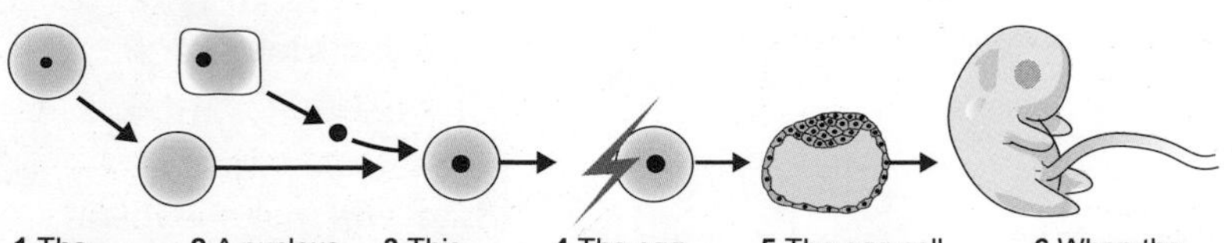

FIGURE 5: Adult cell cloning.

QUESTIONS

5 Suggest why, in adult cell cloning, the egg cell has to be given an electric shock before it will start to divide.

6 Explain why the embryo produced by adult cell cloning does not have the characteristics of the animal from which the egg cell was obtained.

Genetic engineering

Engineering the tomato

Tomatoes like this are science fiction, but genetic engineering is not. Tomatoes which have been genetically modified to be resistant to moulds are now commonly available. They have a longer shelf life which means less wastage.

FIGURE 1: Could genetic engineering make square tomatoes? If it could, would we want them?

You will find out:

> genetic engineering involves transferring genes from one organism to another

> genetically modified crops often produce higher yields

> there are concerns about the effects of genetically modified crops on the environment

How genetic engineering is done

Genetic engineering means taking a gene from one organism, and putting it into another.

Here is one example.

Some people do not make enough of a protein called **insulin**. Insulin stops blood glucose level rising too high. If your body does not make enough insulin, you have diabetes. A person with diabetes may need to inject themselves with insulin several times each day.

Bacteria have been genetically engineered to make human insulin:

> The gene for making insulin is taken out of human cells and put into the bacteria.

> The bacteria use the instructions on the gene to make insulin.

> The bacteria are grown in huge vats.

> The insulin is extracted, purified and sold for use by people with diabetes.

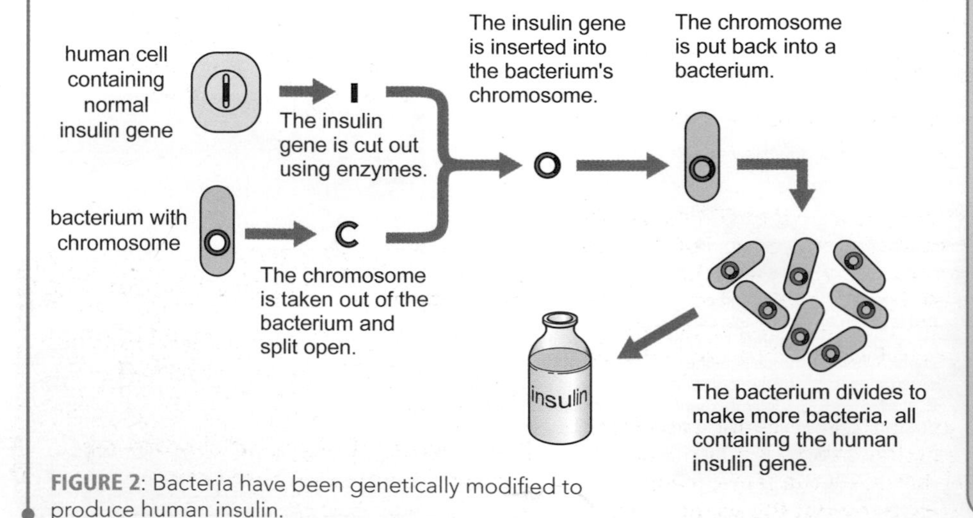

FIGURE 2: Bacteria have been genetically modified to produce human insulin.

Remember

New genes can be transferred to plants. Crops with modified genes are called genetically modified.

QUESTIONS

1 What is genetic engineering?

2 The bacteria that make insulin are said to be 'genetically modified', or GM, bacteria. What do you think this means?

3 Explain why the genetically modified bacteria make insulin.

4 Insulin used to be extracted from the bodies of dead animals. Suggest why using genetically modified bacteria, to make insulin, is better.

... genetic engineering diabetes

Genetically modified (GM) soya

Soya beans are a very important food. Farmers spray bean fields with **herbicides** to kill weeds that compete with soya plants. The spray contains a chemical called glyphosate.

Some soya bean varieties have had a gene that makes them resistant to glyphosate inserted into their cells. When a farmer sprays the field with glyphosate, the weeds die but the bean plants do not.

This is how glyphosate-resistant soya beans are produced:

1 Find an organism that has genes that make it able to withstand glyphosate.
2 Use enzymes to cut out the gene that makes it resistant.
3 Transfer the gene into the cells of embryo bean plants and let them grow.
4 Spray the adult plants with glyphosate to check if they are resistant.
5 If they survive – great, they are resistant. If they die – start again!

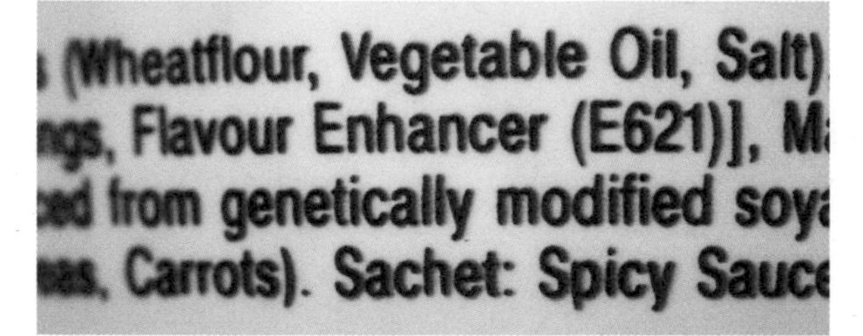

FIGURE 4: Some food manufacturers use GM soya in their products but prefer not to say so on the label. Suggest why.

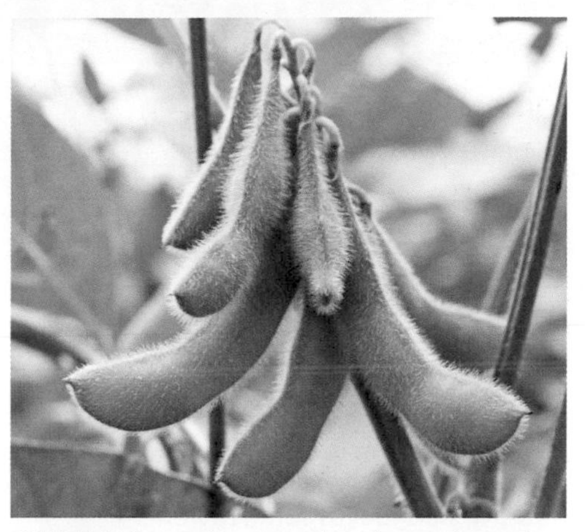

FIGURE 3: Soya plants produce beans in pods.

Did you know?

More than 60% of soya beans grown in the USA have been genetically modified.

QUESTIONS

5 Explain why some GM soya bean plants are not killed by weedkiller.

6 Suggest why the gene is transferred into embryo bean plants, and not fully grown ones.

Genetic modification – good or bad?

There are many examples of GM crops. For instance, crop plants may be genetically modified to make them resistant to attack by insects and other pests. This can greatly increase the yields that the farmers get from their crops. It keeps prices down and reduces the amount of pesticide that has to be sprayed.

GM crops have to be thoroughly tested before they are allowed to be grown on a large scale. All the same, many people are worried about them.

As an example of the concern, imagine a field of GM maize. The maize produces a toxin to kill insects feeding on it. Pollen from this maize could land on the stigma of a wild plant that is closely related to maize. Then wild plants could grow containing the 'extra' gene that had been added to the GM maize. They would be poisonous to insects, which could disrupt natural food chains.

Another concern is the effect, on humans, of eating food from GM plants. Many tests have been done; there is no evidence that eating GM plants does any harm. All the same, many people do not like the idea, no matter what the science says.

Remember

Ideas about GM crops on internet websites are not always written by people who understand the science.

QUESTIONS

7 Make a list of potential advantages of growing GM crops. Try to find at least five.

8 Make a list of potential problems caused by growing GM crops.

Preparing for assessment: Analysing and evaluating data

To achieve a good grade in science, you not only have to know and understand scientific ideas, but you need to be able to apply them to other situations and investigations. These tasks will support you in developing these skills.

Comparing sun and shade leaves

Hypothesis

Jake noticed that the leaves on the sunny side of a tree in his garden looked smaller than the ones on the side that was always shaded by the house. However, when he picked the leaves and compared their weights, the sunny-side leaves felt heavier.

He wanted to test the idea that leaves grew bigger and heavier on the sunny side of the tree.

Plan

Jake decided to pick 10 leaves from the sunny side of the tree. He wanted to make sure all the leaves were about the same age, so he always picked the fourth one back along the twig.

He then picked another 10 leaves from the shady side of the tree, again always choosing the fourth one along a twig.

Jake decided to make two measurements on each leaf – its surface area and its thickness.

Method and results

- First, Jake measured the surface area of the leaf. He put the leaf on paper divided up into squares with sides of 1 cm and drew around it. He counted up all the full squares inside the shape he had drawn. Next, he counted up all the part squares and divided this by two. Then he totalled up the squares.

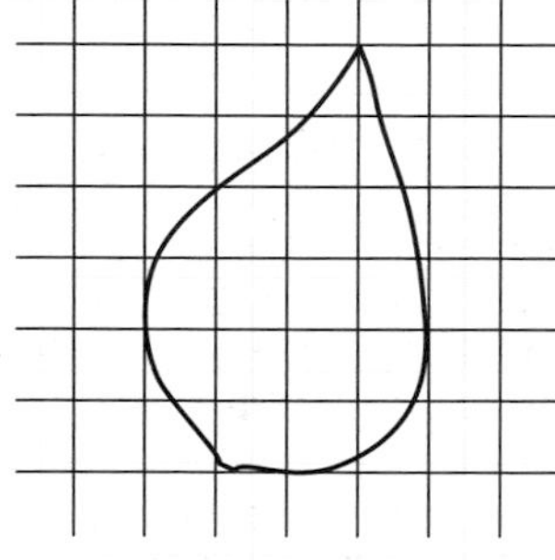

The diagram shows Jake's drawing for the first leaf from the sunny side of the tree.

These are Jake's results for the other 9 sunny-side leaves.

Leaf	1	2	3	4	5	6	7	8	9	10	mean
Area (cm^2)		18.0	13.5	21.5	15.5	12.0	16.0	20.0	19.5	16.0	

These are Jake's results for the 10 shady-side leaves.

Leaf	1	2	3	4	5	6	7	8	9	10	mean
Area (cm^2)	29.0	19.5	28.0	29.5	23.0	32.5	21.5	20.0	22.5	24.0	

- Next, Jake measured the mean thickness of the sunny-side and shady-side leaves. He picked another 10 leaves from each side of the tree, so that he had 20 of each kind. Then he stacked each set up and measured the total thickness of the stack. He divided this by 20 to find the mean thickness of one leaf. These were his results:

  mean thickness of sunny-side leaves = 0.23 mm

  mean thickness of shady-side leaves = 0.18 mm

Collecting and processing data

1. Jake wanted to find out if leaves in the sun differed from leaves in the shade. What was the independent variable in his investigation?

2. Do you think that he controlled other variables effectively? Explain your answer.

3. Use Jake's drawing to find the area of the first sunny-side leaf.

4. Comment on Jake's technique for finding the area of the leaves. How could he have improved the resolution of his measuring technique?

5. Identify the range of Jake's measurements of area for the sunny-side leaves. Then do the same for the shady-side leaves.

6. Comment on Jake's technique for finding the mean thickness of the leaves. How could he have improved this?

7. Calculate the mean area of the sunny-side leaves. Then do the same for the shady-side leaves.

The independent variable is the one that changes.

Look back at the first bullet point to check you are using the correct way of counting the squares.

Resolution means the smallest change that the technique can measure.

Range is the lowest value up to the highest one.

Think about why measuring the thickness of a lot of leaves was easier than trying to measure just one leaf. What should Jake have done to make his measurements on the two sets of leaves comparable?

Remember to give your answer to the same number of decimal places as the original measurements.

Analysing and interpreting data

8. Jake's teacher pointed out to him that, when he was measuring the thickness of the leaves, he had placed the end of his ruler on the bench surface and then read off where the top of the leaf stack was. But the 0 marker was actually about 5 mm up from the end of the ruler. Would that have made Jake's measurements of the leaf thickness too big or too small?

9. Was Jake's measuring error a random error or a systematic error? Explain your answer.

10. Use Jake's results to explain his original observation that the sunny-side leaves felt heavier, even though they looked smaller than the shady-side leaves.

Connections

How Science Works

- Collect primary and secondary data
- Select and process primary and secondary data
- Analyse and interpret primary and secondary data

Science Ideas

B1.4 Interdependence and adaptation (pages 60–61)

Evolution

You will find out:

- > all life on Earth has evolved from simple life-forms that existed more than three billion years ago
- > Darwin and Lamarck both came up with theories about how this might have happened
- > scientific evidence now supports Darwin's idea

How old is life?

No one knows exactly how, when or where life began. The oldest fossils that have so far been found are traces of microscopic bacteria, found in rocks in Sweden that formed almost 3.5 billion years ago.

Did you know?

Bacteria have been on Earth for 3.5 billion years, but humans have only existed for about 400 000 years.

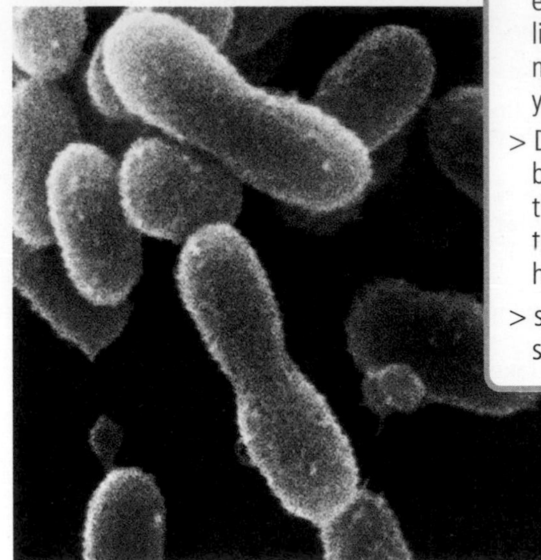

FIGURE 1: The first living organisms on Earth were probably very small bacteria, like these.

What is evolution?

Evolution is how living things change over time. Evolution is going on all of the time.

Life on Earth almost certainly began as very small, single-celled organisms. Since then, over millions and millions of years, new life-forms have gradually evolved from these original ones.

Old and new ideas

People belonging to many different religious faiths believed that living things were created exactly as they are today. Some people still think this, but most now understand that living things have gradually changed over time.

Lamarck's idea

Jean-Baptiste Lamarck (1744–1829) was a French scientist. He suggested that changes in organisms caused by their environment were passed on to their offspring.

Lamarck said that, for example, as a giraffe stretches up to eat leaves from a high tree, its neck gets longer. This characteristic would be passed on to the next generation. So, over the generations, giraffes have grown long necks.

This is not correct. Parents do not pass on changes that have occurred in their lifetimes to their offspring.

They passed on their long necks to their offspring.

FIGURE 2: Lamarck's theory about how evolution happens.

Darwin's idea

Charles Darwin (1809–82) wrote a famous book called *On the Origin of Species*. He suggested that species gradually changed from one form to another by **natural selection**. This is the theory that almost all scientists now believe; there is a lot of scientific evidence to support it.

Darwin thought that, in each generation, only the best adapted individuals:

> survive and reproduce

> pass on their characteristics to the next generation.

QUESTIONS

1 What is evolution?

2 Explain how Lamarck thought that giraffes got their long necks.

3 What did Darwin suggest about the way in which species change?

4 Which theory is now believed to be the correct one – Lamarck's or Darwin's?

Giraffes varied in the length of their necks, because they had different genes.

The giraffes with the longest necks survived.

The giraffes with long necks passed on the genes for long necks to their offspring.

FIGURE 3: Darwin's theory about how evolution happens.

Accepting Darwin's ideas

Darwin's ideas about evolution were based on observation and science. His book challenged the established thinking of the day, so his ideas were not accepted at first.

> Darwin's theory undermined the idea that God made all animals and plants.

> In the late 19th century, there was not much scientific evidence to support the theories of evolution and natural selection.

> At that time, no one even knew that genes existed, let alone the way that they are inherited. This was not discovered until 50 years after Darwin's book was published.

QUESTIONS

5 Make a point-by-point comparison between Lamarck's theory and Darwin's theory.

6 Explain why Darwin's theory was not generally accepted when it was first put forward.

Simple to complex?

The very earliest forms of life on Earth were almost certainly simple, single-celled organisms. Today, many organisms are much more complicated. Does this mean that evolution produces increasingly complex organisms?

There is no doubt that, over the millions of years that life has existed on Earth, more complex species have evolved. Many very simple forms of life are still in existence – and doing very well. Some bacteria living today are almost the same as bacteria that lived billions of years ago. They were, and still are, supremely well adapted to their environment. Bacteria are easily the most numerous and widespread organisms on Earth today.

QUESTIONS

7 Discuss the idea that evolution tends to produce more complex organisms from simple ones. Carry out some research to find examples that support your arguments.

Natural selection

You will find out:

- > mutations produce new forms of genes
- > if the new form of the gene makes an organism better able to survive and reproduce, it may become more common in the next generation

Artificial selection

People breed animals to produce varieties that we want. When people choose a feature, it is called 'artificial selection'. When Nature does it, it's called 'natural selection'.

FIGURE 1: A very rare, miniature dwarf horse.

How natural selection works

This is how natural selection happens:

1 Living organisms produce many offspring.
2 The offspring vary from one another, because they have differences in their genes.
3 Some of them have genes that give them a better chance of survival. They are most likely to reproduce.
4 Their genes will be passed on to their offspring.

FIGURE 2: How natural selection happens in the wild. All the baby rabbits in this family are different. If one has poor hearing, it will not hear a fox creeping up. Why will the gene for poor hearing not be passed on to the next generation?

QUESTIONS

1 Explain why some individuals are more likely to survive than others.

2 Wild gerbils live in dry deserts. Some gerbils are better at managing without drinking water than others. Their genes enable them to do this. Suggest why these genes have a better chance of being passed on to the next generation.

Genes and survival

Occasionally, unpredictable changes to chromosomes and genes happen. These changes are called **mutations**.

Mutations make new forms of genes. These are usually less 'good' than the old ones – they produce features that make it less likely that an organism with a mutation will survive.

Just occasionally, the new form of the gene increases an organism's chances of surviving and reproducing. It is therefore very likely to be passed on to the next generation. Over time, the new feature, produced by this gene, becomes more common in the species.

Did you know?

There are 157 officially recognised breeds of dog, all developed from wolves by artificial selection.

The peppered moth

The peppered moth is a light colour with speckles. One hundred and fifty years ago there was very little pollution in the UK. Peppered moths, resting on lichen-covered tree trunks, were almost invisible to predatory birds.

During the Industrial Revolution, air pollution in cities made buildings and tree trunks black with soot. Lichens could not grow. Birds could easily spot moths resting on these trees.

However, there had always been a few individual peppered moths with a form of the wing-colour gene that made their wings dark (instead of speckled). Very few of these moths had been able to survive: they could be seen easily by birds and be eaten.

These dark moths were well camouflaged. In polluted areas, the dark moths had the best chance of survival. They reproduced and produced more black moths.

In the countryside, the speckled moths were still the best camouflaged. So, this colour remained the most common.

FIGURE 3: Which of these two moths is most likely to pass on its genes to the next generation? Why?

FIGURE 4: Which form of the wing colour gene gives the best chance of survival in a polluted city?

QUESTIONS

3 What is a mutation?

4 Explain why new forms of a gene that make an organism better adapted to its environment are likely to become more common than the 'old' form, as time goes by.

5 Moths with the form of the gene that made their wings dark were extremely rare before the Industrial Revolution. Explain why.

The randomness of mutations

It is important to realise that no organism can make a mutation happen 'on purpose'. For example, some forms of bacteria have become resistant to antibiotics. This happened as a result of mutation in the bacteria producing a form of a gene that helped them survive even when the antibiotic was present in their environment. This was just chance. The bacteria did not purposefully mutate to become resistant.

QUESTIONS

6 A student wrote: "Peppered moths changed their colour so they would be better camouflaged on polluted trees." Explain what is wrong with that sentence.

7 A poison called warfarin is used to kill rats. Some populations of rats have become resistant to warfarin. Use the ideas of random mutation and natural selection to explain how that could have happened.

Evidence for evolution

You will find out:

> evidence shows life began on Earth at least 3.5 billion years ago
> comparing organisms enables us to classify them
> comparing organisms can show their evolutionary relationships
> evolutionary trees show relationships between organisms

Is there life on Mars?

Since 2004, two six-wheeled robotic vehicles, called Spirit and Opportunity, have been exploring the surface of Mars. One of the things that they have been searching for is evidence that water existed. Several pieces of evidence that they have found do suggest that there has been water on Mars. So, we may discover that very simple life-forms once existed there, though it is unlikely that there is any life on Mars today.

FIGURE 1: Mars is much colder than Earth and has a much thinner atmosphere with no oxygen – but it is possible that there was once life on Mars.

Evolving life

How did life start?

Scientists still do not know how, when or where life on Earth began.

Evidence, from fossils, shows that there was life on Earth at least 3.5 billion years ago. At that time, conditions on Earth were very different from today. There was no oxygen in the atmosphere. Volcanoes erupted continuously. Meteors bombarded the surface.

> Some scientists think that life may have been brought to Earth on meteors.

> Others think that life may have evolved deep in the oceans.

> Everyone agrees that the very earliest life-forms were similar to bacteria. They would have had just one simple cell.

Did you know?

Evolution often takes thousands of years. It can also happen very quickly. MRSA evolved from normal *Staphylococcus* bacteria in just a few years.

Comparing living organisms

You can get clues about evolution by looking carefully at organisms that are alive today.

Some organisms look quite different from one another. They are actually very similar when you look more closely. For example, your arm, a bat's wing and a bird's wing all have the same bones in the same places.

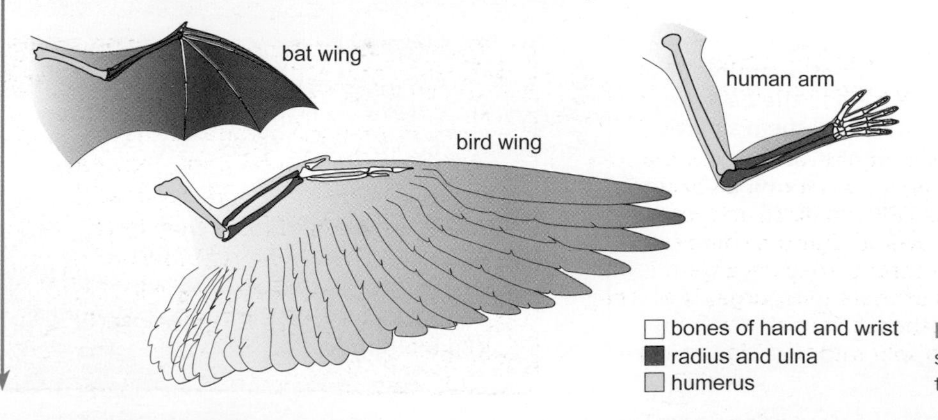

FIGURE 2: What similarities can you see between the structures of these three limb skeletons?

Similarities like this suggest that humans, bats and birds are quite closely related. Long, long ago an animal lived from which humans, bats and birds have all evolved. Humans, bats and birds have a common ancestor.

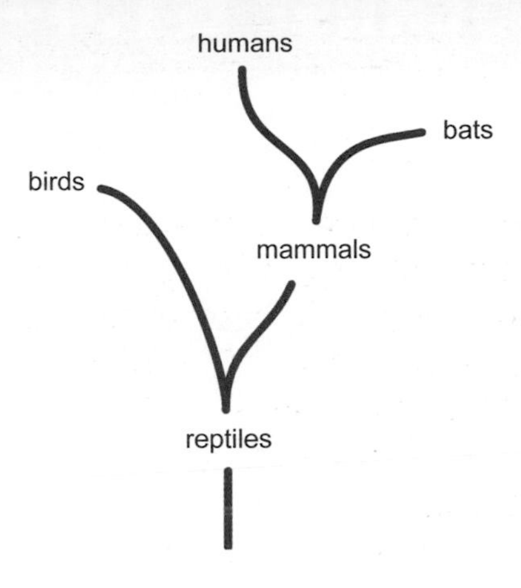

FIGURE 3: An evolutionary tree showing how humans, bats and birds are related.

QUESTIONS

1 When did life on Earth probably first appear?

2 Explain what 'common ancestor' means.

3 Explain how looking carefully at modern-day organisms give clues about evolution.

Classification

Evolutionary trees, like the ones in Figure 3 and Figure 4, show the pathway along which different kinds of organisms may have evolved. Organisms that lived longest ago are at the bottom of the tree. The different groups of organisms that we think evolved from them are the branches on the tree. Models like this help to show how different groups of organisms might be related.

Scientists classify living organisms according to how closely they think they are related. Figure 5 shows how the five main classification groups may have evolved.

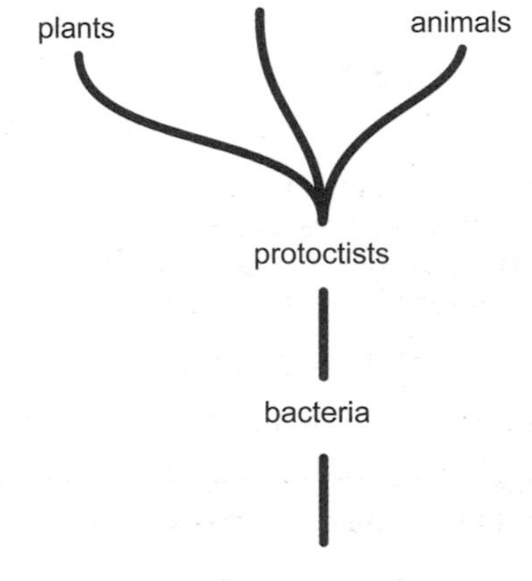

FIGURE 4: Evolutionary trees indicate relationships between organisms.

QUESTIONS

4 Explain how classification and evolution are related to one another.

A new major group?

You would think that, by the 20th century, we would have at least sorted out all living organisms into the major groups.

Science, however, is always making new discoveries that can turn long-held beliefs upside down. Although all bacterial cells look as though they have very similar structures, researchers have recently discovered that the chemistry that goes on inside them is very different. They have found that there are two distinct groups of these microorganisms. They are as different from one another as animals are from bacteria. They have now been split into two big groups – the bacteria and the archaea.

QUESTIONS

5 Current research suggests that the very first organisms on Earth may have given rise to bacteria and archaea. Animals, plants and fungi may have all evolved from archaea. Draw an evolutionary tree, like the one in Figure 4, to illustrate this proposed relationship.

Checklist B1.4–1.8

To achieve your forecast grade in the exam you will need to revise.

Use this checklist to see what you can do now. Refer back to the relevant topic in this book if you are not sure. Look across the three columns to see how you can progress.

Remember that you will need to be able to use these ideas in various ways, such as:

- > interpreting pictures, diagrams and graphs
- > applying ideas to new situations
- > explaining ethical implications
- > suggesting some benefits and risks to society
- > drawing conclusions from evidence that you are given.

Look at pages 276–297 for more information about exams and how you will be assessed.

To aim for a grade E	To aim for a grade C	To aim for a grade A
Recall that organisms compete for resources in order to survive.	Describe the key resources competed for within both the animal and plant kingdoms.	
Recall that all organisms have key features (adaptations) that enable them to survive in their normal habitat.	Describe adaptations of plants and animals to extreme environments such as a desert.	Understand that competition wastes energy and that adaptations to very extreme environments aid survival through greatly reducing competition.
Understand that environmental change can affect the distribution of living organisms.	Understand that environmental changes can be caused by living or non-living factors.	Interpret data about environmental change.
Understand that the Sun is a source of energy and that green plants and algae absorb some of it.	Explain how the Sun's energy is used in photosynthesis.	Explain the energy transfer in photosynthesis and calculate its efficiency in leaves.
Recall that a biomass pyramid shows the decrease in biomass at each stage in a food chain.	Describe how material and energy in biomass reduces at each stage in a food chain.	Explain in detail the material and energy transfer and reduction in a food chain.
Recall that the decay of dead organisms returns useful substances to the environment.	Explain the conditions when decay-causing microorganisms are more active.	Understand that, in a stable community, materials are constantly cycled.
Understand that not all dead organisms decay and some become buried deep underground.	Understand that compressed dead material will be changed into fossil fuel.	

To aim for a grade E	To aim for a grade C	To aim for a grade A
Recall that carbon dioxide is removed from the air by photosynthesis and returned by respiration.	Explain that, through photosynthesis, feeding and respiration, carbon is constantly cycled.	Explain that carbon from carbon dioxide forms various chemicals in plants and animals.
Recall that burning fossil fuels releases carbon dioxide into the air.	Understand that combustion is part of the carbon cycle.	Explain how the formation and combustion of fossil fuels fits into the carbon cycle.
Recall that an individual's characteristics depend on the genes passed on from their parents.	Understand that genes are passed on via sex cells and are carried by chromosomes. Understand that different characteristics depend on different genes and the environment.	Explain how characteristics are influenced by genetic and environmental causes.
Recall that there are two forms of reproduction (sexual and asexual).	Explain the key difference between sexual and asexual reproduction.	Explain how the type of reproduction affects variety in the offspring.
Recall that a clone organism is genetically identical to its 'parent'.	Understand examples of cloning.	Explain the techniques of embryo transplant and adult cell cloning.
Recall that genetic engineering is the process of moving genes from one organism to another.	Describe advantages of transferring genes to crop plants.	Discuss the genetic modification of crop plants.
Understand that evolution describes changes over billions of years and that it is still ongoing. Recall that Charles Darwin developed the theory of natural selection. Understand that there have been many theories for evolution.	Understand how the theory of natural selection works. Explain why Lamarck's theory is not thought to be correct.	Explain the theory of natural selection. Explain why Darwin's theory was only slowly accepted.
Recall that mutations are changes to genes.	Understand that mutations are unpredictable and may affect the chances of an organism surviving.	Explain how a feature from a 'beneficial' mutation can eventually become common in a species.
Recall that living things are classified as animals, plants or microorganisms.	Understand that scientists classify living organisms by looking at their similarities and differences.	Understand the idea of using models such as evolutionary trees to suggest the relationships between organisms.

Exam-style questions: Foundation B1.4–1.8

1. The table shows the number of species of lichens that were found at different distances from a coal-burning power station.

Distance from power station (m)	50	100	500	1000	2000	3000
Number of lichen species	3	6	10	19	24	25

AO3 **(a)** Describe the relationship between the number of lichen species and the distance from the power station. [1]

AO3 **(b)** Suggest an explanation for this relationship. [2]

AO1 **2 (a)** Name the process in which plants transfer light energy to chemical energy. [1]

AO2 **(b)** The diagram shows a pyramid of biomass for this food chain:

grass ⟶ voles ⟶ owls

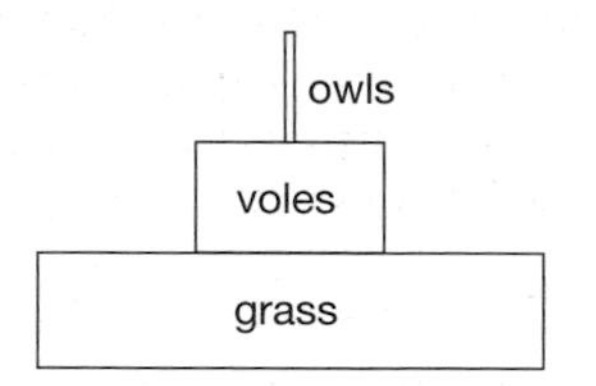

Explain why the pyramid of biomass is this shape. [3]

3. The diagram shows a plant reproducing in two different ways.

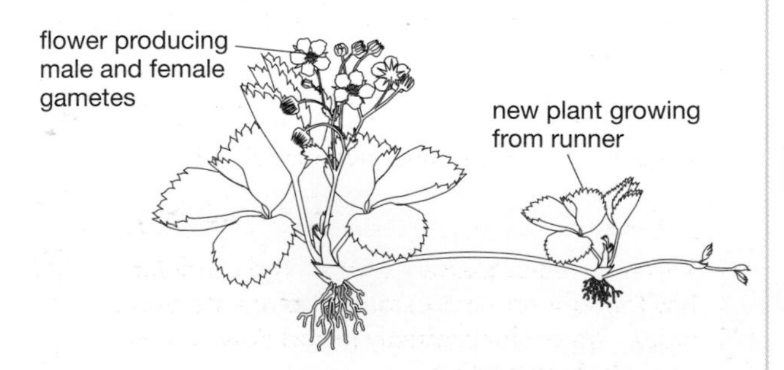

AO2 **(a)** Which method of reproduction, the flower or the runner, will produce genetically identical offspring? Explain your answer. [2]

AO2 **(b)** A gardener planted five new plants, all grown from runners from the same parent plant. She planted them in a different part of the garden to the parent plant. They did not grow as large as the parent plant. Suggest why. [1]

4. A bacterium called *Thermus* lives in hot compost heaps. It feeds on the dead plant material. *Thermus* reproduces faster at temperatures above 50 °C.

The graph shows the temperature of a compost heap over 30 days, and the population of *Thermus* bacteria in the heap.

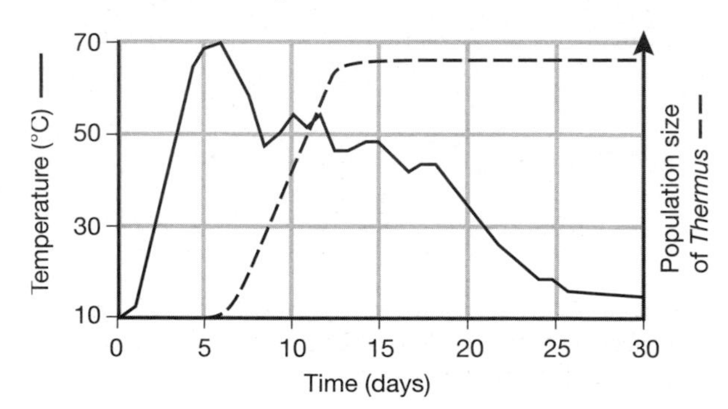

AO3 **(a)** What was the highest temperature reached in the compost heap? [1]

AO2 **(b)** Suggest why the temperature rose. [2]

AO2 **(c)** Suggest why the population of *Thermus* did not start to rise until day 6. [1]

AO2 **(d)** The gardener spreads the compost on the soil in her garden. Explain why the compost will help her plants to grow well. [2]

5. *In this question you will be assessed on using good English, organising information clearly and using specialist terms where appropriate.*

AO2 A population of green lizards lived where they were well camouflaged from predators against green leaves. A few lizards floated on a log to an island where there were no other lizards. The ground was covered by brown rocks. After 30 years, most of the lizards on the island were brown.

Use Darwin's ideas about natural selection to suggest how and why the lizards on the island evolved to become brown. [6]

WORKED EXAMPLE – Foundation tier

In this question you will be assessed on using good English, organising information clearly and using specialist terms where appropriate.

The diagram shows a plant that is adapted for living in a desert.

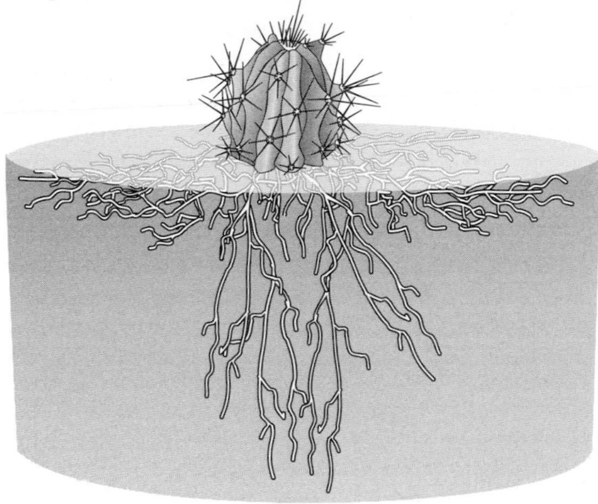

Explain how this plant is adapted to help it to survive in a desert. You should refer to at least three different features of the plant. [6]

It's difficult to live in a desert, because there isn't very much water. So plants that live in deserts have to be adapted to get as much water as they can, and to make sure they don't lose too much water.

It's got a round fat body, so it doesn't have too much surface that water might be lost from.
It doesn't seem to have any leaves. If it had big leaves then it would lose a lot of water from them and might dry out.

It has big roots, so they can get lots of water from the soil. The ones that go deep down can get water even if it hasn't rained for a long time. The ones that go sideways can get a lot of water when it rains, before the rainwater runs off or evaporates away.

It has spines that might stop animals eating it. Animals in a desert are short of food and water, so they will eat plants that aren't protected, and they can get water from the plants they eat.

How to raise your grade!
Take note of these comments – they will help you to raise your grade.

That is a nice start to the answer. It sums up what the problems are for the plant, before it starts to describe how the plant is adapted to solve them.

This is the first feature described in the answer. Notice that the question asked the candidates to 'Explain', not just 'Describe' the features. The candidate first described the feature ('a round fat body') and then explained how this helps it to survive in the desert (it reduces the surface from which water can be lost).

This is the next feature – and once again, the candidate explains how the feature helps the plant.

Here is the third feature and it is well explained. It would be better to say 'long roots', rather than 'big roots', as big could also mean wide.

Here is a fourth feature, well explained.

This is a well-organised answer. The candidate has thought about it carefully before starting to write – he or she probably jotted down a few notes before starting.

The candidate has also read the question carefully, and has made sure that they did 'explain', as well as 'describe'.

The spelling and grammar are good. This answer would get all six marks.

Exam-style questions: Higher B1.4–1.8

AO1 **1 (a)** Explain why all organisms need carbon. [2]

(b) How do each of the following obtain their carbon?

(i) algae **(ii)** animals. [2]

(c) Outline how the carbon in an animal's body is returned to the atmosphere after the animal dies. [3]

2. Researchers carried out an investigation to test the hypothesis that worker ants species that live in warmer places have smaller bodies than those that live in colder places.

They measured the body lengths of hundreds of worker ants from each of six species. They calculated the mean body length of the ants in each species. They recorded the latitude (distance from the equator) at which the ants from each species were collected. Their results are shown in the table.

Species	A	B	C	D	E	F
mean length (mm)	4.2	4.4	4.6	4.9	5.1	5.4
latitude (°N)	50	54	59	63	67	70

AO3 **(a)** Explain why the researchers measured the lengths of hundreds of ants, rather than a few ants, from each species. [2]

AO3 **(b)** Do the results support or disprove the researchers' hypothesis? Explain your answer. [2]

AO2 **(c)** Suggest reasons for the relationship between body length and latitude that is shown by these results. [3]

3. In a marsh, algae grow on the surface of the mud. Small flies eat the algae. Spiders eat the flies.

All the algae, flies and spiders in 2 square metres of the marsh were collected, and their biomass was measured. The results are shown in the table.

Organism	Biomass (g)
algae	200
flies	30
spiders	4

AO3 **(a)** On graph paper, use these results to draw a pyramid of biomass, to scale. Label the pyramid. [3]

AO3 **(b)** Calculate the percentage reduction in biomass between the algae and the spiders. Show clearly how you work out your answer. [2]

AO2 **(c)** Biomass decreases along a food chain because energy is lost at each transfer. Explain why not all the energy in the algae is transferred to spiders. [2]

4. The ground finch, *Geospiza fortis*, lives in the Galapagos islands. It eats seeds, which it cracks with its strong beak. Larger birds are able to eat harder seeds.

Between 1975 and 1977, there was a drought. The graphs show the changes in the seeds during the drought, and the sizes of the ground finches during and just after the drought.

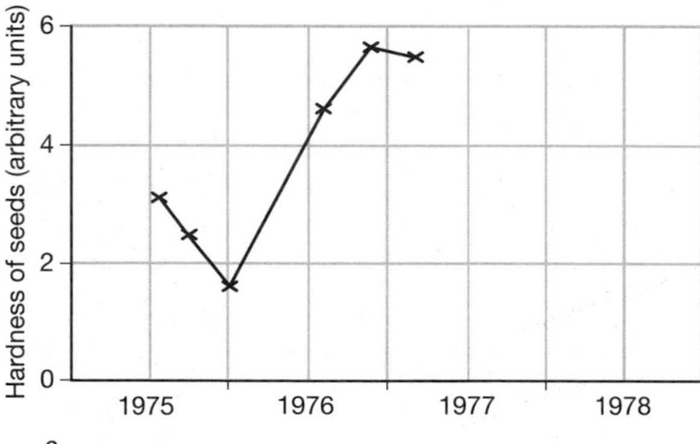

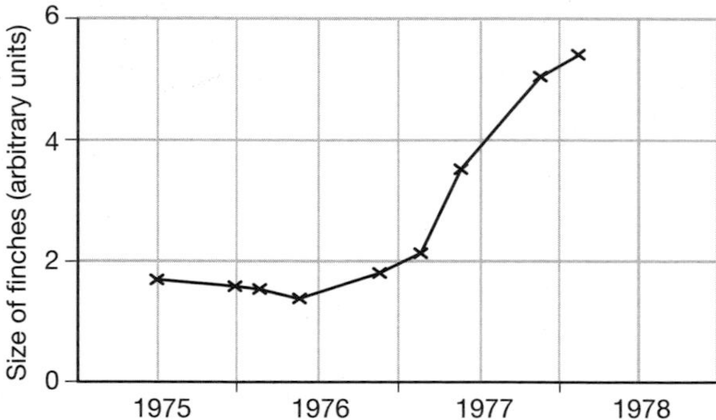

AO3 **(a)** Describe the changes in the hardness of the seeds during the drought. [3]

AO2 **(b)** Use the ideas of natural selection to suggest an explanation for the changes in the size of the finches during the drought. [4]

AO3 **5** *In this question you will be assessed on using good English, organising information clearly and using specialist terms where appropriate.*

Genetic engineering has been used to produce crop plants that are resistant to attack by insects.

Discuss the advantages and disadvantages of growing genetically modified (GM) crops. [6]

AO1 recall the science AO2 apply your knowledge AO3 evaluate and analyse the evidence

WORKED EXAMPLE – Higher tier

In this question you will be assessed on using good English, organising information clearly and using specialist terms where appropriate.

A farmer noticed that fish were dying in a river that ran past his fields.

He decided to investigate the concentration of dissolved oxygen in the river, in case it had got very low and was causing the fish to die.

Describe **two** different methods he could use to do this. Suggest the advantages and disadvantages of these two methods. [6]

He could use an oxygen metre to tell him the concentration of oxygen in the water.

Or he could check what animals were living in the river because there are some animals that can live in water that hasn't got much oxygen in it, so if those animals are there then he knows there isn't much oxygen and if there are other animals like mayfly larvae in it then there is a lot of oxygen.

Using an oxygen metre is best because it just gives you a read-out so you can tell straight away exactly how much oxygen there is.

How to raise your grade!

Take note of these comments – they will help you to raise your grade.

Incorrect spelling – it should be 'meter'.

To get good marks, this should describe *how* he would use an oxygen meter to measure the oxygen concentration in the water.

Again, a good answer would describe *how* he would check the animals – for example, how he would collect a sample from the river and look at the animals in it. It would also be better if the answer said that *invertebrate* animals would be used.

It would be good to name some of the animals that are indicators of water that is low in oxygen.

This sentence is too long. It needs splitting into at least two sentences.

This is an advantage of using an oxygen meter, but no disadvantage is mentioned. There is no mention of advantages or disadvantages of using animals as indicator organisms.

This answer would probably get three marks.

The answer could be better organised. It would be a good idea to start a new paragraph after the first two sentences, and a separate paragraph at the end about advantages and disadvantages. Alternatively, each method could be dealt with separately in two paragraphs.

The answer would be improved if there was more detail. It also needs to address the last sentence in the question much more fully. It should give an example of at least one advantage and one disadvantage for each method.

This candidate almost certainly did not plan the answer before he or she began to write.

Chemistry C1.1–1.4

What you should know

The particle model and properties of materials

The particle model describes how particles are arranged in solids, liquids and gases. It also explains some of their properties and behaviour.

How does the arrangement of particles differ in a solid, a liquid and a gas?

Atoms, elements, compounds and mixtures

Everything is made from tiny particles called atoms. An element consists of one type of atom. Each element has a chemical symbol.

Compounds are made from atoms of two or more elements joined together by chemical bonds. Each has a name and chemical formula.

Each element or compound has its own properties and behaviour. Compounds of the same type show patterns in their behaviour.

Mixtures consist of elements and compounds simply mixed together. They can be separated easily.

List the names of five elements and five compounds.

Acids and bases

Acids and bases are two types of chemical compounds.

Alkalis are bases that dissolve in water.

Give the names of three acids and two alkalis.

Chemical reactions

Chemical reactions happen when substances swap or rearrange their atoms to form different chemical substances.

The reactivity series lists metals in order of their reactivity.

Describe three chemical reactions that you have seen.

You will find out

The fundamental ideas in chemistry

> Substances are made of atoms, and atoms contain sub-atomic particles (protons, neutrons and electrons).

> The arrangement of electrons in an atom affects its reactivity.

> Elements are arranged in the periodic table.

> Chemical bonds connect atoms together to form molecules.

> Equations with chemical formulae describe what happens in a chemical reaction.

Limestone and building materials

> Limestone is important in the building industry.

> Limestone is used to make lime, cement, mortar and concrete.

> Exploiting limestone has environmental, social and economic effects.

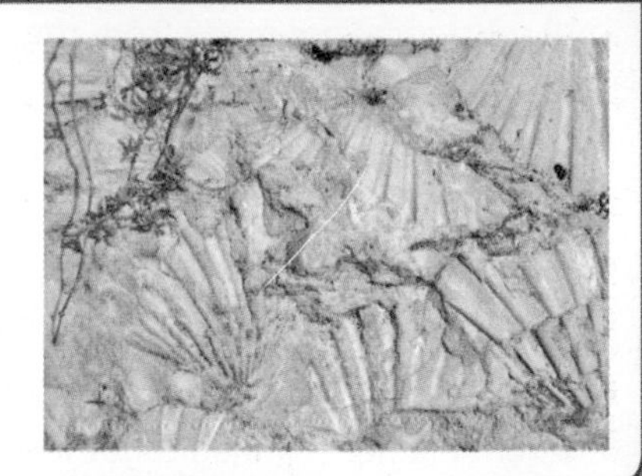

Metals and their uses

> Metals are extracted from rocks.

> Metals and alloys have properties that make them useful.

> Iron and steel, copper, aluminium and titanium are extracted from ores.

> Extracting and using metals has environmental effects.

Crude oil and fuels

> Burning causes air pollution, but there are ways to reduce it.

> Crude oil can be separated into fractions.

> Most of the hydrocarbons in crude oil are alkanes.

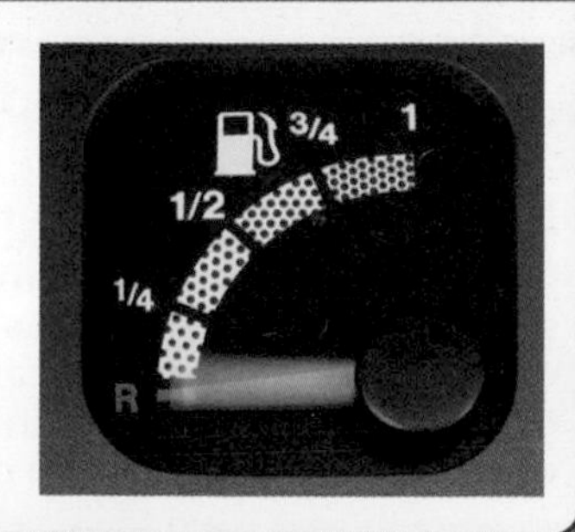

Atoms, elements and compounds

> **You will find out:**
> - all substances are made of atoms
> - what is meant by an element and a compound
> - some elements' symbols that you need to know

Nature's building blocks

Looking around, nature and the world appear very complicated. There are so many different materials – natural and manufactured. Some are living, some used to be alive. Others never lived at all. However, this amazing variety is assembled from only a small kit of parts – the elements.

FIGURE 1: Only a few elements were known in 1806 – and some of these turned out not to be elements.

Chemical substances, symbols and formulae

Elements and compounds

- All substances are made from tiny **particles** called **atoms**.
- An **element** has just one type of atom.
- Atoms of one element are different from the atoms of all other elements.
- Only ninety elements occur naturally. A few more are manufactured. These basic building blocks make up everything on Earth and the rest of the universe.

Compounds are substances that have atoms of two or more different elements joined together.

hydrogen chloride

methane

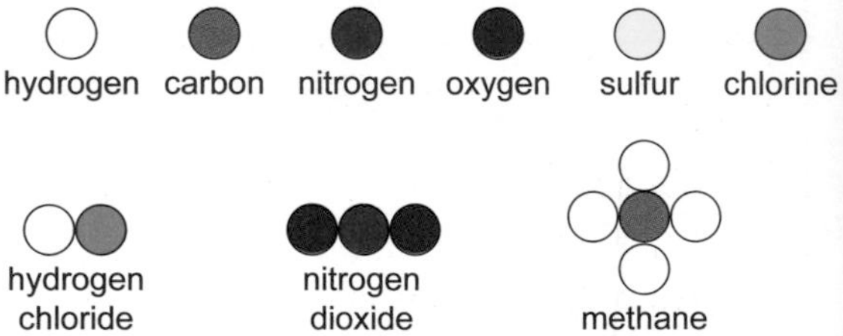

FIGURE 2: Each circle represents an atom – a different colour for each element. The circles are joined together to show compounds.

Chemical symbols and formulae

Each element has a **symbol**. For example, C stands for carbon and O for oxygen.

Symbols for some elements have two letters. The first is always a capital and the second is always a lowercase letter, for example, Ca for calcium, not CA, and Zn for zinc, not ZN. Co is the symbol for cobalt, an element, but CO is carbon monoxide, which is a compound.

A **formula** shows the types and proportions of atoms in a compound. Here are two examples:

- CO_2 shows that carbon dioxide contains carbon atoms and oxygen atoms in the proportion 1:2.
- H_2SO_4 shows that sulfuric acid contains twice as many hydrogen atoms as sulfur atoms, and four times as many oxygen atoms.

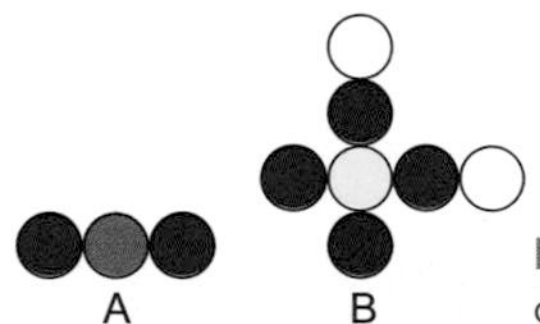

FIGURE 3: What compounds do A and B represent?

> **QUESTIONS**
>
> 1 What is special about an element?
>
> 2 How many different types of element are there in nature?
>
> 3 What is the difference between an element and a compound?

Composition and structure

Is a substance a compound or an element?

Mixtures can be separated into simpler mixtures with fewer parts. Keep on separating and you end up with pure substances (which can be elements or compounds). A compound can be broken down into simpler compounds, but there comes a point at which the product cannot be broken down further. The compound has been broken down into its elements.

... elements compounds GCSE

Structures of elements and compounds

Structures (arrangement of atoms) fall into two categories: **giant structures** and **molecular structures**.

In a giant structure, **atoms** pack together in a huge three-dimensional structure. Metal elements have giant structures, as do a few non-metal elements such as carbon. Many compounds have giant structures, for example, sodium chloride (formula NaCl) and silicon dioxide (formula SiO_2).

In a molecular structure, a small number of atoms cluster to together to form a particle called a molecule. The table shows some of the elements with molecular structures.

Remember

An element is a substance that cannot be broken down by chemical reactions (decomposed) to form any simpler substances.

Element	hydrogen	oxygen	nitrogen	chlorine	phosphorus	sulfur
Molecular formula	H_2	O_2	N_2	Cl_2	P_4	S_8

The symbol for oxygen is O but, in oxygen gas, the smallest particle is a molecule made from two oxygen atoms – shown by writing O_2

Many compounds have molecular structures. Examples are water (H_2O), ammonia (NH_3) and methane (CH_4).

QUESTIONS

4 When limestone is heated, it gives off a gas. Does this show that limestone is an element or a compound?

5 How many atoms are in a molecule of (a) nitrogen (b) phosphorus (c) sulfur?

6 How many atoms of carbon and of hydrogen form a molecule of methane?

7 The formula of ammonia is NH_3. What does this tell you?

Modelling structures

Odd symbols

Some chemical symbols appear not to match the name. This is because they come from the Latin name. For example, Na represents sodium (Latin *natrium*) and Fe represents iron (Latin *ferrum*).

Modelling structures

Chemists use models to show the three-dimensional arrangement of atoms in giant and molecular structures. Coloured spheres represent atoms.

Giant structures can be represented by stacking marbles or polystyrene spheres together.

Molecular structures can be represented using 'ball and stick' models or space filling models.

Ball and stick models can be misleading. A molecule does not look like a number of spheres joined by a rod or stick. A space filling model is better, but for big complex molecules it can be difficult to see all the atoms.

ammonia NH_3

methane CH_4

water H_2O

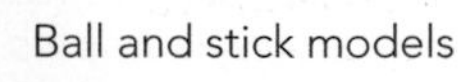

Ball and stick models

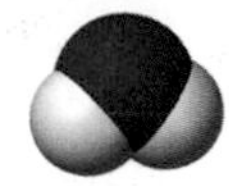

water H_2O

Space filling model

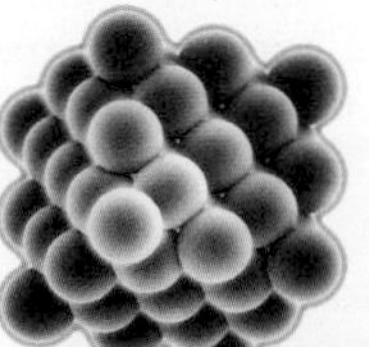

FIGURE 4: Platinum atoms are arranged in a giant structure.

QUESTIONS

8 Try to describe, in words, the shapes of water, ammonia and methane molecules. Can you see any relationship between their shapes?

Inside the atom

You will find out:

> atoms are made up of even smaller particles

> they have a central nucleus surrounded by layers of electrons

> how atoms of different elements differ from each other

Having a smashing time

Hadrons are parts of atoms. The best known are protons and neutrons. The Large Hadron Collider (LHC), at CERN in Geneva, smashes protons into one another at almost the speed of light. By studying these high-energy collisions, scientists hope to learn more about the laws of physics and chemistry.

FIGURE 1: Studying tiny particles needs big equipment.

What is inside an atom?

Atoms are incredibly small. 30 million carbon atoms would only stretch about 1 centimetre. Until the early 1900s, atoms were thought to be the smallest particles. Then scientists discovered even smaller particles inside atoms.

At the centre of every atom is a **nucleus**. It contains **protons** and **neutrons**. The nucleus is surrounded by **electrons** that move around it.

Protons, neutrons and electrons are called **sub-atomic particles**.

What is the difference between elements?

> All atoms of a particular element have the same number of protons as each other.

> Atoms of one element have a different number of protons from every other element.

> The number of electrons in an atom is always the same as the number of protons.

> The number of protons is called the atomic number. Each element has its own unique **atomic number**. The table shows a few examples.

Element	hydrogen	carbon	oxygen	iron	gold
Atomic number	1	6	8	26	79

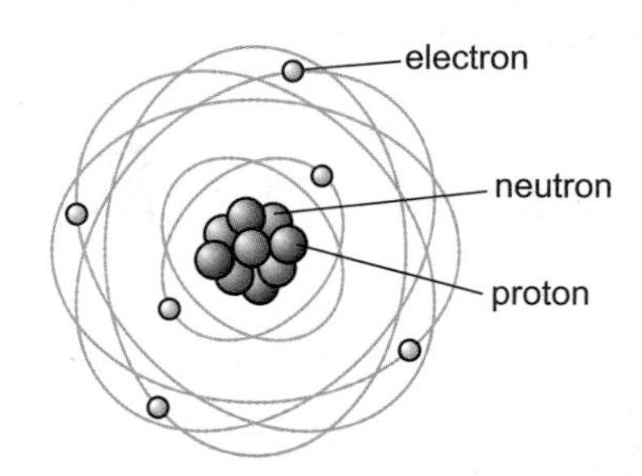

FIGURE 2: This is a 3D drawing of a carbon atom. You can only see four of the protons. How many more are hidden at the back?

QUESTIONS

1 What is at the centre of an atom and which sub-atomic particles are found there?

2 The atomic number of iron is 26. How many protons does an iron atom have?

Sub-atomic particles

Protons, neutrons and electrons

Protons and electrons have an electric charge. Their charges are equal, but opposite. Neutrons have no charge and are neutral. Since an atom has equal numbers of protons and electrons, the charges cancel out. So, an atom has no overall charge.

The sum of the number of protons and neutrons in an atom is its **mass number**. A carbon atom with 6 protons and 6 neutrons has a mass number of 6 + 6 = 12.

Particle	Charge	Relative size of charge
proton	positive	+1
neutron	neutral	0
electron	negative	-1

Q ... subatomic particles for students

How are the electrons arranged?

Electrons moving around the nucleus are arranged in layers called **shells** (also known as **energy levels**). Electrons occupy the lowest available energy level.

> The first shell (lowest energy level) can hold up to 2 electrons.

> The second shell (next energy level) can hold up to 8 electrons.

> The third shell (third energy level) can hold up to 18 electrons.

The arrangement of electrons is called the **electronic configuration**. Here are two examples.

The atomic number of oxygen is 8, so each oxygen atom has eight electrons. Two are in the first shell and six in the second. The electronic configuration of oxygen is written as 2.6

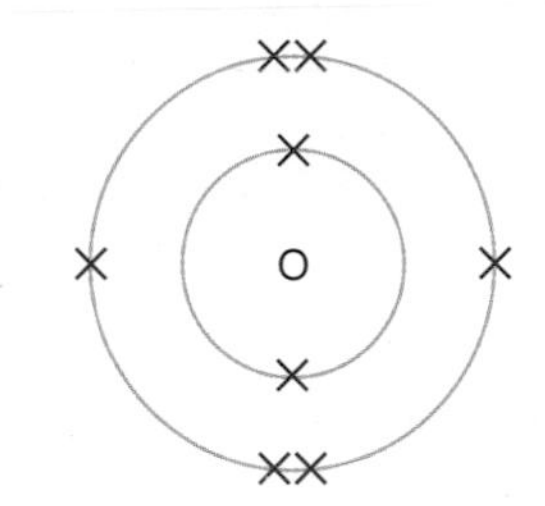

The atomic number of sodium is 11, so each sodium atom has eleven electrons. Two are in the first shell, eight in the second and one in the third. The electronic configuration of sodium is 2.8.1

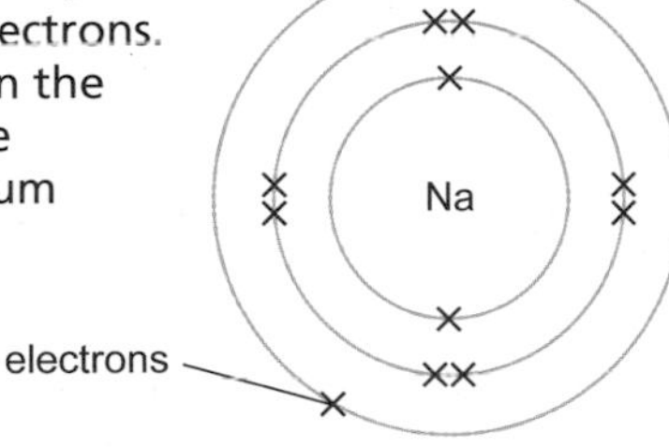

In this type of diagram, electrons may be shown as dots, instead of crosses.

Remember

The higher an energy level in an atom, the further its electrons are from the nucleus.

Did you know?

In the LHC, beams of protons are fired in opposite directions round a 27-kilometres-long circular tunnel. They collide at nearly 300 000 000 (three hundred million) metres per second.

QUESTIONS

3 The mass number of an oxygen atom is 16. How many protons, neutrons and electrons does it contain?

4 Nitrogen atoms have seven electrons. What is its electronic configuration?

5 What is the electronic configuration of aluminium (atomic number 13)?

Why is electronic configuration so important?

Chemical reactions happen when atoms collide. Their outsides make contact, where electrons in the atoms' outer shells are found. Therefore, how an atom reacts depends on the number of electrons in its outer shell.

Elements with the same number of outer electrons react in similar ways. For example, the elements lithium, sodium and potassium react vigorously with air and water. Their other chemical properties are also similar. This is because they all have just one electron in their outer shell (highest energy level).

Although the fourth shell can hold more than eight electrons, potassium's configuration is 2.8.8.1 not 2.8.9. Calcium is similar.

QUESTIONS

6 Work out the electronic configuration of beryllium, magnesium and calcium (atomic numbers 4, 12 and 20). Explain why you would expect them all to react in a similar way.

FIGURE 3: Potassium reacting with water. How can you tell that this is a vigorous reaction?

Element patterns

Precious metals

The elements copper, silver and gold have been known and prized since prehistoric times. This is because they occur 'native' – uncombined with other elements. They were easy to extract from rocks. All three are unreactive and long-lasting. Gold is the most valuable of the three, because it does not tarnish: it retains its shiny golden colour for ever.

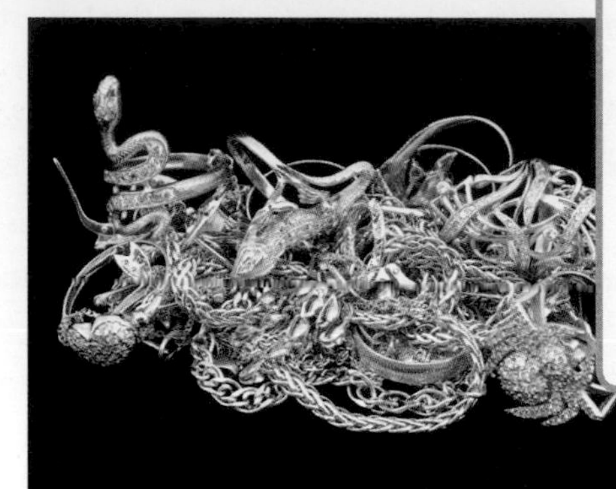

FIGURE 1: Copper, silver and gold were used traditionally for jewellery and coinage.

You will find out:

- > certain groups of elements have similar properties and reactions
- > how and why elements are arranged in the periodic table
- > why the periodic table is a useful tool

Patterns in the elements

Before 1750, only sixteen elements were known. By 1850 there were fifty eight. As more elements were discovered, scientists found that some elements had similar properties. They started to look for patterns.

Lithium, sodium and potassium melt easily and burn with coloured flames. They float on water and react quickly with it, giving off hydrogen gas and forming alkaline solutions.

Copper, silver and gold occur 'native'. They hardly react with water at all.

Scientists tried to show the relationships between elements with similar properties. This was difficult because many elements were still unknown.

The breakthrough came when Dimitri Mendeléev, a Russian chemist, arranged the elements in a table – the **periodic table**.

In the periodic table, elements with similar properties, such as lithium, sodium and potassium, are in vertical columns – called **groups**. The rows are called **periods**. Each period has metals at the left and non-metals at the right, divided by the red zigzag line, shown in Figure 3.

FIGURE 2: Why must these three elements be stored under oil?

QUESTIONS

1 Explain why lithium, sodium and potassium are called alkali metals.

2 What is special about elements in the same vertical column of the periodic table?

Group 1	2											Group 3	4	5	6	7	0
			$^{1}_{1}$H hydrogen														$^{4}_{2}$He helium
$^{7}_{3}$Li lithium	$^{9}_{4}$Be beryllium											$^{11}_{5}$B boron	$^{12}_{6}$C carbon	$^{14}_{7}$N nitrogen	$^{16}_{8}$O oxygen	$^{19}_{9}$F fluorine	$^{20}_{10}$Ne neon
$^{23}_{11}$Na sodium	$^{24}_{12}$Mg magnesium											$^{27}_{13}$Al aluminium	$^{28}_{14}$Si silicon	$^{31}_{15}$P phosphorus	$^{32}_{16}$S sulfur	$^{35}_{17}$Cl chlorine	$^{40}_{18}$Ar argon
$^{39}_{19}$K potassium	$^{40}_{20}$Ca calcium	$^{45}_{21}$Sc scandium	$^{48}_{22}$Ti titanium	$^{51}_{23}$V vanadium	$^{52}_{24}$Cr chromium	$^{55}_{25}$Mn manganese	$^{56}_{26}$Fe iron	$^{59}_{27}$Co cobalt	$^{59}_{28}$Ni nickel	$^{64}_{29}$Cu copper	$^{65}_{30}$Zn zinc	$^{70}_{31}$Ga gallium	$^{73}_{32}$Ge germanium	$^{75}_{33}$As arsenic	$^{79}_{34}$Se selenium	$^{80}_{35}$Br bromine	$^{84}_{36}$Kr krypton
$^{85}_{37}$Rb rubidium	$^{88}_{38}$Sr strontium	$^{89}_{39}$Y yttrium	$^{91}_{40}$Zr zirconium	$^{93}_{41}$Nb niobium	$^{96}_{42}$Mo molybdenum	$^{99}_{43}$Tc technetium	$^{101}_{44}$Ru ruthenium	$^{103}_{45}$Rh rhodium	$^{106}_{46}$Pd palladium	$^{108}_{47}$Ag silver	$^{112}_{48}$Cd cadmium	$^{115}_{49}$In indium	$^{119}_{50}$Sn tin	$^{122}_{51}$Sb antimony	$^{128}_{52}$Te tellurium	$^{127}_{53}$I iodine	$^{131}_{54}$Xe xenon
$^{133}_{55}$Cs caesium	$^{137}_{56}$Ba barium	$^{139}_{57}$La lanthanum	$^{178}_{72}$Hf hafnium	$^{181}_{73}$Ta tantalum	$^{184}_{74}$W tungsten	$^{186}_{75}$Re rhenium	$^{190}_{76}$Os osmium	$^{192}_{77}$Ir iridium	$^{195}_{78}$Pt platinum	$^{197}_{79}$Au gold	$^{201}_{80}$Hg mercury	$^{204}_{81}$Tl thallium	$^{207}_{82}$Pb lead	$^{209}_{83}$Bi bismuth	$^{210}_{84}$Po polonium	$^{210}_{85}$At astatine	$^{222}_{86}$Rn radon
$^{223}_{87}$Fr francium	$^{226}_{88}$Ra radium	$^{227}_{89}$Ac actinium															

FIGURE 3: Modern periodic table. You need to remember the symbols for the highlighted elements.

Q ... periodic table GCSE

Reactive and unreactive

The periodic table and electronic structure

It is the electrons in an atom's highest energy level (outer shell) that determine how it reacts. Elements in a vertical group react similarly because they have the same number of outer electrons. The group number at the top of each column tells you how many.

> The **alkali metals** in Group 1 have one outer electron.

> Magnesium and calcium in Group 2 have two outer electrons.

> Aluminium, in Group 3, has three outer electrons.

> Chlorine in Group 7 has seven outer electrons.

The last number in an electronic configuration, such as 2.8.7 for chlorine, tells you how many outer electrons the atom has.

Noble gases

The **noble gases** are labelled Group 0, although they have eight outer electrons (except helium, which has a full first shell of two electrons). This electron configuration is stable. Because it is stable, noble gases are unreactive. Krypton and xenon react only with fluorine, the most reactive element of all. The others do not react with anything.

Noble gas	Atomic number	Electron configuration
helium	2	2
neon	10	2.8
argon	18	2.8.8
krypton	36	2.8.18.8
xenon	54	2.8.18.18.8

Remember

Vertical columns in the periodic table are called groups. Horizontal rows are called periods.

Did you know?

Platinum is used to make jewellery. It's more expensive than silver and gold, but the most expensive metal is rhodium.

QUESTIONS

3 Explain why elements in the same group have similar reactions.

4 Carbon is in Group 4. How many electrons does a carbon atom have in its outer shell?

5 Why are noble gases unreactive?

Why are the elements arranged in this pattern?

The periodic table lists the elements in order of **atomic number** – the number of protons and electrons in each atom of the element.

Each shell can hold a limited number of electrons. When full, electrons start to fill the next shell. Elements with the same number of electrons in their outer shell occur periodically (at regular intervals) in the list. Since an element's reactivity depends on its electronic configuration, those with similar properties occur periodically – every eighth element at first, then every eighteenth (the number of electrons in the outer shell of stable noble gas atoms).

Cutting the list after each noble gas and starting a new period (row), puts elements with similar properties into vertical columns (groups). The periodic table is emerging.

Period 1								He
Atomic number								2
Electronic configuration								2

Period 2	Li	Be	B	C	N	O	F	Ne
Atomic number	3	4	5	6	7	8	9	10
Electronic configuration	2.1	2.2	2.3	2.4	2.5	2.6	2.7	2.8

Period 3	Na	Mg	Al	Si	P	S	Cl	Ar
Atomic number	11	12	13	14	15	16	17	18
Electronic configuration	2.8.1	2.8.2	2.8.3	2.8.4	2.8.5	2.8.6	2.8.7	2.8.8

Period 4	K							
Atomic number	19							
Electronic configuration	2.8.8.1							

QUESTIONS

6 Explain why the table of elements is called the periodic table.

Combining atoms

> **You will find out:**
> - atoms of one element can join with atoms of another element to form a compound
> - atoms can join together by sharing electrons, or by giving and taking electrons

Atomic bricks

These London landmarks were built with Lego®. The bricks are designed to pull apart easily – so they were glued together to make the models stronger. Atoms are Nature's building bricks, but to make compounds you cannot just push atoms together. Compounds do not fall apart easily, because they are held together by strong forces, called bonds, between the atoms.

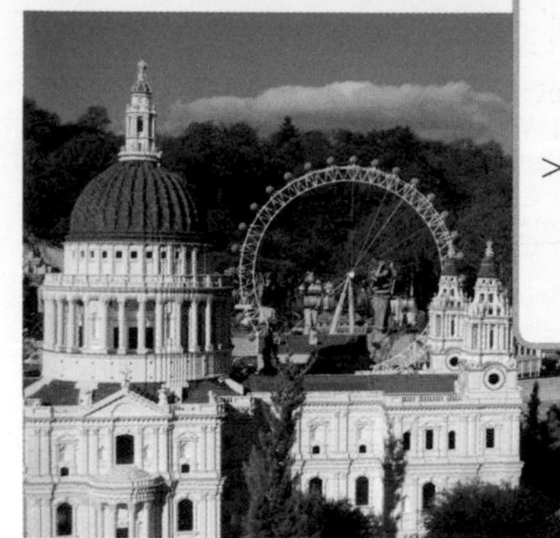

FIGURE 1: 'Building bricks' can be very small.

Two types of bonds

Most elements never exist as single atoms. Instead, two or more atoms join together. A chemical **bond** holds them together. There are two types of bond: **covalent** and **ionic**.

Each molecule of hydrogen, oxygen and nitrogen is made from two atoms. This is why they are written H_2 O_2 and N_2. The atoms are held together by sharing a pair of electrons – one from each. This is a **covalent bond**.

Atoms of two or more different non-metal elements join to form compounds. These are also held together by sharing electrons (covalent bonds).

In a water molecule (H_2O), one oxygen atom shares a pair of electrons with each hydrogen atom. So, a water molecule has two covalent bonds.

When non-metals combine with metals, electrons are not shared. Instead they transfer from the metal atom to the non-metal atom. Both atoms are left with an electric charge. Charged atoms are called **ions**.

Metal ions always have a positive charge, and non-metal ions always have a negative charge. The **attraction** between these oppositely charged ions holds the compound together. This is an **ionic bond**.

Metal compounds are made of ions, not molecules.

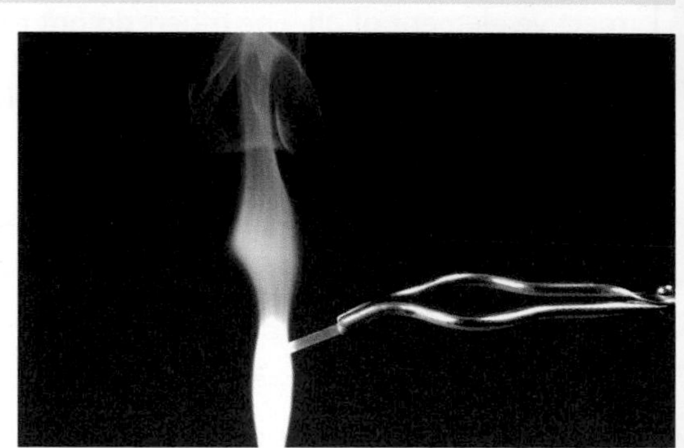

FIGURE 2: When magnesium metal burns, it reacts with oxygen in the air to form the ionic compound magnesium oxide.

QUESTIONS

1 What is a molecule?

2 What is a covalent bond?

3 What charge will an atom have if it (a) gains an electron (b) loses an electron?

Eight in a shell

Noble gases have eight electrons in their outer shell. (Helium has two because that is all the first shell can hold.) They are unreactive elements and form no bonds. They do not gain, lose or share electrons. Eight outer electrons is a stable configuration.

Atoms with fewer than eight outer electrons react in ways that give them a stable group of eight in their outer shell. They may share electrons (covalent bonding), or transfer electrons (ionic bonding).

Did you know?

99% of the air you breathe consists of diatomic molecules (molecules made from two atoms). Less than 1% is single atoms.

... covalent bonds ... ionic bonds

Forming covalent bonds in a water molecule

An oxygen atom needs two more electrons to have eight in its outer shell. A hydrogen atom needs one more to become stable (two electrons in the first shell). Oxygen forms two covalent bonds, one with each of two hydrogen atoms. This way all the atoms achieve a noble gas configuration.

This can be shown using a dot and cross diagram.

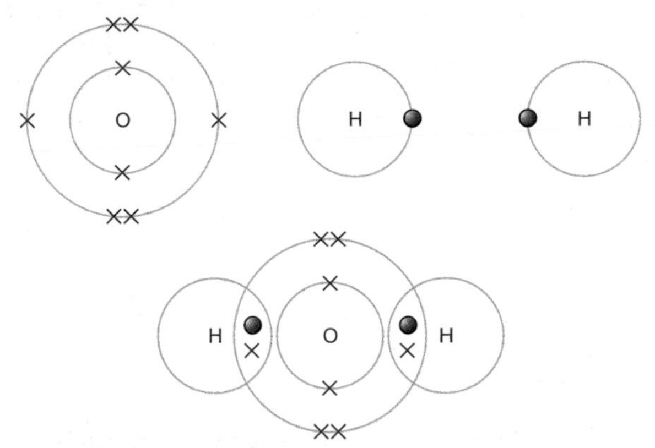

The positive nuclei of both atoms in a covalent bond are attracted to the negatively charged, shared electrons in between.

Note that, although some electrons are drawn as dots and others as crosses, all electrons are identical. The diagrams use dots for electrons from one atom and crosses for electrons from the other atom.

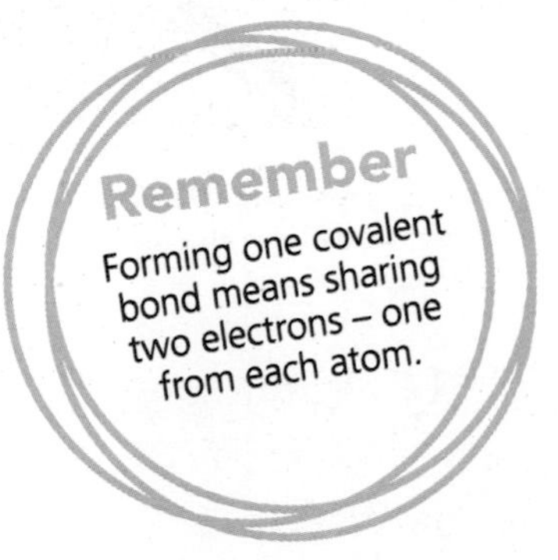

Forming ionic bonds in sodium chloride

The chlorine atom gains an electron to get an outer shell of eight. It is now a negatively charged chloride ion, Cl^-. The electron comes from the sodium atom, leaving the second shell of the sodium atom as its outer shell. This has eight electrons, so it is stable.

The sodium atom becomes a positively charged sodium ion, Na^+.

The oppositely charged ions attract one another.

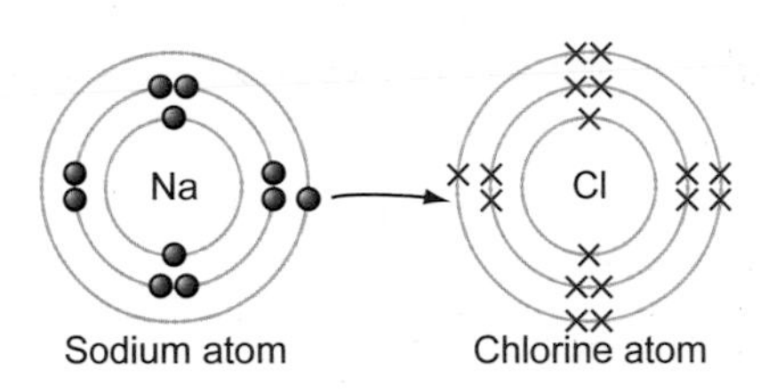

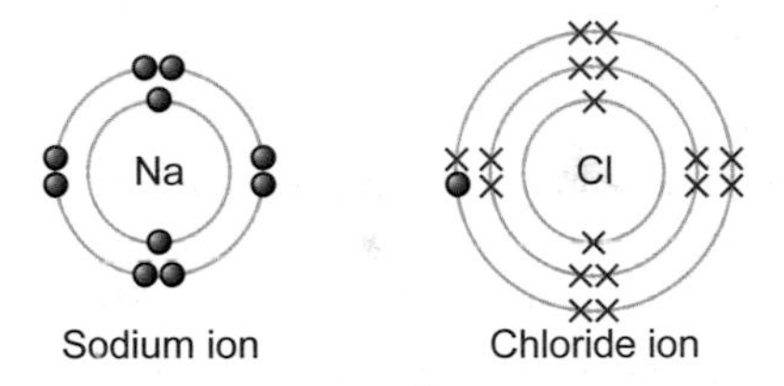

QUESTIONS

4 Does the compound sulfur dioxide have covalent or ionic bonds?

5 Why are metal ions always positively charged?

6 Oppositely charged particles attract one another. What are these particles in (a) a covalent bond (b) an ionic bond?

Representing covalent bonds

Scientists often use a short line drawn between atoms to represent the shared pair of electrons holding the atoms together in a covalent bond.

Using this system, a chlorine molecule (Cl_2) is shown as **Cl–Cl**. An oxygen molecule (O_2) is shown as **O=O**. The double line between the two oxygen atoms represents a covalent **double bond**. The atoms are held together by two pairs of shared electrons.

QUESTIONS

7 A molecule of the compound carbon dioxide (CO_2) can be represented as **O=C=O**. Explain what this shows about its bonding.

8 Draw a dot and cross diagram to show the bonding in an oxygen molecule.

Chemical equations

You will find out:

- > equations describe what happens in a chemical reaction
- > chemical formulae can be used to write equations
- > what a 'balanced equation' means

Simple symbols

Symbols are simple shapes that represent something without using words. They mean the same thing in any language. Medieval alchemists used the same chemical symbols all over Europe. Their symbols showed that certain substances are related – vinegar is an acid, for example. However, the symbols did not show what was in a substance and so it was difficult to describe chemical reactions.

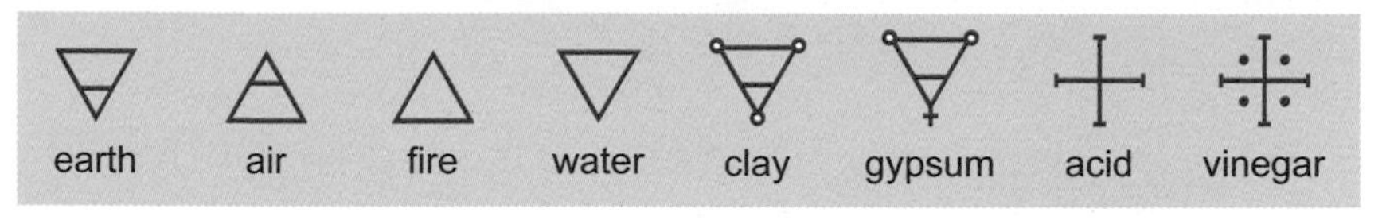

FIGURE 1: How did alchemists show that clay and gypsum are minerals, and vinegar is an acid?

Equations

Chemists use chemical symbols and formulae as abbreviations.

> A formula shows what elements are in a molecule of a substance.

> A number after a symbol shows how many atoms of that element are in the molecule.

You could describe what happens when methane burns:

"A methane molecule contains one carbon and four hydrogen atoms. Each molecule reacts with two molecules of oxygen, each containing two oxygen atoms. This produces one molecule of carbon dioxide, containing one carbon and two oxygen atoms. For each carbon dioxide molecule produced, the reaction also produces two molecules of water, each containing two hydrogen atoms and one oxygen atom."

A chemist would just write a **chemical equation**:

$$CH_4 + 2O_2 \rightarrow CO_2 + 2H_2O$$

That saves a lot of time and space, but you need to understand the abbreviations.

How equations work

> On the left of the arrow are **reactants** and on the right are **products**.

> Formulae replace chemical names: CH_4 for methane, CO_2 for carbon dioxide.

> Signs:

+ means 'reacts with' or 'and'

→ means 'go to' or 'turn into'

> Numbers in front of formulae show how many molecules of a substance react or are produced:

$2H_2O$ means 'two molecules of water'.

No number means 'one molecule of …', so CH_4 means 'one molecule of methane'.

Did you know?

The first chemical equation was written in 1615 by Jean Beguin. It did not use symbols, just words.

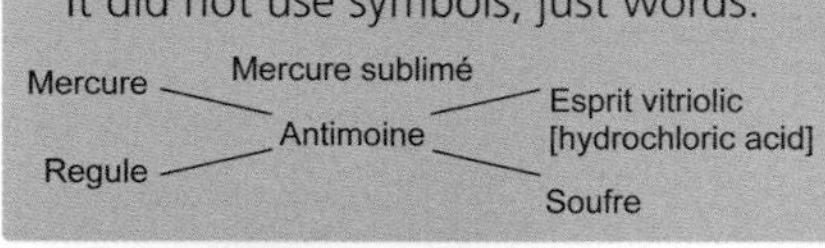

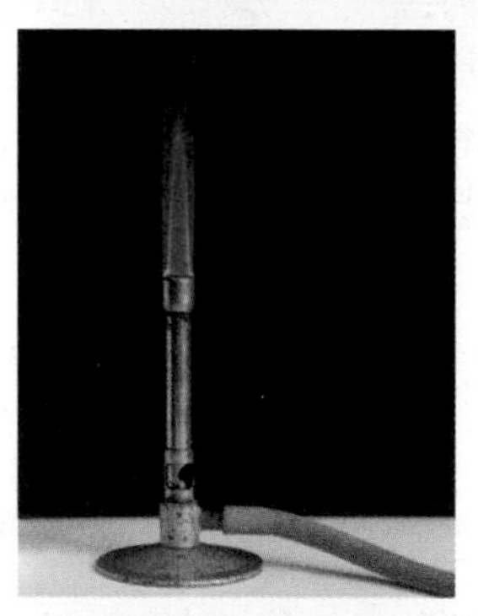

FIGURE 2: Burning methane rearranges the atoms.

QUESTIONS

1 What is the advantage in using chemical equations instead of words?

2 Give the name for substances that are formed during a chemical reaction.

… chemical equations for students

Tracking atoms and molecules

Tracking atoms

During a **chemical reaction**, reactants get used up and new products form. However, no atoms disappear, and no new atoms are produced. They just get shuffled. When methane burns, the atoms in methane (CH_4) and oxygen (O_2) rearrange into carbon dioxide (CO_2) and water (H_2O).

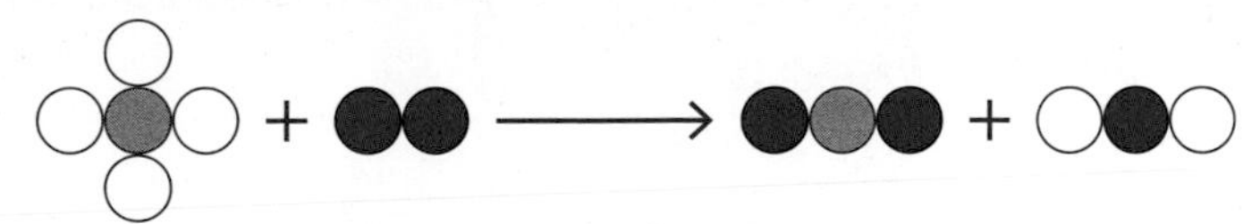

There is a problem here. We appear to have lost two hydrogen atoms and gained an oxygen atom. This is impossible. What really happens is that a methane molecule reacts with two oxygen molecules.

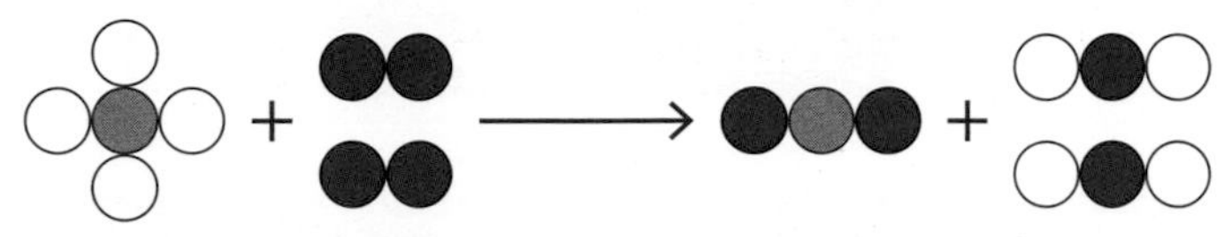

Replacing drawings with formulae, gives the equation:

$$CH_4 + 2O_2 \rightarrow CO_2 + 2H_2O$$

The reactants and products now have the same number of atoms of each element. The equation is **balanced**. The total mass of products is always the same as the total mass of the reactants.

State symbols

The equation above does not show whether H_2O means water, steam or ice. You can specify the physical state of substances by adding **state symbols**:

(s) solid (ℓ) liquid (g) gas (aq) aqueous solution ('aqueous' means 'dissolved in water')

Burning methane produces steam, shown by:

$$CH_4(g) + 2O_2(g) \rightarrow CO_2(g) + 2H_2O(g)$$

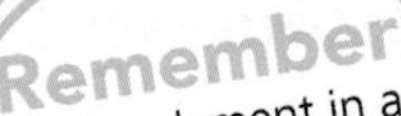

Remember

For an element in a molecule, no number means there is just one atom. If there is a number, it will be a subscript.

QUESTIONS

3 What do equations show, besides the substances involved in the reaction?

4 How do equations show the number of molecules reacting?

5 16 g of methane reacts with 64 g of oxygen to give 36 g of water. Calculate the mass of carbon dioxide produced.

Balancing equations (Higher tier)

To make an equation balance, add numbers in front of formulae as necessary. You must not change any of the formulae, because a different formula would represent a different substance.

Sodium hydroxide solution reacts with sulfuric acid, forming sodium sulfate solution and water:

$$NaOH + H_2SO_4 \rightarrow Na_2SO_4 + H_2O$$

There are two Na atoms in the products, so there must be two in the reactants. Put '2' in front of NaOH.

$$2NaOH + H_2SO_4 \rightarrow Na_2SO_4 + H_2O$$

Count the H atoms: four in the reactants (two in 2NaOH and two in H_2SO_4), but only two in the products. There must be four. Put '2' in front of H_2O.

$$2NaOH + H_2SO_4 \rightarrow Na_2SO_4 + 2H_2O$$

Count the O atoms: six on each side. The equation is now balanced.

QUESTIONS

6 Balance these equations:

(a) $N_2 + H_2 \rightarrow NH_3$

(b) $Na + H_2O \rightarrow NaOH + H_2$

Q ... practice balancing chemical equations

Building with limestone

Rebuilding London

The 1666 Great Fire of London destroyed most of the city because it was mainly built of wood. The city was rebuilt in stone and brick, which are much more robust. Sir Christopher Wren built the new St Paul's Cathedral with Portland stone (a type of limestone). There were no limestone quarries in London, only clay, to make bricks. Buckingham Palace and other famous London buildings are also made from Portland stone, which comes from the Isle of Portland in Dorset.

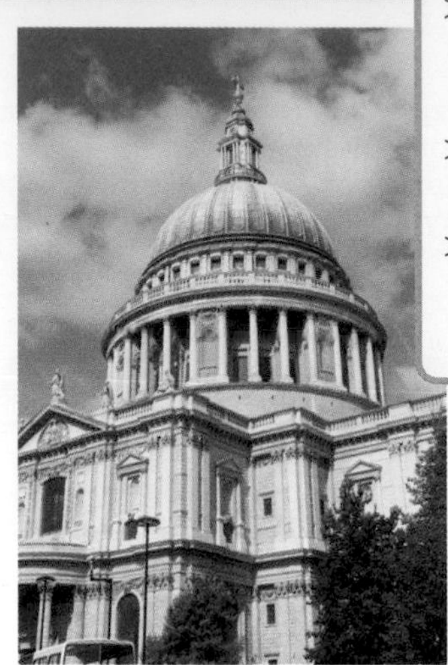

FIGURE 1: St Paul's Cathedral.

You will find out:

> limestone can be quarried and used as a building material

> the composition of limestone

> the importance of limestone to the building industry

Limestone

What is limestone?

Limestone has been, and still is, a popular building material. Limestone started as the shells and skeletons of sea creatures and coral many millions of years ago.

The term 'limestone' covers a range of off-white rocks formed from these skeletons and shells. All are mainly **calcium carbonate** ($CaCO_3$). Some have a yellow–brown colour from the iron oxide in the rock.

Calcium carbonate reacts with acids. Limestone buildings can be badly damaged by acid rain.

Extraction and uses

Limestone is extracted from the ground. After removing the top layer of soil, blasting the rock with explosives dislodges pieces of limestone. This takes place in **quarries**.

As well as its use as a building stone, limestone has other important uses in the construction industry:

- for the foundations of most roads
- in cement, mortar and concrete
- to make iron steel and glass.

FIGURE 2: How can you tell that this limestone formed under the sea?

FIGURE 3: What uses of limestone can you find in this picture?

Did you know?

Limestone, chalk, marble, egg shells, sea shells, stalactites and stalagmites are all made of calcium carbonate.

QUESTIONS

1 How is limestone extracted?

2 What causes limestone buildings and statues to dissolve slowly away?

... limestone definition ... limestone uses

What are the effects of a limestone quarry?

Advantages and disadvantages

The Peak District National Park is an area of outstanding natural beauty. There is high-quality limestone here, which has been extracted since Roman times. We need limestone, but it can only come from where it exists.

There are advantages and disadvantages to an area where a quarry is situated.

Advantages

> Jobs for local people in an area with little industry

> More, better-paid jobs, so more money to boost local economy

> Better healthcare and leisure facilities, as more people move into the area

> Better roads, since needed for lorries

Disadvantages

> Damage to the landscape

> Loss of wildlife habitats

> Noise and vibration from blasting, machinery and vehicles

> Dust pollution in the environment

> Traffic congestion and vibration from heavy lorries

Minimising the disadvantages

Today, all new quarries must be approved by the local planning authority. Although not overcoming the problems, careful planning can reduce them. Methods include:

> allowing a quarry to work only during the day, not at night or at weekends

> where possible, using railways instead of lorries to transport limestone

> restricting the size of a quarry by setting a limit on the amount of limestone extracted

> having fewer, but larger, better-managed quarries.

Although a working quarry scars the landscape, once all the limestone has been extracted it will close. When landscaped, the area can become a nature reserve or used for recreation such as rock climbing.

QUESTIONS

3 If you lived near a limestone quarry, which advantage and which disadvantage would you consider to be the most important? Explain your choices.

4 A public meeting is being held to discuss the opening of a new quarry. What questions would you ask the quarry company?

Limestone caves

Rain dissolves carbon dioxide from the air to form a very dilute acidic solution (carbonic acid). This acid slowly dissolves limestone, forming a solution of calcium hydrogencarbonate, $Ca(HCO_3)_2$.

$H_2O(\ell) + CO_2(g) \rightarrow H_2CO_3(aq)$

followed by

$H_2CO_3(aq) + CaCO_3(s) \rightarrow Ca(HCO_3)_2(aq)$

This is how caves and potholes form in limestone rocks.

As the solution drips from a cave roof, it evaporates. The reaction reverses, producing calcium carbonate again in the form of stalactites and stalagmites.

$Ca(HCO_3)_2(aq) \rightarrow CaCO_3(s) + H_2O(\ell) + CO_2(g)$

FIGURE 4: Limestone cave.

QUESTIONS

5 How do the features in a limestone cave form? Suggest why it takes thousands of years to form such a cave.

... limestone cave formation AND limestone quarrying

Heating limestone

You will find out:

> what happens when calcium carbonate is heated

> how limewater is used to test for carbon dioxide

> how limestone is used to make cement, mortar and concrete

Greener cement

Millions of worn-out tyres are scrapped every year. Amazingly, perhaps, many are used to make cement. Putting flexible rubber in hard, solid cement does not seem right – and it's not. Cement making needs high temperatures. The rubber tyres are used as fuel, not as an ingredient. This reduces the amount of fossil fuel needed and gets rid of the waste tyres.

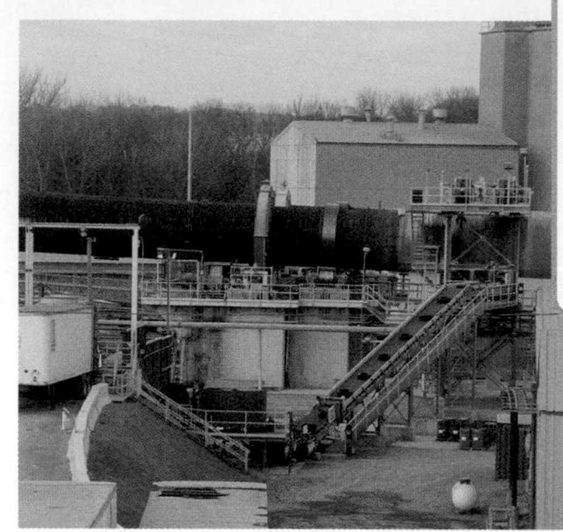

FIGURE 1: Feeding old tyres into a cement kiln.

Decomposing and slaking

What happens when limestone is heated?

Limestone is mainly calcium carbonate ($CaCO_3$). When heated strongly, the compound breaks down (decomposes). Two simpler compounds form: calcium oxide and carbon dioxide.

calcium carbonate → calcium oxide + carbon dioxide

$CaCO_3(s) \rightarrow CaO(s) + CO_2(g)$

Strongly heated limestone glows with a bright, white light. In the 1800s, before electric lighting, theatres used lime, heated by gas, for spotlights. It's where the expression 'being in the limelight' comes from.

Most other metal carbonates, except sodium carbonate and potassium carbonate, also decompose in this way. For example:

magnesium carbonate → magnesium oxide + carbon dioxide

$MgCO_3(s) \rightarrow MgO(s) + CO_2(g)$

This type of chemical reaction is called **thermal decomposition**. It is 'thermal' because the reaction is caused by heat.

When water is added to calcium oxide, it reacts to form calcium hydroxide.

calcium oxide + water → calcium hydroxide

$CaO(s) + H_2O(\ell) \rightarrow Ca(OH)_2(s)$

This compound, known as 'slaked lime', is the lime used by farmers and gardeners. It is slightly soluble in water and a solution of it is called **limewater**.

Uses of slaked lime

Calcium hydroxide (slaked lime) is an alkali.

> Farmers and gardeners spread it on soil that is too acidic – it raises the pH of the soil, which helps to improve crops.

> It is also added to lakes that have been made acidic by acid rain – it neutralises the acid.

FIGURE 2: Blowing powdered lime into a lake in Sweden. This lake needed 8000 tonnes of lime.

QUESTIONS

1 Name the three elements in calcium carbonate.

2 What does 'thermal decomposition' mean?

3 Write equations for the thermal decomposition of zinc carbonate ($ZnCO_3$) and of copper carbonate ($CuCO_3$).

4 Explain why farmers 'lime' their fields.

Q ... thermal decomposition AND slaked lime

Cement

The limewater test

$Ca(OH)_2(aq)$ is calcium hydroxide solution. It can be used to test for carbon dioxide. When carbon dioxide is bubbled through limewater, the clear solution turns cloudy or 'milky' as tiny solid particles of white calcium carbonate form.

$$CO_2(g) + Ca(OH)_2(aq) \rightarrow CaCO_3(s) + H_2O(\ell)$$

Continue bubbling carbon dioxide through the cloudy mixture and it goes clear again. **Insoluble** calcium carbonate reacts to form soluble calcium hydrogencarbonate.

$$CaCO_3(s) + CO_2(g) + H_2O(\ell) \rightarrow Ca(HCO_3)_2(aq)$$

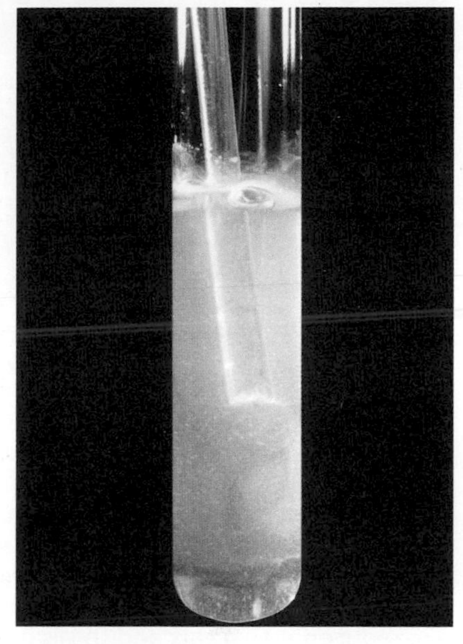

FIGURE 3: Why is this limewater becoming cloudy?

Cement, mortar and concrete

Three important building materials rely on decomposing limestone in kilns. (A kiln is a large oven.)

> **Cement** is made by heating limestone and clay. After adding water to cement, it sets and becomes as hard as stone. Cement is used widely – on its own, and in mortar and concrete.

> **Mortar** is made by mixing cement, sand and water. The mixture slowly sets solid. Mortar binds bricks together in brick walls. Before cement was available, mortar was made using slaked lime.

> **Concrete** is made by mixing cement, sand, gravel (or crushed rock) and water. Concrete is very strong. It is used for the foundations of buildings and for structures such as bridges.

QUESTIONS

5 Why is limestone so important to the building industry?

6 People often think that cement and concrete are the same. Explain the difference.

Did you know?

Calcium carbonate must be heated to 900 °C or more before it decomposes.

The difference between mortar and concrete

Wet cement sets solid, but it does not dry in the same way as washing dries. The water does not evaporate. It takes part in chemical reactions that form 'hydrated cement' crystals. These lock the other ingredients together.

In mortar, it is grains of sand that are locked together.

In concrete, the other ingredients are sand and **aggregate** (gravel or crushed rock). The mix of small sand particles and various sized stones makes concrete much stronger than mortar.

Concrete can be made even stronger by using it with steel, in steel reinforced concrete.

QUESTIONS

7 Find out what steel reinforced concrete is and why it is stronger than plain concrete.

... cement vs mortar vs concrete AND steel reinforced concrete

Preparing for assessment: Applying your knowledge

To achieve a good grade in science, you not only have to know and understand scientific ideas, but you need to be able to apply them to other situations and investigations. These tasks will support you in developing these skills.

Manaccanite

It's 1791 in Manaccan, a small village in Cornwall. The local vicar is the Reverend William Gregor. His hobby is collecting and studying samples of rocks and minerals. He has already collected many local samples and catalogued them, but one day he comes across something that he had not seen before. He shows his wife, Charlotte, a reddish brown substance. "You remember the black sand I found down in the valley?" he explains hurriedly, "You know, that turned out to be magnetic? Well, I reduced it to a calx and this is it. I do believe it is a new material."

Gregor called the new material *manaccanite* after the village. Although he did not succeed in purifying it, his findings were published that year, but aroused little interest. A few years later, without knowing of Gregor's work, another scientist discovered the same material and gave it its modern name.

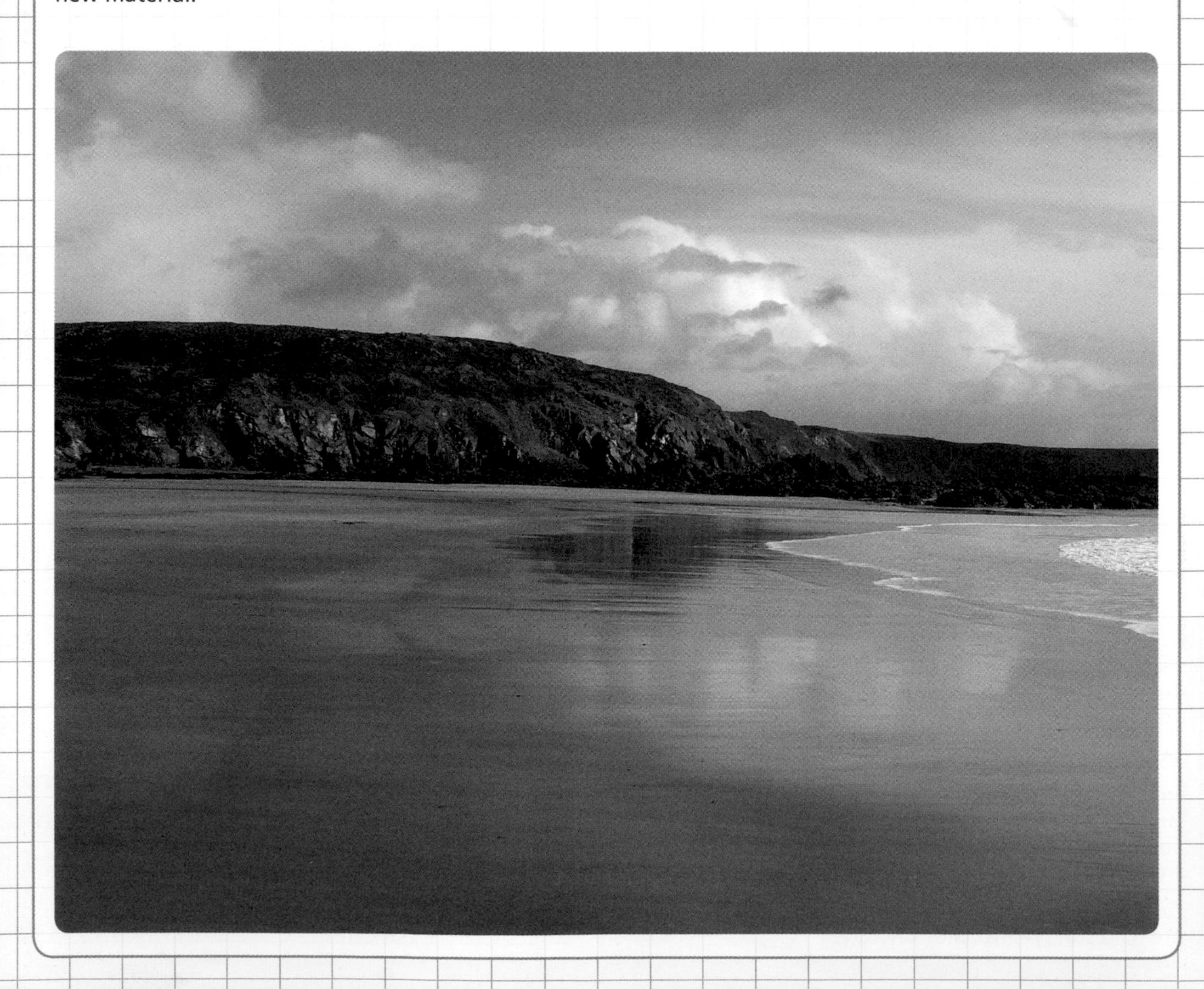

Task 1

(a) Gregor's 'black sand' contained a mineral now called ilmenite. It is an oxide of iron and another metal. Describe how metals are obtained from oxides such as ilmenite.

(b) Two metals can be extracted from manaccanite. One metal, call it Metal X, was unknown at the time that manaccanite was discovered. What was the other metal?

Task 2

(a) Describe where in the periodic table metals are found.

(b) Metal X has a low density and does not react with acids or alkalis (it is resistant to corrosion). Narrow down its likely place in the periodic table.

Task 3

(a) List the characteristic properties of a metal and give some examples of how their properties match their uses.

(b) Metal X has a similar strength to steel, but is 40% lighter. Although it is heavier than aluminium, it is significantly stronger. Its combination of high strength and low density means that it is widely used in aircraft construction – around 77 tonnes are used to build an Airbus A380 'double-decker' airliner. What is the modern name for Metal X and what is its symbol?

Task 4

(a) Atoms of Metal X have 22 protons. What is its atomic number and how many electrons are there in each atom?

(b) The most common isotope of Metal X has a mass number of 48. How many neutrons are there in an atom of this isotope?

Maximise your grade

Grade	Answer includes showing that you...
	know that ores contain metals and metals are often extracted from their reaction with carbon.
E	know where metal elements (including Group 1, Group 2 and transition metals) are found in the periodic table.
	know the difference between properties of metals in Groups 1 and 2 and properties of transition metals.
	know the names of sub-atomic particles and their electrical charges.
C	understand how the properties of metals link to their uses.
	understand why the reaction between metal ores and carbon is called a reduction reaction.
A	understand the meaning of atomic number and mass number.
	can calculate the number of each sub-atomic particle in an atom from its atomic number and mass number.

Metals from ores

You will find out:

- > why some metals were discovered so much earlier than others
- > what ores are, and how we extract metals from them
- > why carbon is important for extracting metals

From stone to metal

The Stone Age gave way to the Bronze Age about 3000BC. People gradually learned how to obtain metals from rocks. Gold and silver were chipped out of the rock and melted. Bronze does not occur naturally. It is an alloy of copper and tin. The metals have to be extracted from rocks using fire and charcoal, and then mixed. Bronze Age people made tools and ornaments from copper, bronze and lead as well as gold and silver. It was not until 1000BC that they learned to extract iron.

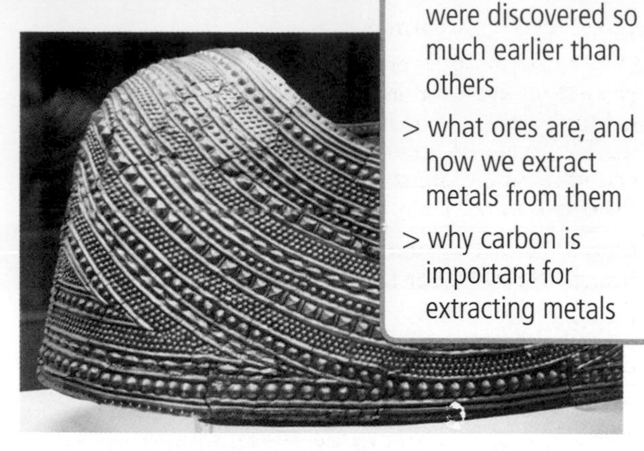

FIGURE 1: This gold shoulder cape from North Wales is nearly 4000 years old and still in good condition.

Metals in the ground

Five metals have been in use for over 5000 years and another three for over 3000 years. The rest (about 60 of them) have been known for less than 350 years. Why?

Unreactive gold and silver occur '**native**', as elements rather than combined in compounds. This means that they were simple to extract from rocks.

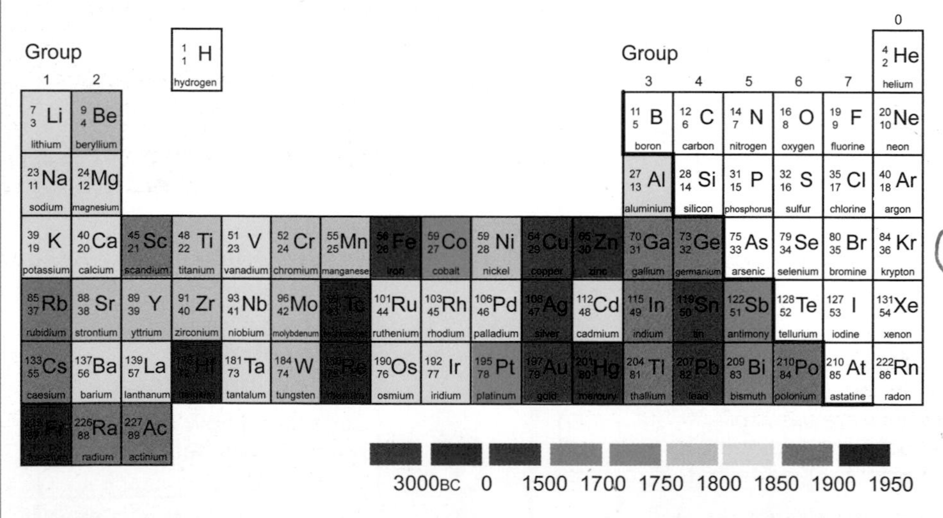

FIGURE 2: Periodic table of discovery dates. Why do you think so few metals were known before 1750?

The more reactive metals occur in rocks as compounds. Rocks that are mined for metals are called **ores**. Copper, zinc and iron ores are examples. The metals are obtained by heating their ores with coke (carbon) in a furnace. This is **smelting**.

Very reactive metals, such as aluminium, sodium and magnesium, need something other than carbon and heat to extract them. They need electricity – and that was not available until the 19th century. The process is **electrolysis**: electricity is passed through a **molten** compound of the metal. This decomposes the compound. Molten means that a substance is so hot that it has melted.

Compare Table 1 and Figure 2. You will see that the least reactive metals were discovered first and the more reactive much later.

TABLE 1: The reactivity series.

Element	Symbol
potassium	K
sodium	Na
calcium	Ca
magnesium	Mg
aluminium	Al
zinc	Zn
carbon	C
iron	Fe
lead	Pb
hydrogen	H
copper	Cu
silver	Ag
gold	Au

Increasing reactivity (upwards)

QUESTIONS

1 Explain why, two thousand years ago, only eight metals were known.

2 What is an ore?

Making use of ores

Ores and economics

Only minerals with enough metal to make it worth extracting are used as ores. Unless the metal is very valuable, it is not economic to process **low-grade ore** (rock containing a small amount of metal). However, economics change; some metals become scarcer or there are new, cheaper methods of extraction.

Metal ores are mined, often on a huge scale, and transported all over the world.

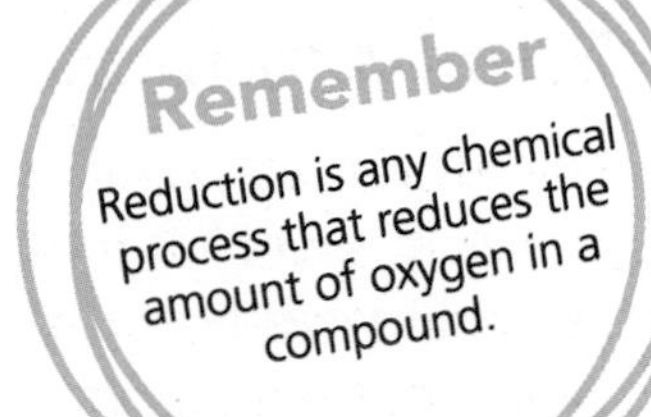

Smelting

At the smelter, the ore is crushed. It may also be concentrated, to remove rock with little or no metal.

Some ores are metal oxides. These can be smelted directly. Other ores are converted to the metal oxide before or during smelting. To convert the metal oxide to the metal, the oxygen must be removed. This is called **reduction**.

The metal oxide is reduced by heating it in a furnace, with carbon. Originally, the type of carbon used was charcoal, but now it is coke (a nearly pure form of carbon, from coal). The overall reaction is:

metal oxide + carbon → metal + carbon dioxide

The furnace must be hot enough to melt the metal. Other materials, such as limestone, are added, to remove impurities in the ore. They form **slag**.

FIGURE 3: The rust coloured ore in this mine is iron oxide. What else has to be mined, to extract the iron from this ore?

Did you know?

There are nearly 70 different metals in Earth's rocks. Apart from a few, they are present as chemical compounds such as metal oxides.

QUESTIONS

3 Suggest reasons for calling a certain type of rock an ore.

4 Why do they add coke to the furnace, at a smelter?

5 Explain what happens to impurities in the ore.

Extracting more reactive metals

Smelting with coke only works for metals less reactive than carbon. Aluminium and other reactive metals, such as sodium and magnesium, are extracted by electrolysis. It is why these metals were not discovered until the 19th century, after electricity had been discovered and the first battery, or voltaic pile, invented.

Electrolysis involves passing an **electric current** through a molten metal compound. This decomposes the compound, splitting it into its metal and non-metal elements. For example:

molten aluminium oxide $\xrightarrow{\text{electricity}}$ molten aluminium + oxygen gas

6 Describe the process to obtain sodium metal from common salt (sodium chloride). What else is produced?

FIGURE 4: In 1807–8 Humphry Davy discovered how to extract sodium, potassium, calcium, magnesium and barium. How did the newly invented voltaic pile enable him to do this?

Extracting iron

You will find out:

> some uses of iron
> how iron is extracted from iron ore in the blast furnace
> about oxidation and reduction reactions

The fiery furnace

In the orange glow, stands a hooded, silvery figure wearing a mask. Beside him flows a river of white-hot metal. He dips a 'cup' on a long pole into the river. No, it's not science fiction. It is everyday science fact. He takes samples of molten iron as it pours from a blast furnace at about 1500 °C. The protective clothing, from head to toe, is coated with aluminium to reflect the heat.

FIGURE 1: Why is this man wearing a silver suit and hood?

Iron

The importance of iron

More iron is extracted from ores than all other metals put together. Most of it is made into **steel**. It is the most important metal, because it is cheap and strong. Steel is in a huge range of modern products including cars, trains, ships, buildings and bridges. It is also used to reinforce concrete. Historians say that the Iron Age began about 3000 years ago... and we are still in it.

Making iron

Iron is extracted in a **blast furnace**. This requires four raw materials:

> Iron ore – such as haematite, which is mainly iron(III) oxide (Fe_2O_3) – the same compound as rust. There are other iron ores which contain different iron compounds.

> Coke – coal that has been heated but not allowed to burn. It is mainly carbon.

> Limestone – mainly calcium carbonate ($CaCO_3$). It removes impurities in the ore, by reacting to form slag.

> Hot air – blasted into the furnace. This gives the 'blast furnace' its name.

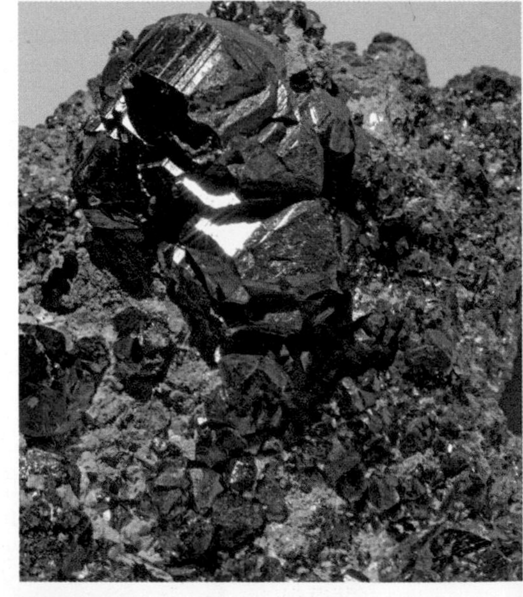

FIGURE 2: This is haematite, iron ore, not iron. Why does it look rusty?

QUESTIONS

1 What solid raw materials are added to a blast furnace?

2 Give the chemical name for the main compound in the ore, haematite.

3 Explain why limestone is added to a blast furnace.

Did you know?

Iron ore production is about one billion metric tonnes a year. That's 1 000 000 000 000 kilograms!

Inside a blast furnace

Measured amounts of iron ore, coke and limestone are fed in at the top of the furnace. Molten iron and slag collect at the bottom. In between, a series of chemical reactions takes place.

Air is blown in at the base. The carbon in the coke burns and releases heat, raising the temperature to over 1700 °C.

carbon + oxygen → carbon dioxide

$C(s) + O_2(g) \rightarrow CO_2(g)$

carbon dioxide + carbon → carbon monoxide

$CO_2(g) + C(s) \rightarrow 2CO(g)$

The carbon monoxide gas reduces the iron oxide to iron. Carbon takes oxygen away from iron because carbon is more reactive than iron.

iron oxide + carbon monoxide → iron + carbon dioxide

$Fe_2O_3(s) + 3CO(g) \rightarrow 2Fe(\ell) + 3CO_2(g)$

The temperature is above iron's melting point (1535 °C). Therefore the iron is molten and flows to the bottom. Here, it is 'tapped' (run off) and taken away for steel making.

Removing impurities

The purpose of limestone in the furnace is to deal with impurities in the ore. The high temperature decomposes limestone to form calcium oxide.

calcium carbonate → calcium oxide + carbon dioxide

$CaCO_3(s) \rightarrow CaO(s) + CO_2(g)$

Calcium oxide reacts with acidic impurities to form molten slag. This floats on the molten iron and is tapped off separately.

Remember
Reduction reduces the amount of oxygen in a compound. Oxidation increases it. You cannot have one without the other.

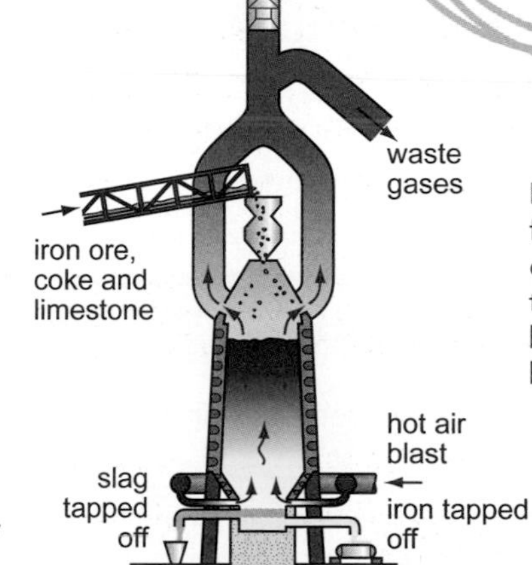

FIGURE 3: In a blast furnace, iron ore and coke are put in at the top and hot air blasted in at the bottom.

QUESTIONS

4 A blast furnace removes oxygen from iron ore. So, why do they add air?

5 Explain why the iron, made in the blast furnace, is molten.

6 Name two waste gases that come out of the top of the furnace.

Oxidation and reduction

Oxidation reactions occur when oxygen is added to a substance.

Reduction reactions occur when oxygen is removed from a substance.

Oxidation is the reverse of reduction and both must always occur together.

In a blast furnace, a series of **oxidation** and **reduction** reactions happen.

Iron(III) oxide is reduced to iron while carbon monoxide is oxidised to carbon dioxide.

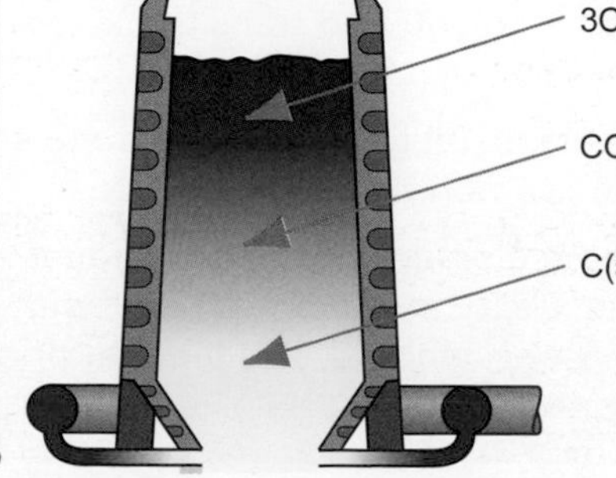

FIGURE 4: Oxidation–reduction reactions occur inside the furnace.

QUESTIONS

7 Explain why reduction cannot occur without oxidation and oxidation cannot occur without reduction.

Metals are useful

You will find out:
- about the properties of metals that make them so useful
- where transition metals are found in the periodic table
- what an alloy is

Tower of steel

The Burj Khalifa in Dubai is the world's tallest building. It was built from 110 000 tonnes of concrete, reinforced with 39 000 tonnes of steel. The Burj opened in 2010 and is 828 metres tall. At this huge height, winds can be very strong – and damaging! Using steel allowed a special spiralling design that minimises damage by wind.

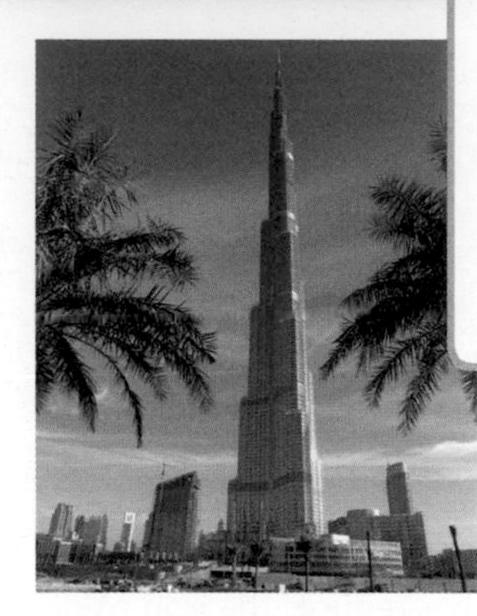

FIGURE 1: The Burj Khalifa – the world's tallest building in 2010.

Important properties and uses of metals

All metals have similar properties, which give them a wide range of everyday uses.

TABLE 1: Uses of metals depend on their properties.

Property	Examples of uses
good conductors of electricity	electrical wiring
good conductors of heat	saucepans, boilers, radiators
not brittle, so can be:	
bent and pressed into shape	car bodies, iron gates, chains
made into tubes	plumbing, bicycle frames
drawn out into thin wires	electrical wiring, wire ropes
shiny surface	jewellery, chrome plating, mirrors

Metals and the periodic table

Group 1	2											Group 3	4	5	6	7	0
			$^{1}_{1}$H hydrogen														$^{4}_{2}$He helium
$^{7}_{3}$Li lithium	$^{9}_{4}$Be beryllium											$^{11}_{5}$B boron	$^{12}_{6}$C carbon	$^{14}_{7}$N nitrogen	$^{16}_{8}$O oxygen	$^{19}_{9}$F fluorine	$^{20}_{10}$Ne neon
$^{23}_{11}$Na sodium	$^{24}_{12}$Mg magnesium											$^{27}_{13}$Al aluminium	$^{28}_{14}$Si silicon	$^{31}_{15}$P phosphorus	$^{32}_{16}$S sulfur	$^{35}_{17}$Cl chlorine	$^{40}_{18}$Ar argon
$^{39}_{19}$K potassium	$^{40}_{20}$Ca calcium	$^{45}_{21}$Sc scandium	$^{48}_{22}$Ti titanium	$^{51}_{23}$V vanadium	$^{52}_{24}$Cr chromium	$^{55}_{25}$Mn manganese	$^{56}_{26}$Fe iron	$^{59}_{27}$Co cobalt	$^{59}_{28}$Ni nickel	$^{64}_{29}$Cu copper	$^{65}_{30}$Zn zinc	$^{70}_{31}$Ga gallium	$^{73}_{32}$Ge germanium	$^{75}_{33}$As arsenic	$^{79}_{34}$Se selenium	$^{80}_{35}$Br bromine	$^{84}_{36}$Kr krypton
$^{85}_{37}$Rb rubidium	$^{88}_{38}$Sr strontium	$^{89}_{39}$Y yttrium	$^{91}_{40}$Zr zirconium	$^{93}_{41}$Nb niobium	$^{96}_{42}$Mo molybdenum	$^{99}_{43}$Tc technetium	$^{101}_{44}$Ru ruthenium	$^{103}_{45}$Rh rhodium	$^{106}_{46}$Pd palladium	$^{108}_{47}$Ag silver	$^{112}_{48}$Cd cadmium	$^{115}_{49}$In indium	$^{119}_{50}$Sn tin	$^{122}_{51}$Sb antimony	$^{128}_{52}$Te tellurium	$^{127}_{53}$I iodine	$^{131}_{54}$Xe xenon
$^{133}_{55}$Cs caesium	$^{137}_{56}$Ba barium	$^{139}_{57}$La lanthanum	$^{178}_{72}$Hf hafnium	$^{181}_{73}$Ta tantalum	$^{184}_{74}$W tungsten	$^{186}_{75}$Re rhenium	$^{190}_{76}$Os osmium	$^{192}_{77}$Ir iridium	$^{195}_{78}$Pt platinum	$^{197}_{79}$Au gold	$^{201}_{80}$Hg mercury	$^{204}_{81}$Tl thallium	$^{207}_{82}$Pb lead	$^{209}_{83}$Bi bismuth	$^{210}_{84}$Po polonium	$^{210}_{85}$At astatine	$^{222}_{86}$Rn radon
$^{223}_{87}$Fr francium	$^{226}_{88}$Ra radium	$^{227}_{89}$Ac actinium															

FIGURE 2: The shaded area highlights the transition metals.

The examples in Table 1 often contain metals such as iron, copper, titanium, chromium, silver or gold. These are **transition metals** and are in the middle block of the periodic table.

Aluminium is not a transition metal, but it is very useful and widely used.

Most transition metals are stronger than non-transition metals. They make excellent structural materials.

- Steel, which is made from iron, is cheap and strong. It is used to build bridges, ships and cars.
- Titanium is strong and lighter than steel, but much more expensive. It is kept for specialist uses.
- Copper is an excellent conductor of electricity and can be pulled into wires easily, making it ideal for electrical wiring. Copper is also a good conductor of heat and does not react with water. Boilers and saucepans are made from copper.

QUESTIONS

1 In which part of the periodic table do you find transition metals?

2 Name (a) five transition metals (b) five non-transition metals.

3 An element does not conduct electricity and is brittle. Say, with reasons, whether it is a metal or a non-metal.

Atoms and alloys

Explaining properties of metals

The properties of metals can be explained by how their atoms pack together. Metal atoms are arranged in giant structures, in regular rows and layers.

> Metals can bend because these layers can slide over each other. The shape can change but the atoms remain bonded together.

> Metals conduct electricity because some of their outer electrons are free to move through the layers.

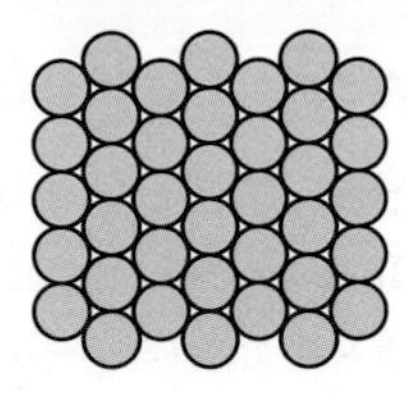

FIGURE 3: Layer of copper atoms. In each layer, each copper atom sits in the gap between the three atoms beneath. Layers slide easily over one another, rather like flexing a pile of paper.

Alloys

Pure metals are too soft for many uses. To make metals harder, they can be mixed to form **alloys**. An alloy is not a compound but is a mixture of elements, mainly metals. In steel, carbon (a non-metallic element) is present in the mix.

TABLE 2: Some common alloys.

Alloy	Elements in the alloy
steel	iron + carbon (often with other metals too)
stainless steel	iron + carbon + chromium + nickel
bronze	copper + tin
brass	copper + zinc
cupronickel for 'silver' coins	copper + nickel
duralumin	aluminium + copper
gold for jewellery	gold + copper

The proportions of each element in an alloy can differ. This affects the alloy's properties and uses:

> A steel containing more carbon is harder than one with less carbon.

> 9 carat gold jewellery ($^{9}/_{24}$ gold + $^{15}/_{24}$ copper) is harder wearing than 22 carat ($^{22}/_{24}$ gold + $^{2}/_{24}$ copper).

QUESTIONS

4 Explain why copper bends easily.

5 How does a metal alloy differ from a metal compound?

6 Name the metals present in (a) brass (b) bronze.

7 Which two elements are common to all types of steel?

Smart alloys

Smart alloys are amazing materials, designed to have a **shape memory**. Like other metals, they can be stretched into a new shape. Unlike other metals, they return to their original shape when heated. A temperature change of 15 °C is enough to trigger the change.

To mend broken bones, strips of smart alloy are cooled, stretched and screwed onto the bone. Inside the body, the smart alloy warms up. As the strips try to shrink to their original size, they hold the bones together, increasing the rate of healing.

QUESTIONS

8 How do smart alloys help broken bones to heal faster? What other uses can you think of, for smart alloys?

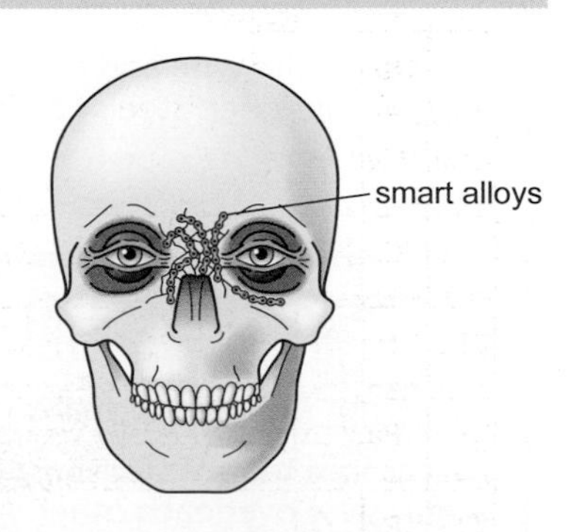

FIGURE 4: Smart alloys can be used to help repair broken bones.

Iron and steel

You will find out:

> the properties of cast iron and and steel
> how iron is converted into steel
> some uses of these metals

Bending metal

When a blacksmith makes gates, railings or horseshoes, the metal must bend without breaking. He uses steel. On the other hand, machinery must not bend. It must keep its shape, or moving parts will get jammed. Foundries make engine blocks, machine tools and much else with cast iron.

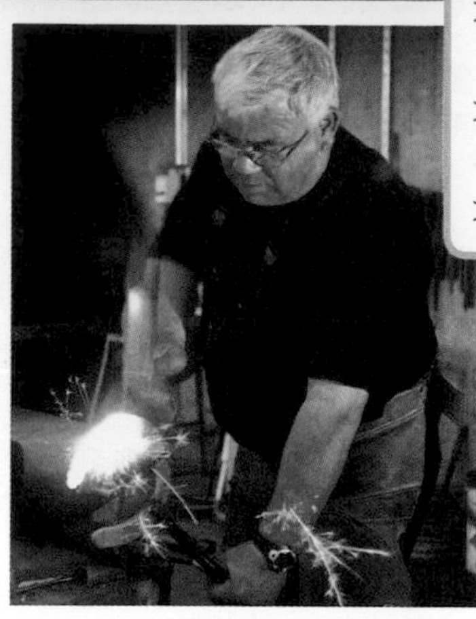

FIGURE 1: Choosing and using the right sort of metal matters.

Types of iron and steel

Blast furnace iron is not very pure. It is about 96% iron and 4% impurities (mainly carbon from the coke used in the furnace). Most iron is turned into **steel**, but not all. Some is used as **cast iron**.

Cast iron is blast furnace iron with some of the impurities removed. Drain covers and machine parts, for example, are made by **casting**: molten cast iron is poured into a mould. It cools and solidifies in the shape needed. Cast iron is strong and does not rust easily but it is very **brittle**. It cracks easily if you try to bend it. It is only used where strength, rather than flexibility, is needed.

You come across many different types of steel everyday. Car bodies, saucepans, cutlery and spanners all need different steels.

Steel is an alloy of iron and carbon (up to 2%). To convert blast furnace iron into steel:

> impurities, including most of the carbon, must be removed

> measured amounts of other metals are added.

By varying the type and quantity of metals and carbon, 'designer' steels can have properties to suit their uses.

FIGURE 2: The Iron Bridge in Shropshire is made of cast iron.

QUESTIONS

1 Give two uses for cast iron.

2 What gives different types of steel their different properties?

Designer steels

Iron is not suitable for many purposes.

Pure iron is too soft for most uses. It bends because layers of iron atoms slide over each other. Adding carbon makes sliding more difficult. It makes the metal stronger and more difficult to bend.

Cast iron has 3 to 4% carbon, which makes it too brittle for most purposes.

Did you know?

Steel has been known for over 3000 years, but the process used today is based on one invented in 1855.

... uses of steel ... cast iron

The answer is to use steel, with less than 2% carbon. The amount of carbon depends on the steel's intended use. For example, a higher carbon content will give steel greater strength in compression (when being crushed).

Type of steel	% carbon	Properties	Examples of uses
low carbon (or mild)	less than 0.4%	soft, very easy to bend and shape	car bodies; 'wrought iron' gates (used instead of wrought iron)
medium carbon	about 0.8%	hard and strong, less easy to bend and shape	nuts, bolts, screws, tubes, girders
high carbon	1.0–1.5%	very hard (good wear resistance), quite brittle	knives, scissors, railway lines

Adding other metals to molten steel can give it special properties. The choice of metal depends on how the steel will be used, and therefore the properties required.

One of the most widely used steel alloys is **stainless steel**. It is about 70% iron, 20% chromium and 10% nickel. Stainless steel is very resistant to corrosion and does not rust. This makes it useful for items in contact with water or acids.

A few examples of the effects of adding metals to steel:

> Chromium and nickel improve corrosion resistance.

> Manganese increases strength and hardness.

> Molybdenum and tungsten increase strength, hardness and toughness at high temperatures.

> Vanadium increases strength and decreases brittleness.

FIGURE 3: From saucepans to industrial food processing equipment, stainless steel is used. Why?

QUESTIONS

3 What is meant by 'designer' steels?

4 Why is manganese added to steel for railway points and tracks where busy lines cross?

Why is steel harder than iron?

The atoms in pure iron, and other pure metals, are arranged in regular rows and layers. When the metal is bent, the layers slide over each other, like sheets of paper in a stack. In steel, carbon atoms disrupt the regular arrangement of iron atoms, pushing them out of line, as in Figure 4. This stops the layers sliding easily over one another. As a result, it is more difficult to bend, and therefore stronger.

Atoms of other metals have a similar effect, when making different alloys.

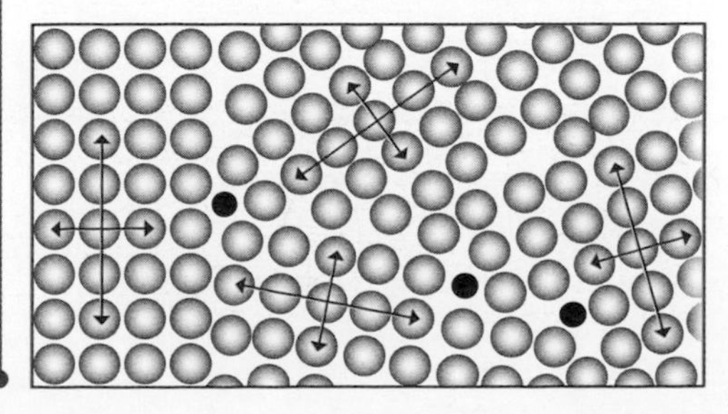

FIGURE 4: Steel is harder than iron because carbon atoms upset the regular layers of iron atoms.

QUESTIONS

5 Explain why steel is harder and more brittle than pure iron.

6 Would you expect pure copper to be harder or softer than bronze (an alloy of copper and tin)? Give your reasons.

Copper

You will find out:

> why copper is a useful metal
> how copper is extracted and purified
> copper-rich ores are limited
> mining these ores causes environmental problems

The Statue of Liberty

The Statue of Liberty is famous all over the world. It is a symbol of freedom and democracy, given to the people of the USA by the people of France, to celebrate the centenary of the American Revolution. Copper covers the outside of the statue. Over time, the statue has changed colour as the copper has been weathered, by naturally occurring acid rain.

Did you know?

Brass door handles disinfect themselves. This is because copper is an antibacterial agent.

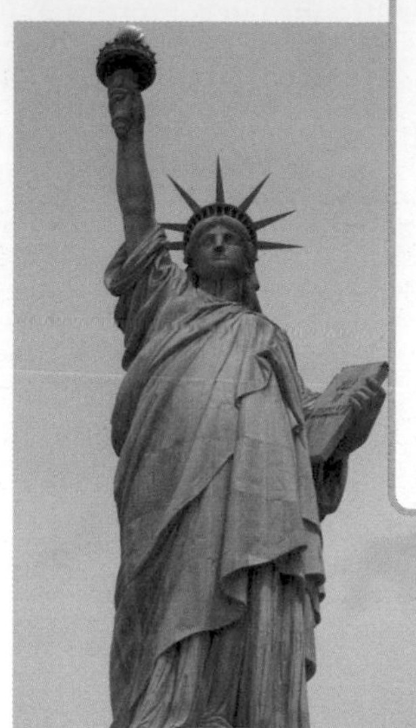

FIGURE 1: Liberty's robes are copper.

Why is copper important?

Copper is an excellent **electrical conductor**. Only silver is better. It is used for electrical wiring. It is also an excellent **thermal conductor**. The best saucepans have copper bottoms.

Copper is unreactive. As it does not react with water, copper is used for boilers, hot water tanks and pipes.

Because it combines low reactivity with excellent conduction of heat and electricity, copper is important for many purposes.

Copper alloys

Like other metals, copper bends without breaking. It keeps its new shape, allowing copper wires and pipes to be bent around corners. Pure copper is too soft for some uses, so copper alloys, such as **bronze** and **brass**, are used instead.

Extracting copper

Good quality copper ores have nearly run out. Most now contain less than 2% copper compounds. The rest is waste rock. Mining the amounts needed has huge environmental impacts.

So does extracting copper from the ores. The process depends on which ore is used. About 50% of copper now comes from **chalcopyrite** ($CuFeS_2$).

Extracting the copper involves several stages.

> Concentrating: separating copper minerals from unwanted rock

> Roasting: heating the minerals in air

> Smelting: heating in a furnace to make impure copper

> Refining: purifying the impure copper using electrolysis

Roasting and smelting both produce poisonous sulfur dioxide.

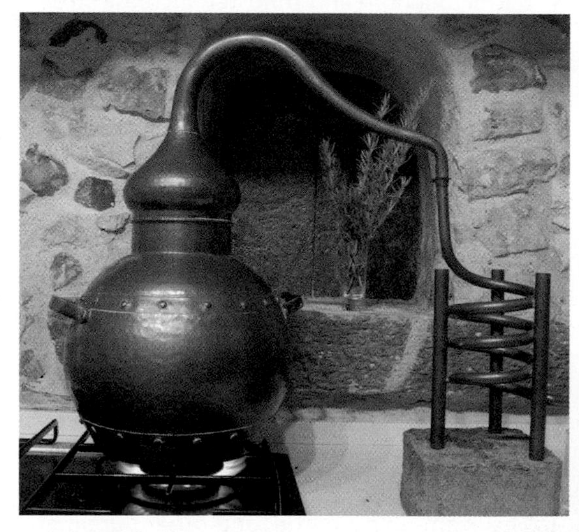

FIGURE 2: Distilling flower petals and water, to make perfume. Give two reasons why the still is made from copper.

QUESTIONS

1 Give two reasons why copper is more expensive than iron.

2 Why is it worth processing ores that contain so little copper?

... uses of copper ... copper extraction

Physical and chemical changes during extraction

Smelting copper

Copper-rich minerals are separated from crushed ore, then dried and roasted in air. This oxidises some of the sulfur to sulfur dioxide.

The roasted ore is smelted in a furnace with limestone and sand (not coke). The reactions are complicated, but the process gives molten copper and more sulfur dioxide.

The iron in the $CuFeS_2$ is not wanted. It is removed by reacting with limestone and sand to form slag.

The copper contains small amounts of other metals, so must be purified before it can be used.

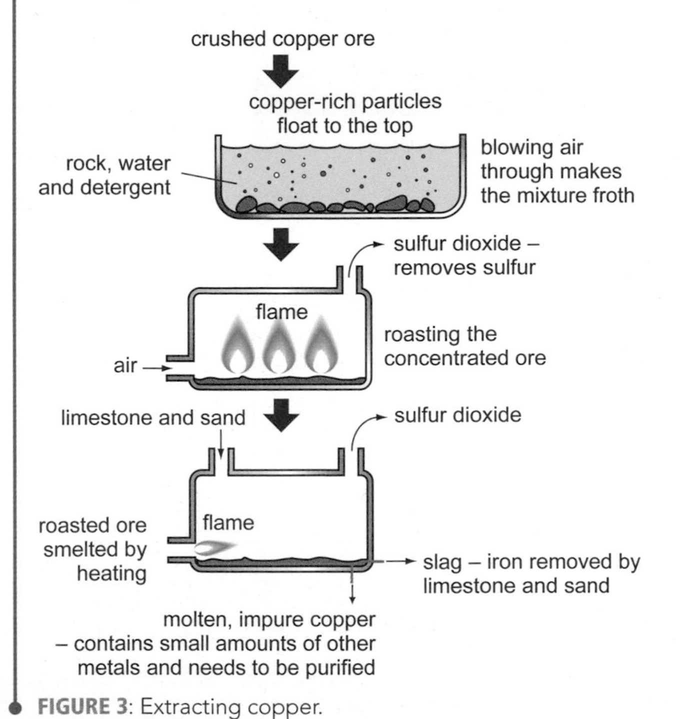

FIGURE 3: Extracting copper.

Refining copper using electrolysis

Copper is purified by electrolysis. Electrolysis means passing an electric current through an electrolyte (a liquid that conducts electricity) to make a chemical change happen.

The electrolyte is copper sulfate solution.

During electrolysis, copper atoms in the impure **anodes** (positive **electrodes**) lose electrons. They become copper ions ($Cu2^+$) and dissolve into the solution. The impurities fall to the bottom.

Meanwhile, at the pure copper **cathodes**, copper ions in the solution gain electrons. They become copper atoms, building up more pure copper on the cathodes.

Copper refined in this way is 99.99% pure.

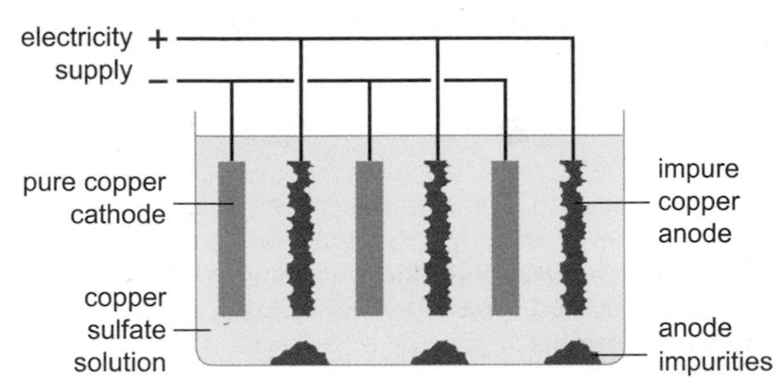

FIGURE 4: Refining copper.

QUESTIONS

3 How does smelting copper differ from smelting iron?

4 Describe what happens to impure copper when it is purified by electrolysis.

Extracting copper from low-grade ores

Scientists are developing methods for extracting copper from low-grade ores (those with very low percentages of copper in them). They give solutions of copper compounds. Copper metal is then obtained by:

> electrolysis, collecting copper on the cathode, or

> adding scrap iron, which displaces the copper from solution.

This causes less environmental damage and uses less energy than smelting.

New methods to extract copper compounds from ores include:

> **Leaching** – Chemical solutions dissolve out the copper compounds.

> **Bioleaching** – Bacteria convert insoluble compounds, into soluble ones.

> **Phytomining** – Plants absorb metal compounds through their roots. After harvesting, the copper can be extracted.

QUESTIONS

5 The waste tip of an old copper mine still contains a little copper ore. Describe a method to recover the remaining copper.

Aluminium and titanium

You will find out:

- some properties of aluminium and titanium
- some uses of these two metals
- how they are extracted from their ores

High-flying metals

The space shuttles' external fuel tanks were made from an alloy of aluminium and lithium. This alloy combines low density and high strength. Both properties are vitally important in the aerospace industry. The F22 fighter aircraft is built from a titanium alloy for the same reasons. Titanium is expensive, but light, strong and can withstand very high temperatures. In contrast, the shuttle's fuel tank must withstand the very low temperature (-217 °C) of the liquid hydrogen inside.

FIGURE 1: Aerospace metals need low density.

Properties and uses

Aluminium and titanium have the usual properties of metals. Apart from kitchen foil, pure aluminium is not used very much. Usually it is made into alloys with other metals, to make it stronger. Aluminium alloys have many useful properties:

- low density (lightweight)
- easy to shape
- strong
- weather resistant
- good electrical conductor
- good thermal conductor.

Compared with aluminium, titanium is:

- more dense, but much less dense than most other metals
- stronger, but still easy to shape
- more resistant to corrosion – not affected by acids or alkalis
- able to withstand much higher temperatures (melting points: Ti, 1660 °C; Al, 660 °C).

FIGURE 2: Cycle frames are often made of an aluminium alloy or titanium alloy. Which would you choose, and why?

Did you know?

Titanium is used for body piercing jewellery.

TABLE 1 : Uses of aluminium and titanium

Aluminium and its alloys
aircraft bodies and wings, bicycle frames, train carriages
window frames, greenhouses
food and drink cans, foil trays, bottle caps
overhead power lines
cooking pans

Titanium and its alloys
military aircraft and missiles
rockets
bicycle frames
replacement joints and other bones
casing of some laptops

QUESTIONS

1 List the properties that make aluminium and titanium suitable for aircraft.

2 Suggest why overhead power lines are aluminium, even though copper is a better conductor.

3 Explain why titanium withstands higher temperatures than aluminium.

... compare aluminium titanium

Extraction

Extracting aluminium

There is more aluminium than iron in Earth's **crust**, so why is aluminium more expensive? Most aluminium is in clay and cannot be extracted. Secondly, aluminium oxide cannot be reduced using carbon. Electrolysis is needed and electricity is expensive.

Bauxite, impure aluminium oxide, is an important aluminium ore. After removing impurities, the aluminium oxide is melted. To decompose the oxide, an electric current is passed through:

aluminium oxide ➡ aluminium + oxygen

$2Al_2O_3 \rightarrow 4Al + 3O_2$

Aluminium oxide contains aluminium ions, Al^{3+}. Unlike in the solid, the ions in molten aluminium oxide are free to move. During electrolysis, the positive Al^{3+} ions move towards the negative electrode (cathode). They pick up electrons to become aluminium atoms:

$Al^{3+} + 3e^- \rightarrow Al$

Each aluminium ion needs three electrons to become an atom. This means that a lot of electricity is needed.

Molten aluminium collects in the bottom of the electrolysis cell and is run off.

Aluminium oxide has a very high melting point. So, for electrolysis, it is dissolved in molten cryolite, another aluminium compound.

Extracting titanium

Rutile, an ore, is a form of titanium dioxide, TiO_2. Titanium cannot be extracted from rutile by reduction with carbon or by electrolysis. Instead, titanium dioxide is converted into titanium tetrachloride and reacted with magnesium, an even more reactive metal.

titanium tetrachloride + magnesium ➡ titanium + magnesium chloride

$TiCl_4 + 2Mg \rightarrow Ti + 2MgCl_2$

This method is very costly, which makes titanium an expensive metal.

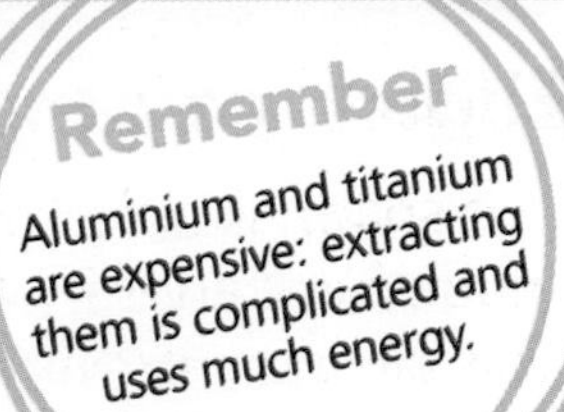

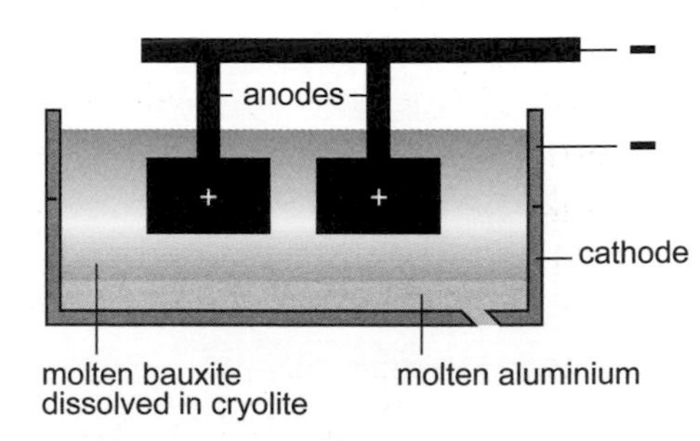

FIGURE 3: Extracting aluminium by electrolysis

QUESTIONS

4 Aluminium and titanium are very common in Earth's crust, yet they are expensive. Explain why.

5 When extracting aluminium, why must the aluminium oxide be liquid, not solid?

Corrosion resistance

Aluminium is a very reactive metal but it does not seem to be affected by air and water. The reason is that aluminium reacts rapidly with oxygen to form a tough surface layer of aluminium oxide, which prevents the aluminium from reacting further. This makes aluminium resistant to corrosion by air and water. Acids and alkalis do attack aluminium, because they react with the oxide layer.

Titanium is also reactive, yet it is used to make replacement joints in the human body. Like aluminium, titanium reacts rapidly with oxygen. It forms a tough layer of titanium dioxide, which prevents the titanium reacting further. Titanium dioxide is not affected much by acids and alkalis. Objects made from titanium are very resistant to corrosion.

QUESTIONS

6 Suggest why titanium is more resistant than aluminium to corrosion by acids and alkalis.

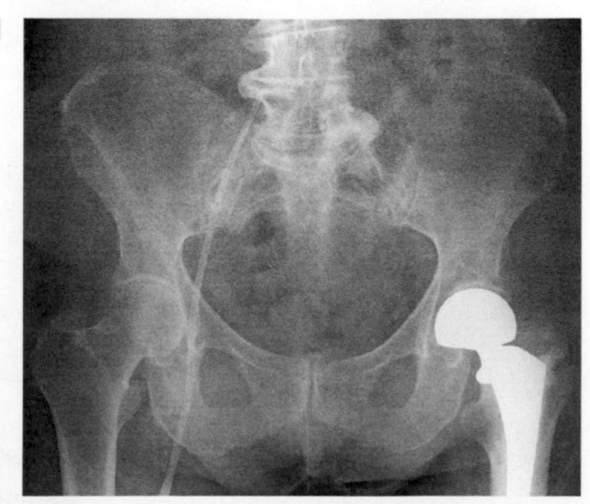

FIGURE 4: An X-ray of a titanium hip joint. Why is titanium a suitable material for replacing worn-out joints or badly broken bones?

Metals and the environment

You will find out:

- > some social, economic and environmental impacts of extracting metals
- > the benefits and problems of recycling metals
- > one way of removing toxic metals from contaminated soil

Billions of cans

Aluminium is used to make cans for fizzy drinks. Worldwide, we use six billion aluminium cans every year. End to end, the cans would reach to the Moon and back. All those cans use about 200 000 tonnes of aluminium.

Did you know?

Recycling an aluminium can saves enough energy to run your computer for three hours.

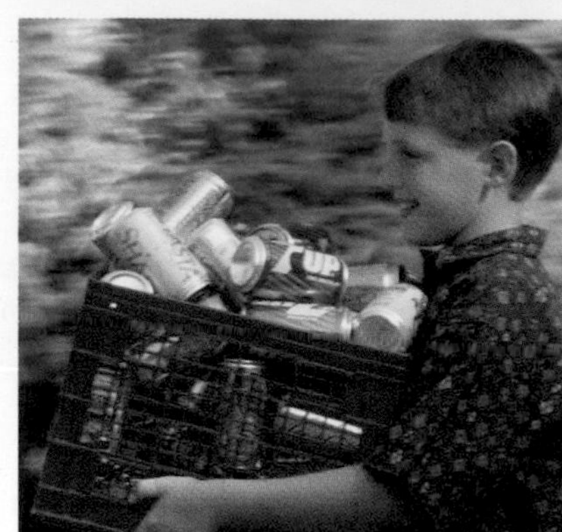

FIGURE 1: Do you recycle your aluminium cans or just throw them away?

Metal extraction

Problems with metal extraction

Nearly everything we do relies on metals – even sleeping (as mattresses often contain metal springs). Obtaining these metals causes many environmental problems:

> Extracting metals from ores uses fossil fuels, to provide energy and to make coke for smelting. This results in pollution.

> Ores are often found in environmentally sensitive areas. For example, bauxite is mined in the Amazon Rainforest in Brazil, an important ecosystem.

> **Opencast mining** literally moves mountains. It turns them into holes in the ground, and creates a mountain of waste rock somewhere else.

Ores and fossil fuels are finite resources. There are limited supplies. Once used, they cannot be replaced.

FIGURE 2: What was once Bingham Mountain in Utah is now Bingham Canyon copper mine. It is over 4 kilometres wide and 1 kilometre deep – and getting bigger. What has happened to all the rock?

Why recycle metals?

By recycling metals, there is less need to mine ore.

Recycling also helps save energy. The energy needed to make one new aluminium can from ore could make twenty cans from recycled aluminium.

Recycling scrap iron and steel saves coke and limestone, as well as iron ore.

Recycling reduces the amount of waste put into landfill.

FIGURE 3: Aluminium and steel cans. What shows that you can recycle both types of can? How can they be separated, if a recycling skip contains both?

QUESTIONS

1 Describe two of the problems with mining metal ores.

2 Explain what a 'finite resource' means.

3 Give two reasons why recycling metals is a good idea.

Q ... environmental impact metal extraction AND recycling for kids

Effects on the environment

Effects of mining ores

Metals ores are often mined from big opencast pits. These, and new access roads, destroy the landscape, including wildlife habitats. They also displace the local people, affecting, or even destroying, their way of life.

Every time a new bauxite mine opens in the Amazon jungle, large numbers of trees are cleared and burned. Native tribes, who used to hunt and fish for food, and farm the land, must move on. They will never be able to return.

Some ores contain very little of the metal mineral that is wanted – see Figure 4.

Metal smelting

Using carbon to reduce metal oxides produces carbon dioxide. The carbon dioxide adds to levels in the atmosphere:

metal oxide + carbon ➞ metal + carbon dioxide

However, amounts are small compared with burning fuels in power stations, vehicles and aeroplanes.

Smelting ores that contain sulfur, such as chalcopyrite, produces sulfur dioxide. Sulfur dioxide causes acid rain, if released into the atmosphere. However, if this sulfur dioxide is collected, it can be made into useful sulfuric acid.

Recycling aluminium and steel

Recycling cans is straightforward and economically worthwhile. Steel cans are separated from aluminium cans.

> Aluminium cans are shredded, cleaned, melted and the metal reused.

> Steel cans, and any other scrap steel such as old cars, may be added back into steelmaking furnaces.

Remember

Recycling means that less ore needs to be extracted from Earth's crust.

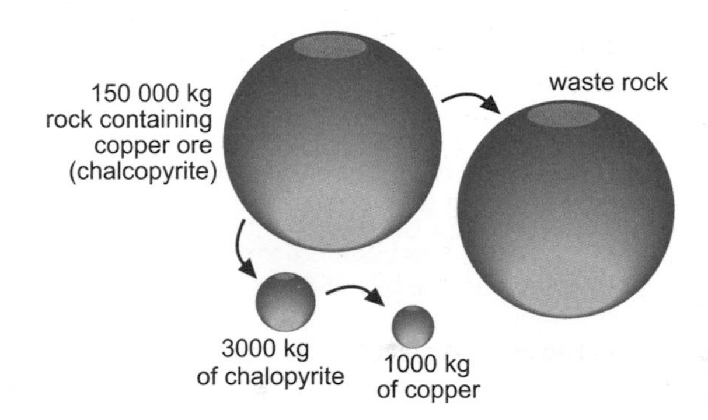

FIGURE 4: How much waste rock is mined, and later dumped, for each 1000 kg copper produced?

QUESTIONS

4 Why are new metal ore mines needed?

5 List some advantages and disadvantages of opening a new mine.

Recovering metals from brownfield sites (Higher tier)

Former industrial sites are often redeveloped for new uses. These **'brownfield sites'** may be contaminated with compounds of toxic metals, such as cadmium, nickel and cobalt. Developers must remove any contamination, especially if the land is to be used for housing.

Conventional methods involve digging out the soil and burying it – a difficult and expensive process.

More recently, **phytoremediation** has been used. *Phyto* means concerned with plants; remediation means improvement or repair. Plants remove metal compounds from the soil by absorption through their roots.

A further stage is **phytomining**: after absorbing the metal compounds, plants are burned and metals recovered from the ash.

FIGURE 5: Alyssum absorbs metal compounds from soil.

FIGURE 6: Water hyacinth cleans contaminated water.

QUESTIONS

6 Explain why it is important to decontaminate brownfield sites which are to be used for housing.

Preparing for assessment: Planning an investigation

To achieve a good grade in science, you not only have to know and understand scientific ideas, but you need to be able to apply them to other situations and investigations. These tasks will support you in developing these skills.

Investigating the strength of concrete

A student wanted to investigate the hypothesis 'the thickness of concrete affects its strength'. The student knew that other people would have already carried out some work into the hypothesis, and that before planning an investigation it would be a good idea to carry out research to find out what was already known.

The student looked on the internet and found this method of investigating the strength of concrete.

Remember that not everything published on the internet is accurate.

Method 1 A steel ball is dropped onto a concrete block. Each time it is dropped from 10 cm higher than the previous height, until the concrete block cracks. This height then indicates the strength of the block.

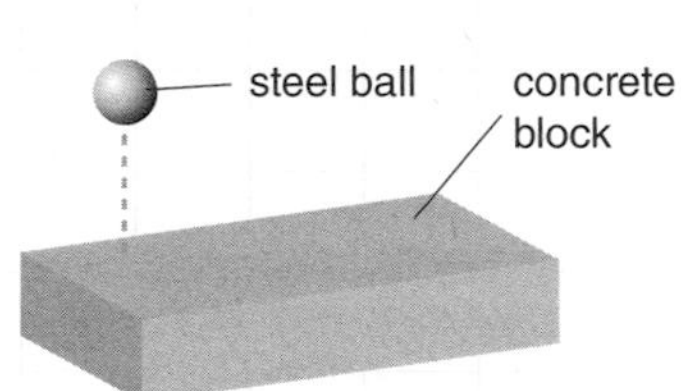

Looking in an old science textbook, the student found an alternative method.

When you have found out as much as you can, think about what is possible. You may not have a lot of time, you may not have some of the equipment, or there may be another reason why you should use one method in preference to another.

Method 2 A concrete block is placed on two supports on the bench. By placing weights on top of the unsupported part of the block, as shown, it is possible to measure the force in newtons needed to crack the block.

masses
concrete beam
supports

Planning

1. Which method – steel ball dropping or the weights on top of the block – would be the better method of obtaining the data to confirm or reject the hypothesis? Explain your choice.

Think about the two methods. Which will give the most accurate results? Does either method have a flaw that might make the results unreliable?

2. Identify the independent and dependent variables in the investigation.

Remember that the dependent variable is the variable that changes as a result of the changes to the independent variable.

3. Suggest two variables that you should keep the same in the method you selected as your answer to question 1.

Remember that control variables are factors that have to be kept the same in an investigation to make it a fair test.

4. Write a clear description of the method that you selected as your answer to question 1.

Make sure that you put your description into a clear order, and suggest values for the size of the concrete blocks or the steel ball.

5. Think about the method you have written. There are some hazards in the method. Suggest two hazards, and how you would reduce the risks to students carrying out the investigation.

Hazards are things that might possibly go wrong, and cause harm or damage. You should be realistic; the hazard should have a reasonable chance of happening.

Processing and evaluating secondary data

The student made two sets of blocks of varying thickness. One set was investigated by Method 1. Here are the results.

1 cm thick: 20 cm high, 1.5 cm thick: 30 cm, 2 cm thick: 50 cm, 2.5 cm thick: 80 cm, 3.0 cm thick: 90 cm

The other set was investigated using Method 2. Here are the results.

1 cm thick: 40 N, 1.5 cm thick: 60 N, 2 cm thick: 80 N, 2.5 cm thick: 100 N, 3.0 cm thick: 120 N

7. Construct and complete a results table for the data from Method 1.

Make sure that the headings of the rows and columns are complete, and that you state the units.

8. Use your table to draw a graph of the results.

The independent variable should be placed on the *x*-axis, and a smooth curve drawn through the points.

9. Construct and complete a results table for the data from Method 2.

10. Add the data to your graph from question 8. Your *y*-axis has the same figures for both sets of data.

This allows you to compare both trends on one graph.

11. Look at the two lines and describe each trend.

Try to use terms such as directly proportional, if the line is straight. A curve indicates that the relationship between the independent and dependent variable is more complex.

12. Explain why the student did not repeat each test to obtain a mean value.

Sometimes it is not possible to repeat tests to get reproducible results. With only one set of beams being tested to destruction, could you repeat the readings?

Connections

How Science Works

> Planning an investigation

> Assess and manage risks when carrying out practical work

> Select and process primary and secondary data

> Analyse and interpret primary and secondary data

Science ideas

C1.2 Heating limestone (pages 114–115)

A burning problem

Invisible enemies

The enemies are tiny, polluting molecules. However, the huge amounts released into the atmosphere every day cause problems worldwide. They are produced by burning fuels. We need energy from fuels, but we pay the penalty. The invisible enemies are all around us.

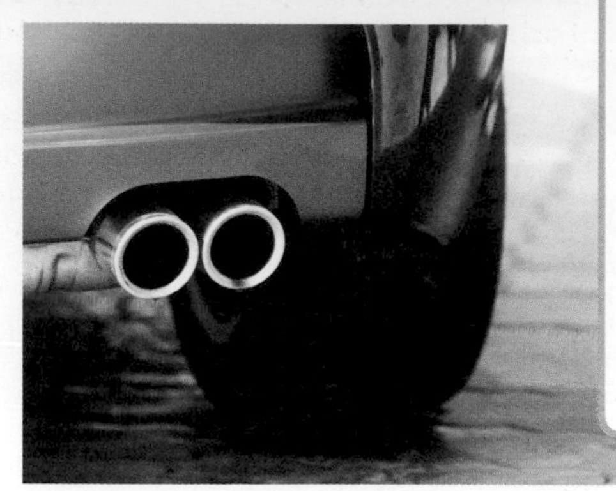

FIGURE 1: Invisible pollutants escape into the air.

You will find out:

> most fuels contain carbon and hydrogen, and many contain some sulfur

> burning fuels releases heat and the oxides of the elements in the fuels

> about some of the problems that these oxides cause

What happens when a fuel burns?

Our main fuels are **coal**, **oil** and **natural gas**. These are **fossil fuels**. Coal is mainly carbon. Oil and gas are mainly hydrocarbons (compounds of carbon and hydrogen).

Biofuels, such as wood and biodiesel, are compounds of carbon, hydrogen and oxygen. They are not fossil fuels.

When a fuel burns, **combustion** reactions oxidise the carbon and hydrogen into carbon dioxide (CO_2) and water (H_2O) in the form of steam.

hydrocarbon + oxygen → carbon dioxide + water

How does burning fossil fuels affect the environment?

Modern life depends on energy from burning fossil fuels, but there are problems:

> Fossil fuels are **finite resources**. There is only a limited amount in Earth's crust. Once they have been used, there are no more.

> Carbon dioxide contributes to **global warming**. Rising sea levels is one possible effect of global warming.

> If there is insufficient oxygen, poisonous carbon monoxide (CO) and carbon (as soot) are produced.

> Some fuels produce tiny smoke particles that contribute to **global dimming**.

> Fossil fuels contain small quantities of sulfur compounds. These burn to form **sulfur dioxide** (SO_2), a poisonous gas. Sulfur dioxide causes breathing problems and acid rain.

FIGURE 2: Why might burning fuels cause more floods around the world?

QUESTIONS

1 Name three fossil fuels.

2 Name the gas produced when carbon is burned in a good supply of oxygen.

3 Which gas causes acid rain?

Did you know?

In 2007, the UK produced 564 000 000 000 kg of carbon dioxide.

Global effects

What is global warming?

Earth's surface is warmed by the Sun, and then emits infra red radiation back into space. Carbon dioxide absorbs some of this energy. Increasing levels of carbon dioxide absorb more infrared radiation. This means that less is escaping, and Earth gets warmer.

What is global dimming?

Krakatau, a volcano in Indonesia, exploded in 1883, forcing millions of tonnes of ash into the atmosphere. This prevented sunlight getting through, so the region was shrouded in darkness for several days.

Smoke from burning fuels has a similar, though less drastic, effect. Sunlight becomes dimmer. Some scientists believe global dimming may lead to Earth's surface cooling down.

Both global warming and globing dimming may affect weather patterns around the world. However, their combined effects are uncertain.

FIGURE 3: Krakatau 1883 – the world's biggest recorded explosion. How did it turn day into night?

FIGURE 4: Suggest what killed these trees.

How is acid rain caused?

Rainwater is slightly acidic, with a pH around pH 5.5. One reason is because carbon dioxide, in the atmosphere, dissolves in the rain.

As well as carbon dioxide, burning fossil fuels produces sulfur dioxide. Oxides of nitrogen are produced in hot car engines. Each of these gases reacts with water to form acids.

$$CO_2(g) + H_2O(\ell) \rightarrow H_2CO_3(aq)$$

carbon dioxide   water   carbonic acid

$$SO_2(g) + H_2O(\ell) \rightarrow H_2SO_3(aq)$$

sulfur dioxide   water   sulfurous acid

Sulfurous acid reacts with oxygen to form sulfuric acid H_2SO_4.

In a similar way, oxides of nitrogen react with water to form acids.

All these acids dissolve in the rain, making it more acidic, with a lower pH.

Acid rain attacks limestone buildings much more quickly than normal rain. It can also cause serious damage to trees and to aquatic life in affected lakes.

QUESTIONS

4 Explain how increased levels of carbon dioxide lead to global warming.

5 Explain why normal rainwater is slightly acidic.

6 Describe how acid rain is formed.

Other problems with combustion

During combustion, each carbon atom in a fuel needs two oxygen atoms to form carbon dioxide, CO_2. If there is not enough air, some carbons get only one oxygen, or none at all.

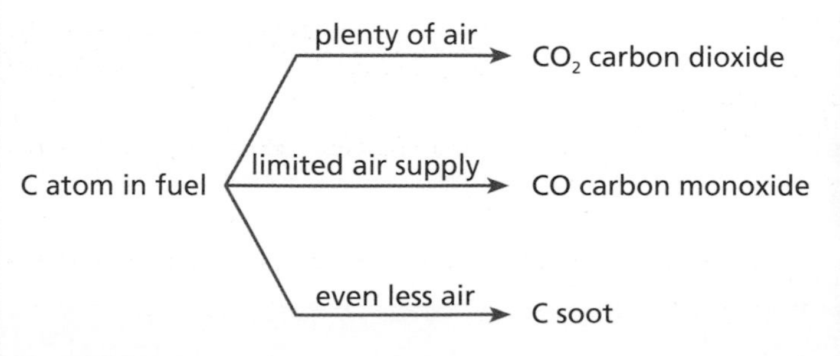

If you have a gas fire, you should have a carbon monoxide alarm, because carbon monoxide is poisonous, but invisible and **odourless** (has no smell). You will not notice that you are breathing it in, but it may still kill you.

FIGURE 5: Explain why a Bunsen flame alters when you adjust the air hole.

QUESTIONS

7 If your home has gas appliances, a smoke alarm may not provide enough protection. Explain why and how you could be better protected.

Reducing air pollution

You will find out:
- > ways to reduce sulfur dioxide pollution from burning fuels
- > why modern cars are fitted with catalytic converters
- > alternative ways of powering vehicles

Lunar exploration

In 1971, the Apollo 15 astronauts explored the surface of the Moon. They used a lunar buggy powered by silver–zinc batteries. There were two reasons. There is no air on the Moon, so a fuel cannot burn, and it was important not to pollute the Moon with exhaust fumes.

FIGURE 1: Non-polluting lunar buggy.

What are the problems?

Burning fuels causes **air pollution**. The main sources are power stations and road vehicles – mostly cars.

Reducing sulfur dioxide pollution

Sulfur compounds can be removed from oil and gas. When burned, these 'cleaner' fuels produce little or no sulfur dioxide. Removed sulfur can be sold and used to make sulfuric acid.

Sulfur compounds are not removed from coal – it is too difficult. However, sulfur dioxide produced in coal-fired power stations can be removed by reaction with limestone.

What other fuels could we use?

Biofuels such as ethanol and biodiesel can be used. These are sulfur-free and made from plants, so are **renewable resources**.

Hydrogen is another alternative. Burning hydrogen produces only water vapour – there is no pollution. However, making hydrogen needs electricity, and therefore power stations, which do cause pollution.

FIGURE 2: What is the advantage of using low sulfur fuel?

QUESTIONS

1 Suggest which human activities produce the most air pollution.

2 Give two advantages of removing sulfur compounds from oil and gas.

3 Explain why biodiesel and ethanol are renewable fuel resources.

Alternative solutions

Reducing pollution from vehicles

Vehicles that burn fossil fuels produce poisonous carbon monoxide. They also produce several different nitrogen oxides, known as NOx. This is because engine temperatures are very high, causing nitrogen (from the air) and oxygen to react and form nitrogen oxides.

All new cars have a **catalytic converter** ('cat') in their exhaust pipe. The 'cat' contains precious metals such as platinum. These catalyse chemical reactions that convert pollutant gases into harmless ones.

Did you know?

Air pollution can be reduced by removing impurities from the hydrocarbon fuel and the pollutants formed when it burns.

... biodiesel ethanol for students

Carbon monoxide and unburnt hydrocarbons react with nitrogen oxides. They are **oxidised** to carbon dioxide. Nitrogen oxides lose oxygen, for example, by reacting with carbon monoxide. They are **reduced** to nitrogen.

Fuel must be unleaded, because lead 'poisons' the catalyst and stops it working.

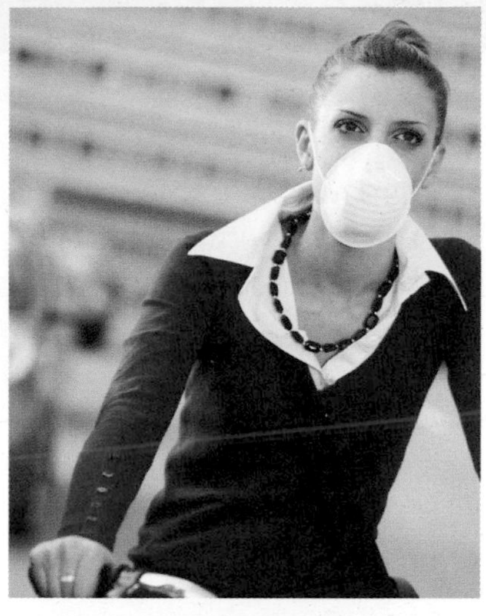

FIGURE 3: Harmful pollutants in vehicle exhaust gases affect air quality and health.

Alternative fuels

Car engines normally burn petrol or diesel, but they could burn ethanol or biodiesel instead.

Ethanol is produced from sugar. It is already popular in countries, such as Brazil, that grow sugar cane. A mixture of ethanol and petrol produces less carbon monoxide than petrol alone. However, ethanol releases less energy per litre than petrol.

Biodiesel is made with vegetable oils, mainly from rapeseed and soya beans. It is usually mixed with diesel, but can be used on its own if the engine is modified.

Ethanol contains no sulfur and biodiesel almost none. When mixed with normal fuels they reduce sulfur dioxide pollution.

Hydrogen is an ideal fuel. It is renewable, releases lots of energy, and is non-polluting as it produces only water vapour. However, it cannot be used in normal engines. One alternative is to use hydrogen **fuel cells** to generate electricity, to power electric vehicles.

FIGURE 4: Why are hydrogen-powered vehicles 'emission-free'?

QUESTIONS

4 Explain how catalytic converters reduce harmful emissions.

5 Why do cars with a 'cat' need unleaded petrol?

6 Give one advantage and one disadvantage of mixing ethanol with petrol.

Problems with alternatives

Fuels

Using biofuels reduces air pollution but creates ethical and environmental problems. Rainforest is cut down to grow sugar and soya bean for biofuels. Elsewhere, land that could grow crops to feed the world's increasing population is, instead, producing fuels.

Using hydrogen as a pollution-free fuel poses technical problems. Hydrogen can be produced by electrolysis of water, but this is pollution-free only if the electricity is generated using renewable energy. At present, such methods are expensive. Secondly, liquid hydrogen boils at -253 °C. You cannot store it in a normal fuel tank.

Vehicles

Electric vehicles do not emit harmful pollution. However, their range is limited and their batteries must be recharged regularly. Generating electricity at conventional power stations does produce pollution.

QUESTIONS

7 Explain how an increase in the number of electric vehicles could lead to an increase in the levels of air pollution.

... alternative fuels ... electric cars

Crude oil

Can't live without it?

Only two liquids occur naturally on Earth. All living things depend on water. In contrast, only humans depend on the other liquid – oil. We depend on it for transport, clothing, plastics, medicine and much more. We're addicted. It is going to be difficult to give up, but we will have to. Sooner or later crude oil will run out.

FIGURE 1: People drill for oil in remote places in unpleasant conditions.

You will find out:

- > crude oil is a mixture of different compounds
- > most of the compounds in crude oil are hydrocarbons
- > how crude oil can be separated by fractional distillation

Crude oil fractions

Crude oil is a mixture

Petrol is just one of many useful products made from **crude oil (petroleum)**. Crude oil is a mixture of many different compounds. Most are **hydrocarbons** – compounds containing only hydrogen and carbon. Some hydrocarbons have molecules made from a few atoms. Others are larger. Some have hundreds of atoms.

Fractional distillation

Because crude oil is a mixture, it can be separated by a physical process such as **distillation**.

Fractional distillation separates the crude oil into **fractions**. Each fraction is a mixture of different-sized hydrocarbon molecules with similar boiling points. Fractions with larger molecules boil at higher temperatures.

Uses of fractions

TABLE 1: Uses of fractions depend on boiling points.

Name of fraction	Carbon atoms per molecule	Boiling point range (°C)	Uses
petroleum gas	1 to 4	below 20	heating, cooking, LPG fuel
petrol	5 to 9	20–100	fuel (cars and lorries)
naphtha	6 to 10	40–170	to make other chemicals
kerosene	10 to 16	170–240	jet fuel, paraffin
diesel	14 to 20	240–330	fuel (cars, lorries and trains)
fuel oil	20 to 50	330–400	fuel for ships, factories and heating
bitumen	more than 50	over 400	tar for road making

FIGURE 2: Models of methane, a gas, and octane, a liquid. What is the chemical formula for octane?

QUESTIONS

1 Give the name for compounds made only from hydrogen and carbon atoms.

2 What is a crude oil 'fraction'?

3 Explain how crude oil is separated into fractions.

Did you know?

Petra is Greek for rock and *oleum* is Latin for oil, hence 'petroleum' because it is found in rocks underground.

... crude oil facts

How does fractional distillation work?

The atoms that form a molecule are held together by strong **covalent bonds**. However, there are also attractions (bonds) between molecules which are very much weaker. The boiling point of a compound depends on the strength of attraction between its molecules. The stronger the attractions, the higher the boiling point (though they are still very weak). Small hydrocarbon molecules have weaker attractions, so have lower boiling points.

Fractional distillation separates the hydrocarbons according to their boiling points. This happens in a **fractionating column** (Figure 3). Crude oil is heated to about 400 °C. Most of it boils and **vaporises**.

> The hydrocarbons with highest boiling points do not boil. They remain liquid.

> Vapours consisting of the other hydrocarbons rise up the column, gradually cooling. When cooled below their boiling point, they **condense** back to liquid and are run off.

> Hydrocarbons with high boiling points condense first. The remaining gases continue upwards, cooling further. The lower their boiling point, the higher up the column they rise before condensing.

> Hydrocarbons with different size molecules condense at different levels, separating the crude oil mixture into a series of fractions. Each fraction contains hydrocarbons with a limited range of boiling points.

Remember

Bonds within a molecule (covalent bonds) are much stronger than bonds between one molecule and another.

QUESTIONS

4 Explain why the four smallest hydrocarbons are gases at room temperature.

5 Pentane's chemical formula is C_5H_{12}. Dodecane's is $C_{12}H_{26}$. Which boils at the higher temperature? Explain why.

6 Does dodecane condense higher up the fractionating column than pentane, or lower down? Explain why.

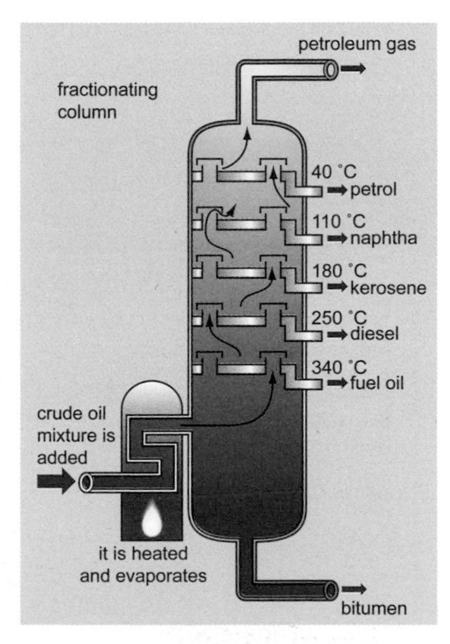

FIGURE 3: What happens to the oil vapours as they pass up the column?

Why do boiling points depend on size?

A liquid boils when its molecules gain enough energy to break the attractions between each other and become a gas (vapour). When a liquid is heated, energy is transferred to its molecules. Small molecules need to gain only a little energy to overcome the very weak bonds between them. Hence, hydrocarbons with small molecules boil at low temperatures.

Larger hydrocarbon molecules have stronger bonds between them, need to gain more energy to break their bonds and, therefore, boil at higher temperatures.

QUESTIONS

7 Explain the relationship between the number of carbon atoms and the boiling point of the hydrocarbon.

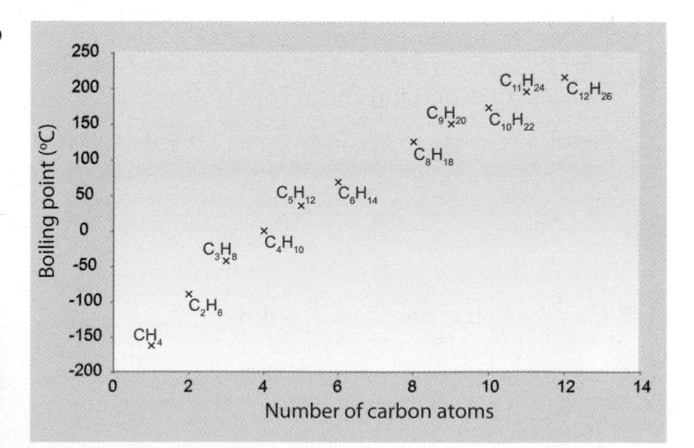

FIGURE 4: Boiling point increases with the size of the molecules. Estimate the boiling point of the hydrocarbon with seven carbon atoms. Check your answer in a data book.

Alkanes

You will find out:

> most of the hydrocarbons in crude oil are alkanes
> how to recognise an alkane
> how members of the alkane family are related to each other

Fuels for racing cars

Petrol is mainly a mixture of hydrocarbons called alkanes with 5 to 10 carbon atoms per molecule. The petrol used by Formula 1 cars must not contain anything that is not in normal petrol. It's just blended differently to give the maximum performance under race conditions. Petrol for normal cars is very similar, but blended to give good economic performance rather than maximum power.

FIGURE 1: Formula 1 cars use specially blended mixtures of alkane hydrocarbons.

What are alkanes?

Methane

Natural gas is used for cooking and heating, and at school in Bunsen burners. It is mainly **methane**, the smallest and simplest hydrocarbon. Methane molecules are made up of one carbon atom joined to four hydrogen atoms. Methane's formula is CH_4.

Methane belongs to a family of hydrocarbons called **alkanes**. Most hydrocarbons in crude oil are alkanes. Their names all end in 'ane'. The next members of the family are ethane (C_2H_6), propane (C_3H_8) and butane (C_4H_{10}).

Can you see a pattern in their formulae? Try to work out how many carbon and hydrogen atoms the next member has.

Alkanes

All alkanes are similar to one another. Their formulae follow a regular pattern. They have similar chemical properties.

All alkanes are hydrocarbons, but not all hydrocarbons are alkanes. There are other hydrocarbon families with different names, properties and formula patterns.

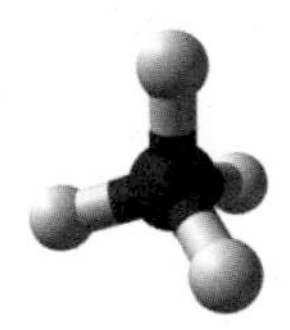

FIGURE 2: What does this model represent?

QUESTIONS

1 Name the elements in alkanes.

2 What is the formula of methane?

3 Is ethanol an alkane? How can you tell?

Did you know?

'Octane rating', usually 95 or 97, shows how well a petrol performs, compared with octane (C_8H_{18}).

Patterns and properties

Why is there a pattern in their formulae?

Each alkane has one carbon atom and two hydrogen atoms more than the previous member of the family. They all fit the general formula C_nH_{2n+2}, where n is the number of carbon atoms.

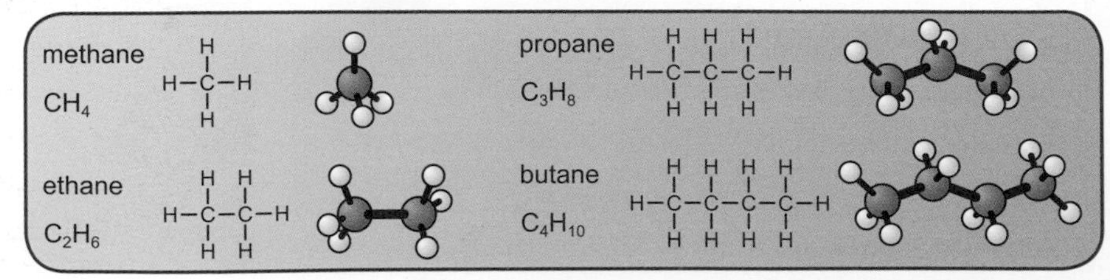

FIGURE 3: The four simplest alkanes. You need to know their names.

To work out the number of hydrogen atoms, double the number of carbons and add two. This is because there are two hydrogen atoms attached to each carbon atom, plus one at each end. The chain increases by CH_2 each time.

You need to recognise alkanes represented in the three ways shown in Figure 3. Each line (–) represents a single covalent bond.

QUESTIONS

4 Hexane has six carbon atoms. How many hydrogen atoms does it have? Draw its structure.

5 Suggest the type of fuel you would expect to contain hexane.

Properties and uses

TABLE 1: These properties influence how crude oil fractions are used to make different types of fuel.

Fraction	Alkanes	Boiling point (°C)	Properties	Uses
petroleum gas	CH_4 to C_4H_{10}	below 20	> flammable gases > liquefy under pressure	household gas supplies bottled gas lighter fuel petrol
petrol	C_5H_{12} to C_9H_{20}	20–100	> flammable liquids > vaporise easily > ignited by a spark	small road vehicles needing good acceleration
kerosene	$C_{10}H_{22}$ to $C_{16}H_{34}$	170–240	> liquids that vaporise and burn at higher temperatures	jet fuel (balancing high energy with good acceleration)
diesel	$C_{14}H_{30}$ to $C_{20}H_{42}$	240–330	> liquids needing higher temperature to ignite > give more energy	transport (lorries, buses and trains) needing high energy but less acceleration
fuel oil	$C_{20}H_{42}$ to $C_{50}H_{102}$	330–400	> slow burning > high energy content	domestic central heating energy for industry ships (very low acceleration)

What makes alkanes similar to each other?

All alkanes contain carbon and hydrogen atoms held together by single covalent bonds. They are said to be **saturated** hydrocarbons. You will meet **unsaturated** hydrocarbons, with double bonds, later.

Alkanes have similar structures to one another, explaining the similarities in their properties. The regular increase in size from one member to the next (an extra CH_2 unit) explains trends in their properties.

As the number of carbon atoms increases:

> molecules become larger and heavier

> boiling point increases

> flammability decreases (catch fire less easily)

> viscosity increases (liquid becomes thicker).

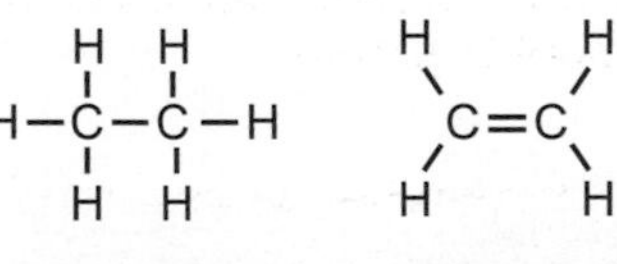

FIGURE 4: Ethane is a saturated hydrocarbon. Ethene is an unsaturated hydrocarbon.

QUESTIONS

6 Explain why alkanes behave in a similar way to one another.

7 Suggest reasons why ethane and ethene react differently from one another.

Checklist C1.1–C1.4

To achieve your forecast grade in the exam you will need to revise

Use this checklist to see what you can do now. Refer back to the relevant topics in this book if you are not sure. Look across the three columns to see how you can progress. **Bold** text means Higher tier only.

Remember that you will need to be able to use these ideas in various ways, such as:

- interpreting pictures, diagrams and graphs
- applying ideas to new situations
- explaining ethical implications
- suggesting some benefits and risks to society
- drawing conclusions from evidence you are given.

Look at pages 276–297 for more information about exams and how you will be assessed.

To aim for a grade E	To aim for a grade C	To aim for a grade A
Recall that metal elements are found on the left of the periodic table and non-metals on the right.	Know that vertical columns are groups and horizontal rows periods. Name and position the three sub-atomic particles.	Know charges of sub-atomic particles. Be able to work out the structure of the first twenty elements.
Recall that electrons orbit a central nucleus of the atom.	Know that electrons are arranged in energy levels or shells. Know how many each shell can accept.	Describe the electron arrangement of the first twenty elements.
Know that noble gases are very unreactive because they have full outer shells of electrons.	Know that elements in the same group have the same number of outer electrons and that this determines chemical reactivity of the elements.	
Recall that elements are made of atoms, and that atoms join together to form compounds.	Understand that in non-metal compounds the atoms share electrons to form molecules, and that in metal and non-metal compounds the atoms transfer electrons to form ions. Know that the sharing of electrons between atoms is called covalent bonding.	
Write word equations from the names of the reactants and products. Know that the total mass of reactants will always equal the total mass of the products.	Use a symbol equation to determine the number of atoms. Calculate the mass of a reactant or product from the masses of the other reactants and products.	**Be able to balance a symbol equation.**
Recall that limestone is mainly calcium carbonate ($CaCO_3$). Know that limestone is quarried, and used as a building material, and to make cement and concrete.	Understand that limestone decomposes on heating, producing calcium oxide and carbon dioxide, and that other metal carbonates decompose on heating. Understand that limestone quarrying has environmental implications.	Evaluate the use of limestone products in the building industry. Understand that Group 1 carbonates may not decompose at the temperatures reached by a Bunsen burner.
Know that calcium oxide dissolves in water to form calcium hydroxide (limewater); carbon dioxide turns limewater cloudy.	Understand that calcium hydroxide solution is alkaline and neutralises acids. Carbon dioxide in limewater forms calcium carbonate.	

To aim for a grade E	To aim for a grade C	To aim for a grade A
Know that limestone is damaged by acid rain.	Understand that carbonates react with acids to produce carbon dioxide, a salt and water.	
Recall that limestone is heated with clay to make cement.	Know that cement is mixed with sand to make mortar or sand and aggregate to make concrete.	
Recall that ores contain metals. Know that unreactive metals, for example gold, are found as the metal and that others are extracted by carbon.	Know that ores are mined and may be concentrated before extraction. Evaluate the social, economic and environmental impacts of exploiting metal ores, and recycling.	
Recall that more reactive metals than carbon, for example aluminium, are extracted by electrolysis.	Understand that electrolysis needs lots of energy and metals extracted by electrolysis are expensive. Know that copper-rich ores are heated in a furnace and purified by electrolysis. Know that copper-rich ores are limited.	
Explain why we should recycle metals. Recall that most iron is converted into steels and that most metals in everyday use are alloys.	Know that copper can be obtained from low grade ores, by phytomining, or bioleaching. Describe the differences between iron, alloys and steel.	**Know that plants can absorb some metals out of soil.** Describe the differences between low and high carbon steel.
Recall that the central blocks of the periodic table are transition metals.	Describe the uses and properties of copper, aluminium and titanium.	
Know that crude oil is a mixture. Recall that the compounds in crude oil consist of molecules of hydrogen and carbon atoms only (hydrocarbons).	Know that the hydrocarbons in crude oil are saturated hydrocarbons called alkanes, which have the general formula C_nH_{2n+2}	
Name the structural formula of alkanes. Know that crude oil is separated into fractions by fractional distillation.	Understand that each fraction has different properties, and that the properties depend on the size of the molecules. Describe the trends in these properties: boiling points, viscosity and flammability.	
Name the gases released when a carbon and hydrogen fuel burns. Know that solid particles may also be released.	Understand that during combustion the carbon and hydrogen in the fuels are oxidised. Know that the solid particles may contain both soot and unburned fuels.	
Understand that some fuels contain sulfur and that this can produce the gas sulfur dioxide, which causes acid rain. Know how biodiesel and ethanol are produced.	Understand that carbon dioxide causes global warming, and solid particles cause global dimming. Know that sulfur can be removed from fuels before they are burned and that sulfur dioxide is removed from the combustion gases in power stations. Evaluate the economic, ethical and environmental issues surrounding the use of biofuels.	

Exam-style questions: Foundation C1.1–1.4

1. The numbers in this periodic table represent elements.

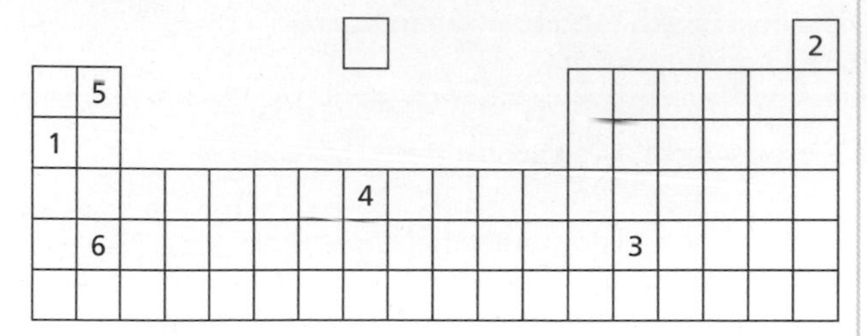

AO1 **(a)** Which numbered elements are in the same group? [1]

AO1 **(b)** Which numbered elements are in the same period? [1]

AO2 **(c)** Which numbered element is a non-metal? [1]

2. Natural gas used to heat our homes is mainly the hydrocarbon methane. Some students used this apparatus to investigate the gases produced when methane burns in oxygen.

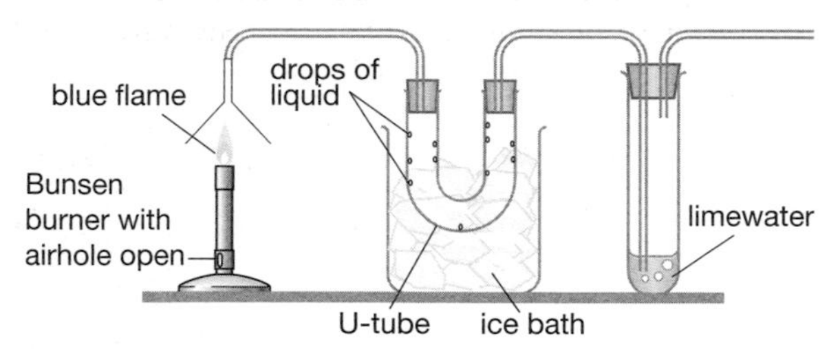

AO1 **(a) (i)** Name the gas that turns the limewater cloudy. [1]

AO3 **(ii)** Name the liquid that collects in the U-tube. [1]

AO3 **(b) (i)** During the investigation the funnel became coated in a black substance.

Name this substance. [1]

AO2 **(ii)** Explain why this substance forms. [2]

3. Iron is a commonly used metal. It is extracted using carbon in a blast furnace.

AO2 **(a)** Explain why carbon is used to extract iron in the blast furnace. [3]

(b) Steel is made from iron. Look at this bar chart showing the composition of iron and several steels, and use the information to help you answer the questions below.

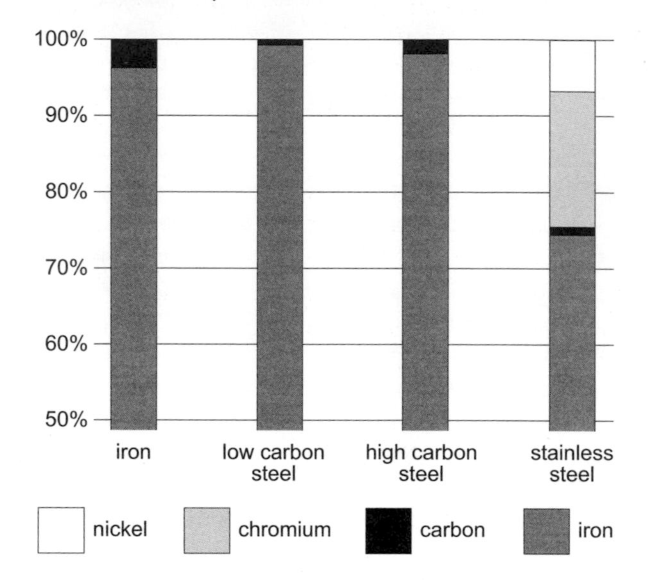

AO2 **(i)** Metal-cutting tools are best made from high carbon steels. Explain why. [2]

AO2 **(ii)** Car bodies are best made from low carbon steel. Explain why. [2]

AO2 **(iii)** Which metals are used to make the alloy stainless steel? [2]

AO2 **4.** *In this question you will be assessed on using good English, organising information clearly and using specialist terms where appropriate.*

Copper is an important metal and has many uses. Copper ores are expensive. Alternative sources of copper include:

- phytomining of plants, such as brassicas
- bioleaching by bacteria
- refining scrap copper.

Use your knowledge and understanding to give both positive and negative environmental impacts of using each of these methods to provide copper for industry. [6]

AO1 recall the science AO2 apply your knowledge AO3 evaluate and analyse the evidence

WORKED EXAMPLE – Foundation tier

Atoms are made up of three main particles called protons, neutrons and electrons.

Use the periodic table on the data sheet to help you to answer these questions.

(a) Calcium is in Group 2 of the periodic table.

(i) Why are calcium and magnesium in the same group of the periodic table? [1]

Both calcium and magnesium have two outer electrons so they are in the same group of the periodic table.

(ii) How many protons are there in an atom of magnesium? [1]

24

(iii) Each magnesium atom has 12 electrons. Complete the electronic structure of magnesium. [2]

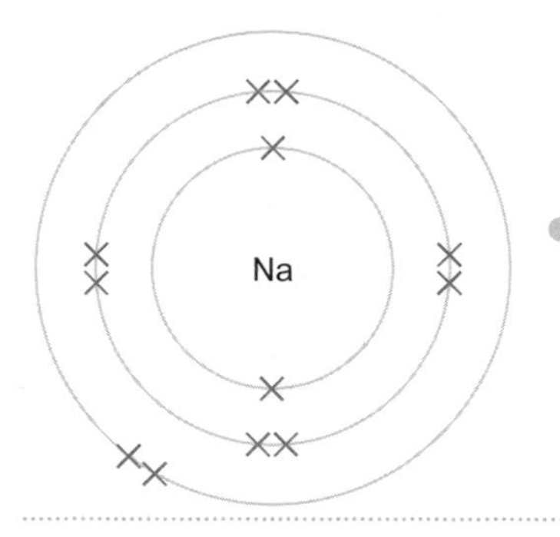

(b) The chemical equation for a reaction of magnesium carbonate is shown below.

$MgCO_3 \rightarrow MgO + CO_2$

Describe this reaction of magnesium carbonate in terms of the names of the substances and the numbers of the atoms involved. [3]

Magnesium carb has 1 Mg atom, 1 C atom and 3 O atoms. It makes 1 molecule of magnesium oxide, and 1 molecule of carbon dioxide.

(c) 84 grams of $MgCO_3$ was heated, and 40 grams of MgO was made. Calculate the mass of CO_2 gas produced. [1]

84 – 40 = 44 grams

How to raise your grade!

Take note of these comments – they will help you to raise your grade.

It would be better to say 'They are both in Group 2. Atoms in this group all have two outer electrons'.

This is wrong – the incorrect number has been picked from the periodic table. If you are not sure which number to use, either remember the proton number is the smaller of the two, or use the key on the periodic table.

Good diagram worth both marks. There is no need to pair electrons, and the outer shell would have been better if the electrons were at opposite sides of the shell.

This answer starts off well, except for the abbreviated form of carbonate, and gives the atoms in $MgCO_3$. It then correctly names both products, but fails to name and count the atoms in each compound produced, so loses one mark here. It is also incorrect to say "a molecule of magnesium oxide", but as this question is about counting atoms, there is no penalty. Watch out! Do not abbreviate/shorten words in questions requiring good English.

A good answer that shows clear working out, with the correct units given.

Exam-style questions: Higher C1.1–1.4

1. Limestone and clay are heated by burning methane to make cement in a rotary kiln.

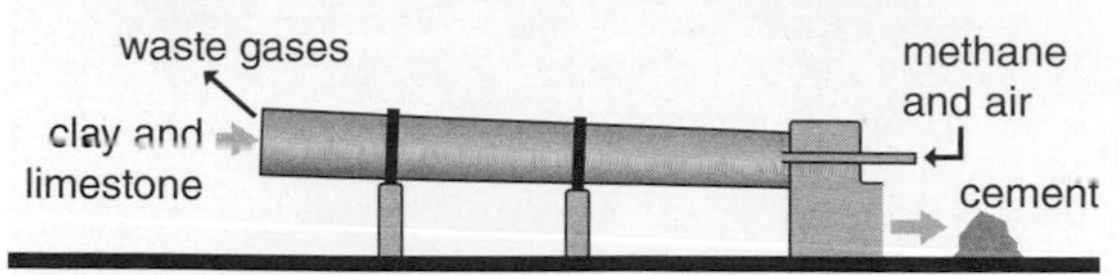

AO2 **(a)** Use information in the diagram to name the two waste gases produced by the process. [2]

(b) Cement is mixed with sand and aggregate to make concrete. A student investigated the strength of six concrete mixtures. Each concrete mixture contained 100 cm^3 of cement, 200 cm^3 of sand and a different volume of aggregate.

To do the investigation, the student:

- added and stirred water into each concrete mixture
- put each mixture into the same sized, beam-shaped moulds
- left each mixture to set hard
- hung weights from the suspended beam, increasing the weights until the concrete cracked
- recorded the results in a table.

Volume of sand and cement (cm^3)	300	300	300	300	300	300
Volume of aggregate (cm^3)	300	400	500	600	700	800
Weight needed to crack beam (N)	98	91	84	72	70	63

AO3 **(i)** What happens to the strength of the concrete as the volume of aggregate increases? [1]

AO3 **(ii)** The student was worried about an anomalous result. Which result was anomalous? Explain why you have chosen this result. [2]

AO3 **(c)** The student repeated the investigation, but this time used six mortar mixtures. Mortar is a mixture of sand and cement. From the results she concluded correctly that concrete was stronger than mortar.

Suggest **one** reason why concrete is stronger than mortar. [1]

AO1 **2. (a) (i)** Draw a diagram to represent the electron arrangement of a fluorine atom. [1]

AO1 **(ii)** Explain how a fluorine atom can change into a fluoride ion, F^-. [1]

AO2 **(b)** Use the periodic table to help you copy and complete this table for a fluoride ion, F^-. [3]

Number of protons	Number of electrons	Mass number

[3]

AO2 **(c)** Use your knowledge of electronic structure to explain why chlorine and bromine are in the same group of the periodic table as fluorine. [2]

3. Copper is a widely used metal. A common ore of copper contains copper sulfide. Copper is extracted from copper sulfide in a three stage process.

(a) In the first stage of extraction the copper sulfide is heated in air.

AO2 **(i)** Balance the symbol equation for the reaction.

Cu_2S + ...O_2 $\longrightarrow$ ....CuO + SO_2 [1]

AO3 **(ii)** Explain why there would be an environmental problem if the gas from this reaction was allowed to escape into the atmosphere. [2]

AO1 **(b)** In the second stage copper oxide, CuO, is reduced using carbon. Describe and explain what happens during this reaction. [2]

(c) In the third stage, the copper is purified like this.

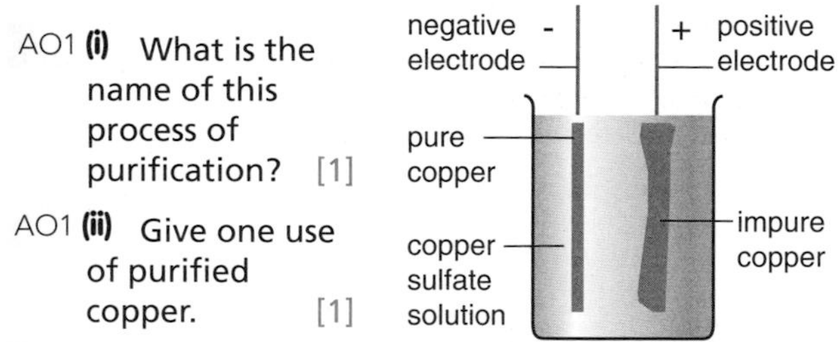

AO1 **(i)** What is the name of this process of purification? [1]

AO1 **(ii)** Give one use of purified copper. [1]

AO1 **(iii)** Copper is used to make alloys. Explain why copper alloys are more useful than pure copper. [4]

AO3 **2.** *In this question you will be assessed on using good English, organising information clearly and using specialist terms where appropriate.*

Copper-rich ores are becoming scarce. Phytomining is a new process being used to produce ores. The plants are grown on mining waste from copper mines. Then the plants are harvested, dried and burned. Finally the ash is used as copper ore.

Use the information above and your knowledge and understanding to give positive and negative environmental impacts of using this method to produce copper ores. [6]

AO1 recall the science AO2 apply your knowledge AO3 evaluate and analyse the evidence

WORKED EXAMPLE – Higher tier

In this question you will be assessed on using good English, organising information clearly and using specialist terms where appropriate.

Copper-rich ores are becoming scarcer. Bioleaching is a new process being used to produce ores:

- Bioleaching uses bacteria.
- Ores and waste with too little copper to be mined are injected with the bacteria.
- The bacteria produce soluble copper compounds from the ores and waste. These compounds dissolve in water and run off.
- The run-off solution is then reacted with scrap iron to obtain copper.

Use the information above and your knowledge and understanding to give positive and negative environmental impacts of using this method to produce copper ores. [6]

Positive impacts. Bioleaching reduces down the need for mining and quarrying of metal ores. It allows copper to be obtained from polluted areas where at the moment few plants can grow. Reducing copper in the area helps wildlife as there are less copper compounds which are toxic in the plants eaten by animals.

Negative impacts. It allows more copper compounds into rivers and streams, harming the fish and plants, particularly after flooding. Previously the amount of copper leaching out of the ground is small, but this is increased by using bacteria designed to produce copper compounds.

How to raise your grade!

Take note of these comments – they will help you to raise your grade.

It is a good idea to organise your answer into positive and negative sections, to make sure both parts of the question are covered well.

The candidate gains several marks for recognising the benefits of reducing toxic copper compounds in terms of plant growth and animal health. Good use of scientific words.

The candidate gets a mark for discussing the negative risk of increasing pollution in rivers and streams. Remember, environmental effects also include other factors, such as noise, dust and increased traffic.

The answer gains five marks as it is well organised, contains several good science points and uses scientific language appropriately.

Chemistry C1.5–1.7

What you should know

Atoms, elements, compounds and mixtures

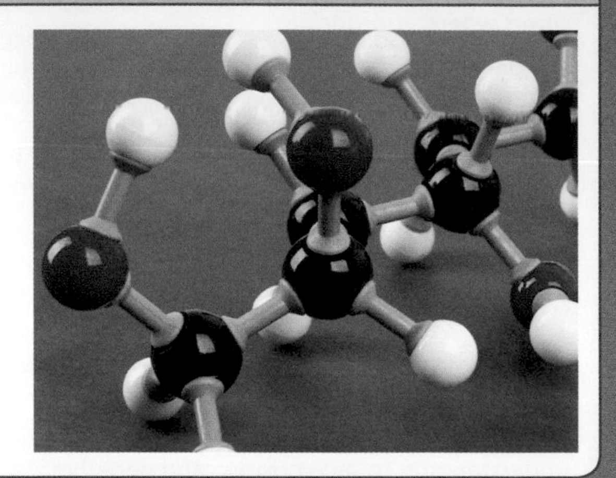

Compounds are made from atoms of two or more elements joined together by chemical bonds. Each has a name and chemical formula.

Each element or compound has its own properties and behaviour. Compounds of the same type show patterns in their behaviour.

Mixtures consist of elements and compounds simply mixed together. They can be separated easily.

 **Write the chemical formulae of three compounds.**

Chemical reactions

Chemical reactions happen when substances swap or rearrange their atoms to form different chemical substances.

The reactivity series lists metals in order of their reactivity.

 **Name as many metals as you can and arrange them in order of their reactivity.**

Rocks, weathering and erosion

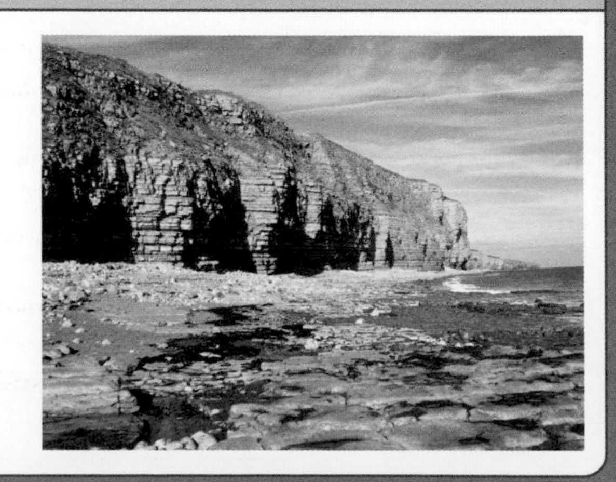

There are three types of rocks: sedimentary, igneous and metamorphic.

Rocks are worn down and broken up by physical, chemical and biological weathering.

 **Give one example of each of the three types of rock.**

You will find out

Other useful substances from crude oil

> Cracking produces the starting materials for plastics.
> There are differences between alkane and alkene hydrocarbons.
> Ethanol is manufactured from ethene and by fermentation of sugar.
> Plastic polymers are made from alkenes and have many useful properties.
> The disposal of waste plastics presents problems.

Plant oils and their uses

> Some plants are a source of oil; these oils have important uses.
> The production and use of biofuels has advantages and disadvantages.
> Oils and fats are important in foodstuffs.
> Emulsions are droplets of one liquid suspended in another. Emulsions have many everyday uses.

Changes in the Earth and its atmosphere

> Earth is a layered structure of crust, mantle and core.
> Earth has changed since it formed, and continues to change.
> Earthquakes are linked to the movement of tectonic plates. Volcanoes form where magma bursts through the crust.
> Air is a mixture of elements and compounds and the mixture has changed since Earth formed.
> The atmosphere influenced the development of life on Earth.
> Changes in atmospheric carbon dioxide levels affect the whole Earth.

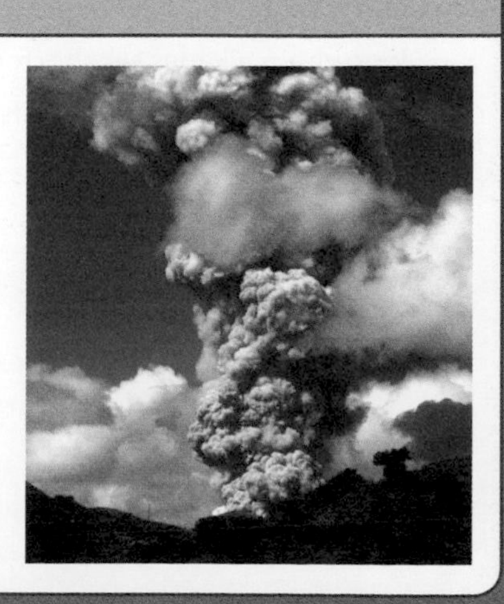

Cracking

> **You will find out:**
> > cracking breaks down large oil molecules to make them smaller
> > cracking produces starting materials for plastics

Oil companies crack the problem

Oil companies have a problem. Crude oils vary in composition, but none contains enough of the petrol fraction needed for cars. Fortunately, the companies have got it cracked – literally. They 'crack' large hydrocarbon molecules into smaller pieces. This uses up fractions that nobody wants – and turns them into the fraction that everybody needs. Brilliant!

FIGURE 1: Without cracking there would not be enough petrol.

What's the problem?

Fractional distillation separates crude oil into fractions. Each fraction consists of hydrocarbons with similar-sized molecules. Hydrocarbons consisting of small molecules are in greatest demand (to make petrol). Unfortunately, there are not enough in crude oil to meet the demand.

On the other hand, there is too much of the larger hydrocarbons. Fewer people want these fractions, so oil companies break up the larger molecules into smaller ones. This is called **cracking**.

> **Remember**
> Cracking an alkane always gives a shorter alkane plus an alkene.

Cracking

The compounds in crude oil are mostly **alkanes**. Their molecules contain chains of carbon atoms. Cracking breaks the chains into shorter lengths.

> Liquid hydrocarbons are vaporised by heating and mixed with a hot catalyst. A **catalyst** is a substance that speeds up a reaction. It does not get used up, so can be used again and again.

> Cracking breaks down long alkane molecules to make a mixture of shorter alkanes. This mixture can be used to make more petrol.

> Cracking also produces another type of hydrocarbon called **alkenes**. These can be made into **plastics** and other **petrochemicals**.

long-chain alkanes → shorter-chain alkanes + alkenes

Cracking is a very useful process. It converts unwanted material into profitable products. It also makes oil companies a lot of money.

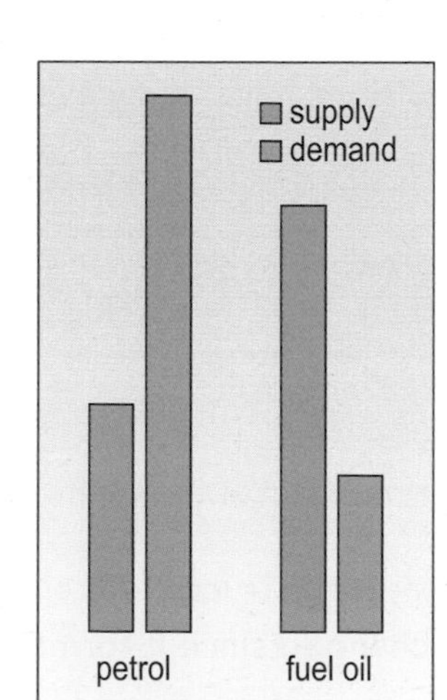

FIGURE 2: Crude oil has more fuel oil than is needed and not enough petrol.

QUESTIONS

1 Explain why cracking is necessary.

2 What else is formed during cracking, besides shorter alkanes?

3 Suggest why alkenes are useful compounds.

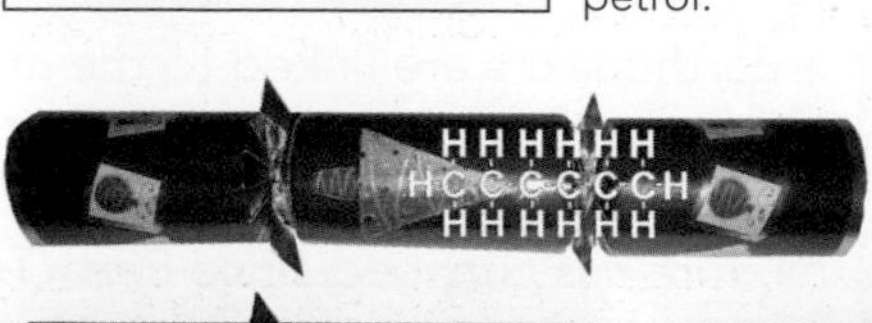

FIGURE 3: Cracking is like a Christmas cracker being pulled apart.

... cracking crude oil

A closer look at cracking

Build a molecular model of an alkane with, say, 10 carbon atoms ($C_{10}H_{22}$). Crack your alkane by holding both ends and waggling it until it breaks. Count the carbons in each of your products. Rebuild the molecule and try again. The chances are that it will crack into different products this time.

```
                                          H  H  H  H  H  H  H  H          H   H
                                          |  |  |  |  |  |  |  |           \ /
                                       H--C--C--C--C--C--C--C--C--H  +     C=C
                                          |  |  |  |  |  |  |  |           / \
                                       ↗  H  H  H  H  H  H  H  H          H   H
   H  H  H  H  H  H  H  H  H  H                   octane                 ethene
   |  |  |  |  |  |  |  |  |  |
H--C--C--C--C--C--C--C--C--C--C--H
   |  |  |  |  |  |  |  |  |  |
   H  H  H  H  H  H  H  H  H  H
            decane                     ↘  H  H  H  H  H  H  H             H H
                                          |  |  |  |  |  |  |              \|/
                                       H--C--C--C--C--C--C--C--H  +         C    H
                                          |  |  |  |  |  |  |              / \   /
                                          H  H  H  H  H  H  H             H  C=C
                                                 heptane                     /  \
                                                                            H    H
                                                                           propene
```

That is how cracking works. Heating makes the molecules waggle until a carbon–carbon bond breaks. They break in different places, giving a mixture of products. So $C_{10}H_{22}$ could crack to form
$C_8H_{18} + C_2H_4$ or $C_7H_{16} + C_3H_6$ or $C_6H_{14} + C_4H_8$ and so on.

In each case, one product is an alkane (general formula C_nH_{2n+2}). The other is not an alkane. It has a **double bond**, shown as C=C, so that each carbon atom still has four covalent bonds. It is a member of a hydrocarbon family called the alkene family. Alkenes have the general formula C_nH_{2n}.

The double bond in alkenes makes them much more reactive than alkanes. They can react to make many different compounds. This makes alkenes extremely useful.

Ethene is a particularly important alkene product of cracking. It is the starting point for making polythene and many other plastics.

Remember

One pair of electrons is shared in a C–C single bond. Two pairs of electrons are shared in a C=C double bond.

QUESTIONS

4 Write the chemical equation for cracking decane to form octane and ethene.

5 Explain why cracking always produces an alkane and an alkene, rather than two alkanes. (Hint: count the atoms.)

6 If cracking dodecane ($C_{12}H_{26}$) produces nonane (C_9H_{20}), give the name and formula of the alkene also formed.

Types of cracking

Oil refineries use two main processes to crack fuel oil to make smaller alkanes for petrol.

Steam cracking: Fuel oil is mixed with steam in a furnace at about 850 °C. The hydrocarbons undergo **thermal decomposition**. That is, the high temperature decomposes them into smaller molecules. By altering the proportion of steam to hydrocarbon, the reactions can be controlled to give the required mix of products.

Catalytic (or 'cat') cracking: Fuel oil is vaporised and mixed with a catalyst at about 600 °C. Using a catalyst allows the cracking reaction to take place at a lower temperature than in steam cracking. The catalyst does not get used up – it is separated from the products and reused.

QUESTIONS

7 Suggest why heating long chain hydrocarbons, to a high temperature, causes thermal decomposition.

Did you know?

Oil companies, in the UK, process about 250 million litres of crude oil every day.

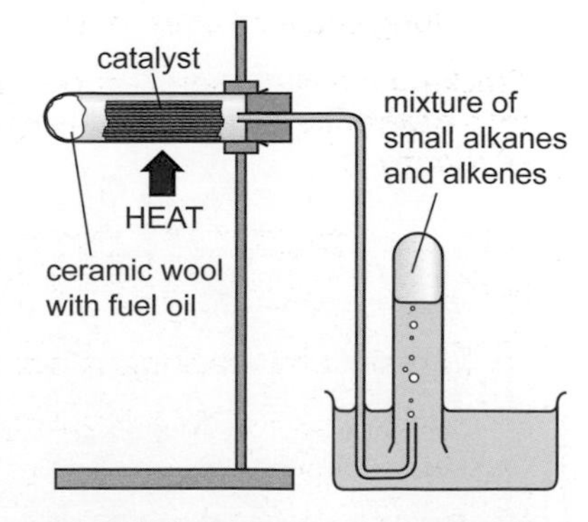

FIGURE 4: A laboratory investigation of catalytic cracking.

Alkenes

Bananas

Hanging bananas on a hook is not just a fashion trend. There is a good chemical reason for it. Ripe bananas give off ethene gas. Yes, the same as produced by cracking petroleum. Ethene causes nearby fruit to ripen more quickly. Hanging them up allows the ethene to blow away. This prevents the fruit from going brown and spotty too quickly.

FIGURE 1: Fashionable and functional. Hanging bananas helps them to keep fresh longer.

You will find out:

> alkenes are another family of hydrocarbons

> all alkenes have a double bond

> why alkenes are so useful

The alkene family

You have already met the alkane family and know that cracking alkanes produces alkenes. **Alkenes** are another family of hydrocarbons. Their names start like the alkanes, but end in 'ene' instead of 'ane'.

> The general formula for alkenes is C_nH_{2n}. Each molecule has twice as many hydrogen atoms as carbon atoms.

> All alkenes have a **carbon–carbon double bond**, shown as C=C. Every carbon atom still has four bonds in total. The double bond is two bonds (two pairs of electrons are shared).

> The double bond makes alkenes much more reactive than alkanes. They take part in more chemical reactions – and that makes them much more useful than alkanes.

> In industry, ethene and propene are known as ethylene and propylene. They are very important industrial chemicals, particularly for making polymers.

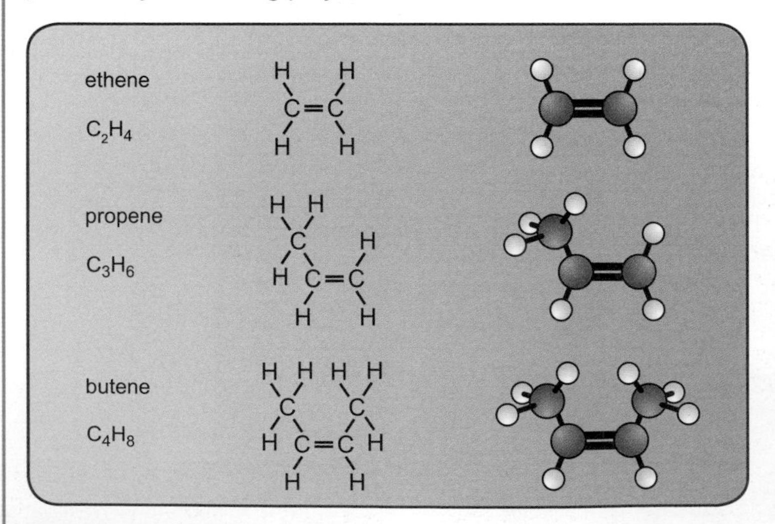

FIGURE 2: The first three members of the alkene family.

Remember

Carbon always forms four bonds. Always count them when you draw structures. Count a double bond as two bonds.

Did you know?

The tangy smell when you peel an orange or lemon is due to limonene. It's not an alkene, but the ending –ene tells you that it does contain a C=C double bond.

QUESTIONS

1 Hexene has six carbon atoms. Give its formula.

2 What is the major difference between alkenes and alkanes?

3 Explain why there is no such thing as methene.

4 What makes alkenes more reactive than alkanes?

Reactive alkenes

Why are alkenes more reactive than alkanes?

Hydrocarbon molecules have covalent bonds. Two atoms (a carbon and a hydrogen, or two carbons) share a pair of electrons. A double bond is just two bonds. The atoms share two pairs of electrons.

One of the two bonds can 'open up', allowing each carbon atom to form a bond with another chemical. This adds extra atoms to the molecule – it is an **addition reaction**.

The reaction of ethene with steam to form ethanol is an example of an addition reaction. The water 'adds across' the double bond.

$$H_2C{=}CH_2 + H{-}O{-}H \Rightarrow H{-}\overset{H}{\underset{}{C}}{-}\overset{H}{C}{-}H + H{-}O{-}H \Rightarrow H{-}\overset{H}{\underset{H}{C}}{-}\overset{H}{\underset{OH}{C}}{-}H$$

$$\underset{\text{ethene}}{C_2H_4} + \underset{\text{water (steam)}}{H_2O} \longrightarrow \underset{\text{ethanol}}{C_2H_5OH}$$

> **Remember**
> One pair of electrons is shared in a carbon–carbon single bond. Two pairs are shared in a carbon–carbon double bond.

NTS	
ed vegetable oils; Salt (2.4%); (E471); Flavourings; Vitamin E; (E160a); Vitamins A and D.	
ORMATION	
FAT	75
of which saturates	27
monounsaturates	32
& polyunsaturates	15
trans	1
FIBRE	r
SODIUM	0.9
f which saturates 2.7g, Salt 0.24	

FIGURE 3: A food label showing the types of fat present. What is the difference between mono-unsaturates and polyunsaturates?

Saturated and unsaturated hydrocarbons

Each carbon atom in an alkane already has bonds to four other atoms. So, unlike alkenes, alkanes cannot react by adding extra atoms. Alkanes are **saturated** – they cannot add any more atoms. Alkenes can, so alkenes are **unsaturated**.

The terms saturated and unsaturated do not just apply to hydrocarbons. You may have come across 'polyunsaturated' and 'saturated' fats in foods. The former have many C=C bonds in the fat molecules. The latter have only single bonds.

QUESTIONS

5 Explain how two carbon atoms form a C=C double bond.

6 Give the word that describes all molecules which contain a C=C bond.

7 What is an 'addition' reaction?

Detecting double bonds

You can use an addition reaction to test whether a compound contains C=C bonds.

1 Add a few drops of bromine water (an orange–brown solution of bromine in water) to a sample of the compound.

2 Shake well to mix the contents. If the bromine turns colourless, the compound contains one or more C=C bonds.

The bromine adds across the double bond, forming a colourless product. It turns the C=C bond into a C–C bond, with a bromine atom attached to each carbon atom.

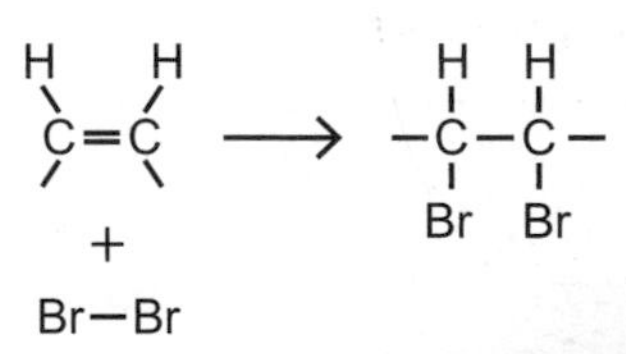

QUESTIONS

8 Hexane and hexene are both colourless liquids. Write a method to identify which is hexane and which is hexene.

FIGURE 4: Name some gases that would give a positive test like this.

Making ethanol

Alcohol fuels

Most cars burn petrol or diesel. Both come from crude oil. One day, crude oil will run out, so we need alternatives. Ethanol is commonly known as 'alcohol'. It has been made for drinking, from sugar, for thousands of years. However, it can also be used as a fuel, as Brazil discovered. It is made from sugar cane on a huge industrial scale now.

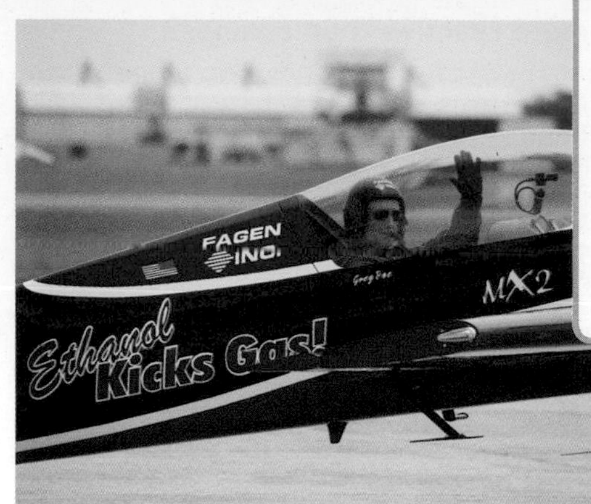

FIGURE 1: Ethanol is not just used in road transport.

You will find out:

- > ethanol is a compound that can be used as a fuel
- > how ethanol is made from ethene and by fermentation
- > other uses for ethanol

Alcohol and ethanol

Alcohols

To most people, alcohol is something that they drink. To chemists, **alcohol** is the name of another chemical family.

Alcohols are compounds with a hydroxyl group (–**OH**). Like other chemical families, such as alkanes and alkenes, there are many different alcohols.

Ethanol is a member of the alcohol family. Like *eth*ane and *eth*ene, *eth*anol has two carbon atoms in its molecule. Its formula is C_2H_5OH. It is written like that, rather than C_2H_6O, to show that it has the –OH group.

This is the structure of ethanol.

```
  H  H    H
  |  |   /
H-C--C--O
  |  |
  H  H
```

Ethanol is not a hydrocarbon because it contains oxygen as well as carbon and hydrogen. It is an alcohol because it contains the –OH group.

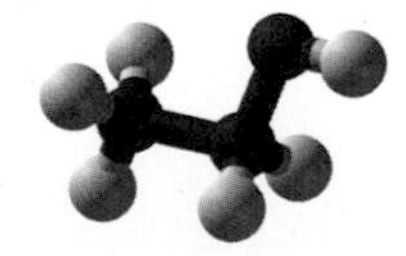

FIGURE 2: A molecular model of ethanol. What does the red ball represent?

Making ethanol

Ethanol is a colourless liquid that boils at 78 °C. It is highly **flammable** (catches fire very easily) and is a good fuel. Ethanol has many other important uses.

Ethanol is manufactured from two sources:

> Crude oil.

Cracking large alkanes produces ethene. Blowing ethene and steam over a hot catalyst makes ethanol.

> Sugar.

Adding **yeast** to sugar dissolved in water causes **fermentation** and changes sugar into ethanol.

Both methods give a solution of ethanol in water. Ethanol is separated by fractional distillation.

QUESTIONS

1 How do alcohols differ from hydrocarbons?

2 Give the chemical name for common alcohol.

3 Name the process that makes ethanol from sugar.

Converting into ethanol

Hydration of ethene into ethanol

When ethene and steam are blown over the hot catalyst, an *addition* reaction takes place. The double bond opens up and a water molecule adds across it: H on one side, –OH on the other. The catalyst speeds up the reaction.

$$C_2H_4 + H_2O \longrightarrow C_2H_5OH$$

ethene + water (steam) → ethanol

... hydroxyl group ... ethene ethanol

Making ethanol by fermentation

Sugars occur naturally in fruit, sugar beet and sugar cane. Fermentation happens when microorganisms, called yeasts, feed on the sugar and convert it into ethanol.

To manufacture ethanol for use as a fuel, a sugar solution made from sugar cane or maize (corn) is prepared. Yeast is added to cause fermentation.

sugar → ethanol + carbon dioxide

FIGURE 3: How is sugar cane used to make ethanol?

QUESTIONS

4 Name the two substances that react to make ethanol from crude oil.

5 Explain the purpose of the catalyst used in this reaction.

6 Explain what biofuels are and why they are renewable.

Ethanol is a biofuel

Fuels obtained from animals and plants are called biofuels. Ethanol made from sugar or maize is a biofuel. After harvesting, new plants are grown and used to make more ethanol. Thus, ethanol and other biofuels are renewable fuels. Unlike crude oil, they will not run out. Ethanol made from ethene (obtained from crude oil) is not a biofuel and is non-renewable.

Ethanol can be used instead of petrol, or mixed with petrol to make **gasohol** (gasoline + alcohol).

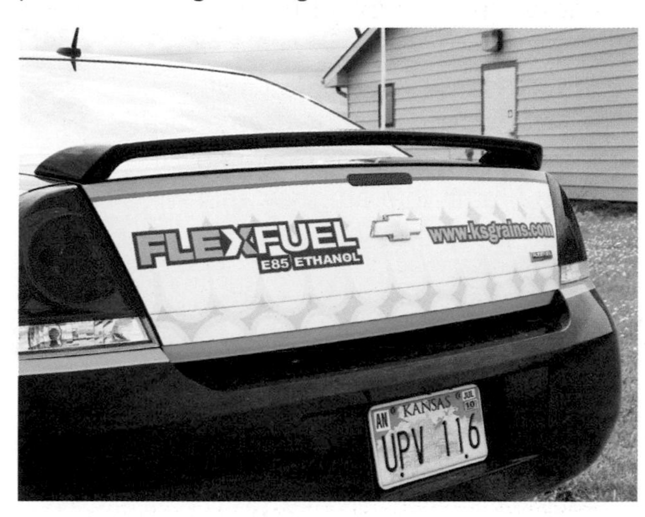

FIGURE 4: In the USA, the ethanol in gasohol is made from maize. Suggest what E-85 means.

Did you know?

Methylated spirits is ethanol with poisonous substances added to make it undrinkable.

Other uses of ethanol

As well as being a fuel, ethanol has many important uses:

> It is a useful solvent – a liquid that dissolves other chemicals. Some substances do not dissolve in water, but do dissolve in ethanol. This is useful when making pharmaceuticals (medicines), perfumes and aftershaves, inks and varnishes.

> It is the major ingredient in surgical spirit (an **antiseptic** that kills pathogens).

> It is used to make a wide range of other chemicals, including flavourings and perfumes.

QUESTIONS

7 Explain why the nurse rubs your skin with surgical spirit, when you have an injection. Suggest what happens to the ethanol left on your skin.

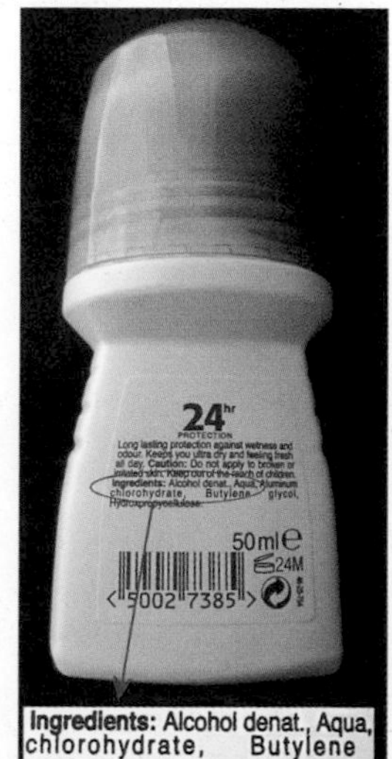

FIGURE 5: Why is ethanol one of the main ingredients in this deodorant?

Polymers from alkenes

You will find out:

> alkenes are used to make plastics
> plastics are polymers
> the differences between types of plastics

Plastics are everywhere!

How many plastic items have you used today? Probably too many to count. Plastics have replaced traditional materials like paper, wood and metal for many purposes. Their uses are almost endless – from carrier bags and credit cards to window frames and lorry cabs. Plastics are incredibly useful, but what are they and where do they come from?

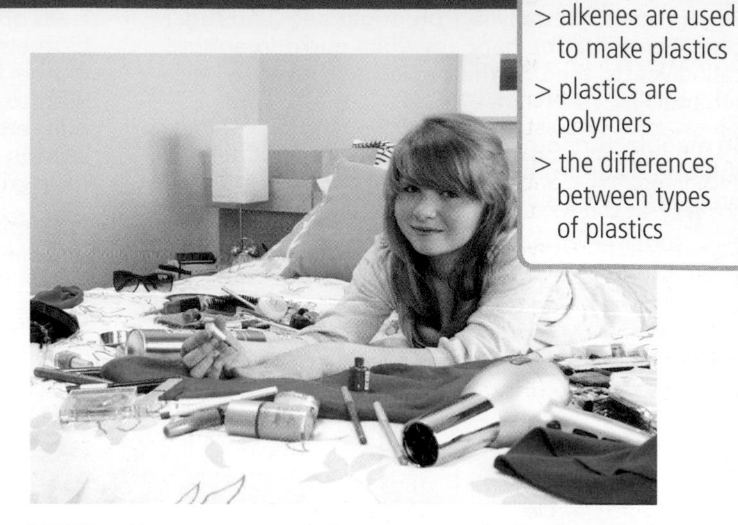

FIGURE 1: How many uses of plastics can you find in this scene?

From crude oil to polymers

Most plastics are made from alkenes obtained by cracking crude oil fractions. Alkene molecules have C=C double bonds. These bonds enable alkene molecules to link together into long chains, known as **polymers**.

Polymer molecules contain up to a million atoms joined together like beads in a necklace. Chemists nearly always refer to plastics as polymers.

Polymerisation is the process that joins together small molecules, called **monomers**. 'Mono' means one; 'poly' means many.

The most important monomer is ethene. Its molecules link up to form the polymer **poly(ethene)**, commonly known as polythene.

```
H     H  H     H  H     H          H   H   H   H   H   H
 \   /    \   /    \   /           |   |   |   |   |   |
  C=C      C=C      C=C     ---> ----C - C - C - C - C - C----
 /   \    /   \    /   \           |   |   |   |   |   |
H     H  H     H  H     H          H   H   H   H   H   H
```

three ethene molecules → poly(ethene)

The diagram shows only a small section of the polymer molecule. The dots mean that the chain continues in both directions. The polymer chain contains many thousands of ethene units. This polymerisation is often shown like this.

```
   H H          / H H \
   | |          | | | |
n  C=C   --->  -+-C-C-+-
   | |          | | | |
   H H          \ H H / n
```

ethene → poly(ethene)

Other alkenes polymerise in the same way. For example, propene forms poly(propene).

```
H     CH3 H     CH3 H     CH3          H  CH3  H  CH3  H  CH3
 \   /     \   /     \   /             |   |   |   |   |   |
  C=C       C=C       C=C      ---> ----C - C - C - C - C - C----
 /   \     /   \     /   \             |   |   |   |   |   |
H     H   H     H   H     H            H   H   H   H   H   H
```

three propene molecules → poly(propene)

FIGURE 2: In what way is a polymer like this necklace?

QUESTIONS

1 Explain what is meant by 'monomer' and 'polymer'.

2 What type of chemical bond does a molecule need, in order to make a polymer?

A variety of polymers

Polymerisation is a chain reaction

To make poly(ethene), ethene is heated under pressure. A catalyst sets off a chain reaction. It makes a C=C bond open up and join onto another ethene molecule. The double bond in this one opens up and joins onto the next, and so on. A polymer chain forms.

The reaction involves only the C=C double bond. It does not matter what other atoms are attached to the carbons. Replacing one or more of ethene's hydrogen atoms with other atoms or groups of atoms gives different monomers. Using different monomers, gives a wide variety of different polymers.

```
H     H  H     H  H     H          H  H  H  H  H  H
 \   /    \   /    \   /           |  |  |  |  |  |
  C=C      C=C      C=C     ➡  ----C--C--C--C--C--C----
 /   \    /   \    /   \           |  |  |  |  |  |
H     H  H     H  H     H          H  H  H  H  H  H
```

three ethene molecules — poly(ethene)

Some common polymers

Most well-known plastics are polymers made from alkenes. Their chemical names are all poly(monomer name). Some are known by more familiar names or abbreviations.

All monomers polymerise in the same way as ethene. They form polymer chains with the repeating pattern of the monomer.

In the equation below and in Table 1, X represents an atom, or group of atoms.

```
H     X  H     X  H     X          H  X  H  X  H  X
 \   /    \   /    \   /           |  |  |  |  |  |
  C=C      C=C      C=C     →  ----C--C--C--C--C--C----
 /   \    /   \    /   \           |  |  |  |  |  |
H     H  H     H  H     H          H  H  H  H  H  H
```

TABLE 1: Some common examples of polymers.

Polymer familiar name	X	Monomer	Polymer chemical name
polythene	H	ethene	poly(ethene)
polypropylene	CH_3	propene	poly(propene)
PVC	Cl	chloroethene	poly(chloroethene)
polystyrene	C_6H_5	phenylethene	poly(phenylethene)
PTFE or *Teflon*®	all four H's replaced by F	tetrafluoroethene	poly(tetrafluoroethene)

FIGURE 3: This pan is coated with PTFE non-stick plastic. How is PTFE made?

QUESTIONS

3 How many ethene molecules polymerise to form a poly(ethene) molecule 100 000 carbon atoms long?

4 The initials PVC stand for poly vinyl chloride. Give the other chemical name for the monomer.

5 Draw part of the PVC polymer chain, showing three monomer units.

Polymers and plastics

All plastics are polymers, but not all polymers are plastics.

> Polymers are molecules consisting of very long chains made from carbon atoms, with various other atoms attached. Many natural materials are polymers, including silk, rubber, starch, proteins and DNA.

> Plastics are **synthetic** (made by people) materials that are soft enough to shape in a mould during manufacture.

Most polymers have low densities, are not brittle (and do not break easily), can withstand corrosion by chemicals and soften when heated (can easily be moulded into shape).

These properties vary from one polymer to another. Chemists can make polymers that have the combination of properties needed for a specific use.

FIGURE 4: Why are these items made from different polymers?

QUESTIONS

6 A plastic is needed to wrap frozen food such as pizza or chicken. List the properties that the polymer should have.

Designer polymers

You will find out:

- plastic polymers have very useful properties
- some everyday uses of polymers that rely on these properties
- some uses of polymers with unusual properties

Protected by polymers

People wear bullet-proof vests in war zones, and wherever there is a risk of being shot. The vests are heavy, bulky and difficult to wear, but will stop a bullet. So, they protect the wearer from serious harm. Bullet-proof vests are made of a special polymer called *Kevlar*®. It's five times stronger than steel.

FIGURE 1: Bullet-proof vests can protect dogs.

Polymers for a purpose

Slippery stuff

Hip joints in some older people start to wear out. Bones grind against each other and wear away, making it painful and difficult to walk. An artificial hip joint has a titanium ball that swivels in a *Teflon*® coated socket (Figure 2). **PTFE** (or *Teflon*) is a slippery plastic polymer. It is also the non-stick coating on cookware.

Tough stuff

A surfboard must be tough and light – supporting the surfer's weight without breaking and having low density so that it floats. Many are made of a polymer called **polyurethane**, produced in the form of rigid foam.

Sofas and car seats contain a flexible type of polyurethane foam. It can be squashed thousands of times without damage.

Remember

Using different monomers, polymers may be designed with the properties needed for different uses.

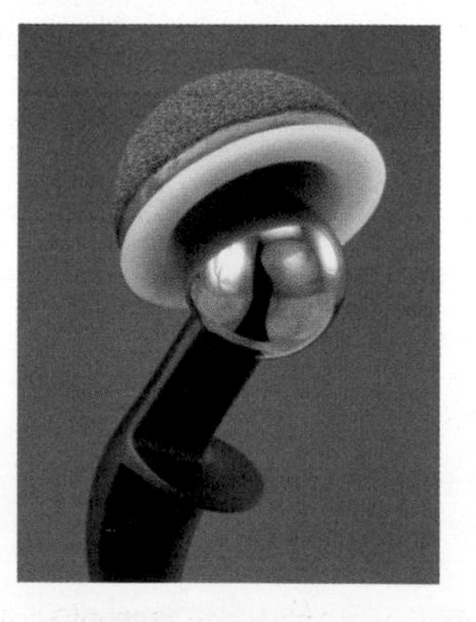

FIGURE 2: An artificial hip joint. The PTFE coating, on the inside of the socket, helps the joint to move freely.

FIGURE 3: Why is this surfboard made from polyurethane?

QUESTIONS

1 What makes PTFE special?

2 Explain why different types of polyurethane foam are used for surfboards and for sofas.

3 Name the very strong plastic used to make bullet-proof vests.

Did you know?

Twenty times more plastic is used now than was used fifty years ago.

Special polymers

Plastic aeroplanes

Aircraft are usually made of aluminium, because of its low density. However, passengers will soon be flying in plastic airliners. The *Boeing 787* fuselage, wings and tail are made from a **composite** material called CFRP (carbon fibre reinforced plastic). It is a polymer strengthened with carbon fibres.

CFRP is expensive, but very strong and less dense than aluminium. Being lighter, these aircraft use less fuel. They can fly further, non-stop. CFRP also needs less maintenance than aluminium.

FIGURE 4: The world's first plastic airliner took off in 2009.

Designing special polymers

Plastic polymers are easily moulded into shape, low density (lightweight), waterproof and resistant to acids and alkalis. Properties such as strength, hardness and flexibility vary. Polymers can be designed to have the specific properties needed for a particular purpose:

> Packaging materials that are light and strong, even when wet. Some need to let gases, such as oxygen, pass through. Others, for fizzy drinks bottles, stop them passing.

> Fabric coatings that repel water, for items such as tents and cooks' aprons.

> Breathable waterproof fabrics, such as *GORE-TEX®* stop rain getting in, but let perspiration (warm air and water vapour) out.

> Hydrogels, spread onto wound dressings, retain moisture. This softens dead tissue so that it can be removed, enabling quicker healing.

> White polymer for tooth fillings is soft and flexible when applied and shaped. Then, shining ultraviolet light on it makes it go very hard. Dentists also use polymers for crowns and dentures (false teeth).

QUESTIONS

4 Explain how plastics strong enough to build airliners are made.

5 Explain why plastic polymers have so many different uses.

6 Some measuring cylinders and beakers are plastic. What properties does a polymer need for this use?

What about the future?

Some new polymers offer exciting possibilities. Smart materials, for example, respond to changes, such as temperature or voltage.

> **Conducting polymers** conduct electricity, unlike other plastics. Possible uses: printing cheap, flexible, metal-free electronic circuits.

> **Light-emitting polymers** give off light when electricity passes through. Unlike light-emitting diodes (LEDs), they are flexible. Possible uses: computer and mobile phone screens.

> **Shape memory polymers** 'remember' the shape of the object. Uses: spectacle frames, dental braces, 'staples' in surgery, aircraft. On some makes of cars, dented car bumpers made of a shape memory polymer are repaired simply by warming the bumper, to make it return to its original size and shape.

> **Biodegradable polymers** rot away, unlike other polymers that cause waste problems. They are made from cornstarch or other plant materials. Uses: plastic bags and other packing that decay naturally: 'soluble' surgery stitches (do not need an operation to remove them).

FIGURE 5: Which type of smart polymer would you need, to make a colour version of a folding screen like this?

QUESTIONS

7 Suggest, and briefly describe, some other applications for smart polymers.

Polymers and waste

You will find out:

- > about the problems caused by throwing away plastics
- > recycling is the best option
- > about other ways to deal with waste plastics

Sailing on plastic

In 2010, the yacht *Plastiki* sailed 15 000 km across the Pacific Ocean. She was built from recycled plastic and 12 500 empty plastic bottles. The purpose of the voyage was to raise awareness of the enormous amount of waste plastics floating around the oceans. One patch of waste plastics that *Plastiki* sailed through is thought to be larger than the USA.

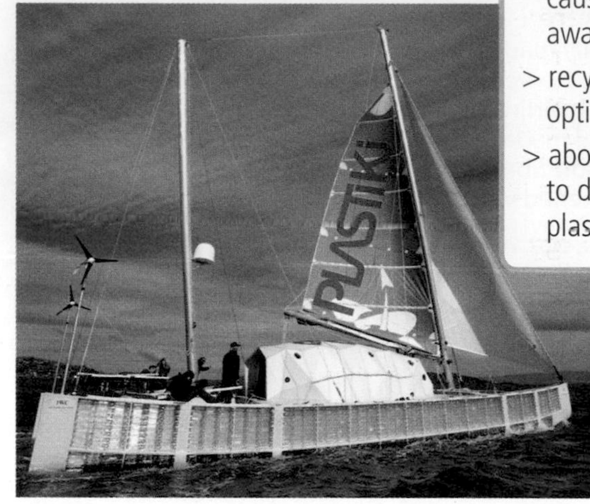

FIGURE 1: The *Plastiki* expedition drew attention to the problem of waste plastics.

A load of rubbish

Every year, people in the UK throw away three million tonnes of plastics. Much of it is packaging, such as food wrappers and bottles. Throwing away plastics causes serious problems.

> Most plastics are not **biodegradable** – they do not rot away. Any plastics dropped as litter just blow around for years, unless someone picks them up.

> Plastics put in rubbish bins mostly end up in big holes in the ground – landfill sites.

> More plastics must be made to replace those used. That takes a lot of energy and crude oil.

> Before long, there will be no landfill holes left, and no crude oil to make more plastics.

What can be done about these problems?

An alternative future – reusing plastics

Like aluminium, steel and glass, most plastics can be **recycled**. They can be melted down and used again. For example, fizzy drinks bottles can be turned into polyester fleece jackets.

Recycling plastics means less litter, less buried in landfills and less oil used to make new plastics. Unfortunately, recycling plastics is not simple.

> There are so many types of plastic, each designed for different uses. Mixed plastic waste is much less useful.

> Storing and transporting waste plastic takes a lot of space.

> Plastics melt at much lower temperatures than metal or glass. Paper labels or printed designs do not burn off, so they must be removed first.

> Some types of hard plastic do not melt at all.

FIGURE 2: What waste product could be used to make fleeces?

FIGURE 3: Waste plastics are lightweight but very bulky.

Did you know?

Used plastics, dumped in landfill rubbish tips, will still be there in 100 years, because they do not decompose.

QUESTIONS

1 Explain why throwing away plastics is bad for the environment.

2 What does 'recycling' mean?

3 List types of plastic that cannot be recycled.

Sorting out waste

Identifying different polymers

Sorting plastics for recycling means knowing what plastics items are made of.

Most disposable plastic items are labelled with a recycling symbol and code number to identify the polymer. Mixed plastic waste could be sorted by numbers, but someone needs to do this. It is time-consuming and, as labour is expensive, it may not be economical.

Recovering energy from polymers

Instead of dumping in landfills, household rubbish, including plastics, can be burned in incinerators. The heat produced may be used to generate electricity or heat local buildings. However, this is a waste of useful polymers and, unless carefully controlled, can cause air pollution.

Symbol	Polymer	Packaging uses
1 PET	poly(ethene terephthalate)	fizzy drinks bottles cooking oil bottles freezer-to-oven food trays
2 HDPE	high density poly(ethene)	supermarket milk bottles squeezy bottles butter / spread tubs
3 PVC	poly vinyl chloride or poly(chloroethene)	cooking oil bottles shampoo and cosmetics bottles cling film
4 LDPE	low density poly(ethene)	plastic carrier bags freezer bags bubble wrap
5 PP	poly(propene)	plastic jars bottle caps milk crates
6 PS	polystyrene or poly(phenylethene)	egg cartons, foam plastic cups and food trays CD cases

FIGURE 4: Which polymer is this made from?

QUESTIONS

4 Give three reasons why it is better to recycle plastic items than to bury or incinerate them.

5 Give two problems with recycling plastics.

6 What changes are needed to encourage people to recycle more plastics?

Biodegradable plastics

Most plastics are not biodegradable. Plastics in landfill do not rot, and the ground will be unusable for agriculture.

A few plastics are water-soluble and/or biodegradable. Here are two important examples.

> Biodegradable plastic carrier bags are made from polythene and corn starch. When buried, soil microorganisms feed on the starch between the polymer molecules, causing the plastic to fall apart. The polythene itself is not decomposed – just broken down into microscopic pieces.

> *Biopol*® is a biodegradable plastic produced by microorganisms. It has medical uses: *Biopol* stitches hold a wound together while it heals, but then dissolve away. The stitches do not need to be removed. Similarly, the material can hold broken bones together, gradually dissolving as new bone forms.

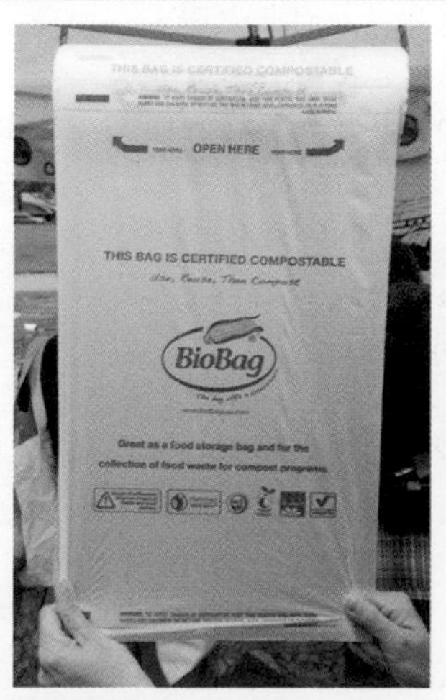

FIGURE 5: What makes this plastic bag biodegradable?

QUESTIONS

7 Explain why it might cause problems if a batch of recycled plastic contains some biodegradable polymers.

Preparing for assessment: Applying your knowledge

To achieve a good grade in science, you not only have to know and understand scientific ideas, but you need to be able to apply them to other situations and investigations. These tasks will support you in developing these skills.

Walking in the Lake District

Will, Sacha, Mike and Nisha are going on a field trip to the Lake District, at Easter. They will do plenty of walking and climb some of the peaks. Weather information says that the average monthly rainfall is about 75 mm in April and the temperature is 3–12 °C. Choice of clothing is important.

Staying dry is important. If you get wet you soon start to feel cold and miserable, making walking seem even harder. However, climbing to some of the peaks is also hard work and you soon start to sweat even on a mild day. Feeling hot and sticky is uncomfortable.

This is a problem when selecting suitable clothing. Outer clothes that are not waterproof would soon get wet through. Ordinary waterproofs, made from nylon, keep rain out but they also keep perspiration in.

A special material called *GORE-TEX®* provides an answer, and many moorland walkers now use hats, coats, trousers and boots made from it. *GORE-TEX* is a membrane made from a substance called PTFE, combined with nylon. It keeps the rain out, but lets water vapour from the body escape.

GORE-TEX works because PTFE is hydrophobic ('fears' water) and causes water to gather in beads on its outer surface. The beads are too large to pass through to the inside. Water vapour (perspiration) from the body can still pass through to the outside since the molecules are around 700 times smaller than the pores in the membrane. There are around 14 million pores per square millimetre of fabric.

Task 1

(a) Why do Will and his friends need coats that are 'waterproof and breathable'?

(b) Why is *GORE-TEX* successful at keeping them dry while not making them sticky with sweat?

Task 2

(a) Describe the difference between how water molecules are arranged in liquid water and in water vapour.

(b) Explain why water vapour can pass through *GORE-TEX*, but liquid water cannot.

Task 3

(a) PTFE and nylon are polymers. Describe how polymers form from their monomers.

(b) Explain why fabrics are made from polymers and not from monomers.

Task 4

(a) Explain the links between: crude oil, hydrocarbons, alkanes, alkenes, cracking and polymers.

(b) PTFE stands for poly(tetrafluoroethene). Explain how PTFE forms from its monomer, tetrafluoroethene (C_2F_4).

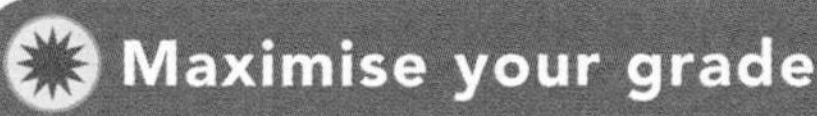

Maximise your grade

Answer includes showing that you...	
	know that suitable coats should be (a) waterproof or (b) breathable.
	know that water molecules are arranged differently in liquid water and in water vapour.
E	know that suitable coats should be waterproof and breathable.
	know how polymers form from their monomers.
	know the relationships between crude oil, hydrocarbons, alkanes and alkenes.
C 	understand why water vapour can pass through *GORE-TEX*, but liquid water cannot.
A 	understand why fabrics are made from polymers and not from monomers.
	understand the processes of cracking and polymerisation.

Oils from plants

You will find out:
- > about oils found in seeds, nuts and fruits
- > how plant oils are extracted from plants
- > plant oils are used in foods
- > why plant oils are good for your health

Oil from sunflowers

North American Indians grew sunflowers 2000 years ago. Sunflower seeds can be crushed to produce oil for cooking and for making butter substitutes and soaps. The seeds can be eaten either raw or cooked; they provide energy, protein, vitamin E and several essential minerals.

FIGURE 1: Sunflower seeds are a useful source of plant oils.

Vegetable oils

Sources of vegetable oils

Vegetable oils are fats that are liquid at room temperature. Plants produce **natural oils** in their seeds or nuts. They provide the energy that seeds need to start growing. Some fruits are also rich in oil.

Plants grown for their oil include:

> sunflower, rape, maize (corn), soya beans, linseed and sesame (from the seeds)

> groundnut (peanut), coconut, almond and walnut (from the nuts)

> olive and palm (from the fruits).

Extracting oils

Fruits, seeds and nuts are crushed to obtain their oil. Pulp is pressed between filters to squeeze out the oil and separate it from solids. These solids still contain some oil, which can be extracted by dissolving the oil in a **solvent**. Oil is separated from the solvent by distillation.

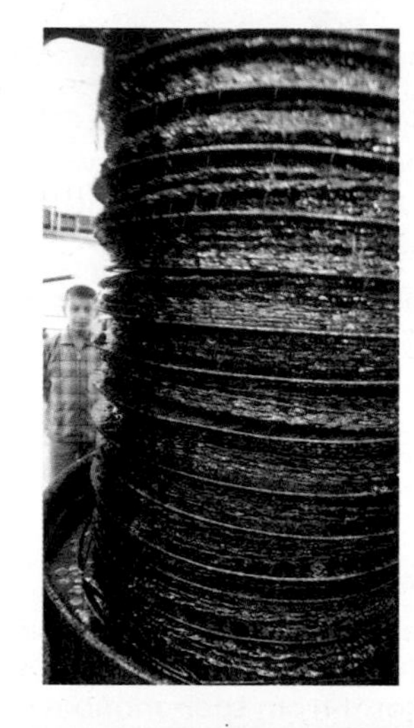

FIGURE 2: Oil drips out from olive pulp. The pulp is being pressed between filter mats.

Using vegetable oils	Some examples
foodstuffs	cooking oils, low-fat spreads, salad dressings, flavourings
cosmetics	soap, skin care, hair care, perfume, aromatherapy
fuels	biodiesel

QUESTIONS

1 Name three parts of a plant that contain oils.

2 Give the main stages in extracting oils from plants.

3 Suggest why *Palmolive*® soap has this name.

Oils in food and fuel

Natural oils in our diet

Natural plant oils store a lot of energy. This energy is released when the oils are eaten. Oils provide more energy than most other foods.

Plant oils have higher boiling points than water, so frying cooks food at higher temperatures. The food cooks faster and develops different flavours and textures. However, the food absorbs some of the oil and this increases the food's energy content.

Did you know?

Chocolate melts in your mouth because it contains cocoa butter, which melts at 37 °C. Cocoa butter is a fat extracted from cocoa beans.

... plant oils ... diet plant oils

Natural oils also contain other nutrients needed in a healthy diet:

> **Essential fatty acids**, for the heart, muscles and nervous system to function properly. Omega-3 and omega-6 acids are particularly important.

> **Vitamins**. A deficiency (shortage) of one or more vitamins causes disease. Seed and nut oils are particularly rich in vitamin E.

> **Minerals**. Minerals are compounds of metals and non-metals such as potassium, calcium, iron and phosphorus.

> **Trace elements**, but only in tiny amounts. For instance, a single Brazil nut provides the whole recommended daily allowance (RDA) of selenium.

Using vegetable oils as fuels

Vegetable oils are energy-rich and flammable, so can be used as fuels. Olive oil and castor oil, easily obtained by crushing, were used in oil lamps for thousands of years. Nowadays, oils from rapeseed, soya beans and other crops are converted into biodiesel fuel.

Omega Mix

Dry roasted sunflower, pumpkin, sesame, linseed, flax, hemp & rape seeds with a dash of savoury sauce

Why they are good for you:

Nutritional information (Typical values per 100 g)	Per 100 g	% RDA
Energy	613Kcal 2544kJ	
Protein	28.4g	
Carbohydrate	13.1g	
of which: Sugars	1.5g	
Fat	49.7g	
of which: Saturates	6.5g	
Mono-unsaturates	12.7g	
Omega 9	12.1g	
Poly-unsaturates	28.4g	
Omega 3	3.0g	
Omega 6	25.2g	
Cholesterol	0.0g	
Dietary Fibre	2.2g	
Sodium	255mg	
Calcium	6.1mg	
Iron	8.0mg	57%
Zinc	20mg	136%
Vitamin E	11.8mg	118%
Salt	0.65g	

FIGURE 3: Seeds are good for you. How much energy would you gain by eating 10 g of seeds?

QUESTIONS

4 Which gives you more energy per 100 g, chips or boiled potatoes? Explain why.

5 Name the vitamin that is present in most natural oils.

6 Name one metal and one non-metal found in natural oils.

Essential oils

Extracting essential oils

Flowers contain **essential oils**. These are quite different from the natural oils in seeds and nuts. They have low boiling points, so evaporate easily, giving flowers their scents. They are used in perfumes.

Essential oils can be extracted using **steam distillation** (Figure 4). Steam vaporises the oils. Steam and oil vapours pass to the condenser and become liquid again. Oils and water do not mix. The essential oils are less dense and float on the water.

Essential oils do dissolve in other solvents. To extract oils that might be damaged by heating, flowers are immersed in a solvent which dissolves out the oils. When the solvent evaporates, the oils are left behind.

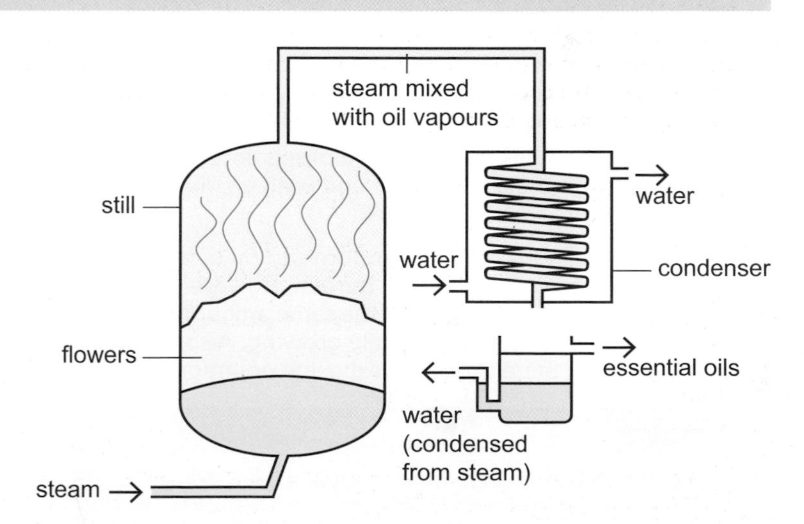

FIGURE 4: Steam distillation of flowers.

Vegetable and mineral oils

Vegetable oils are similar to animal fats, but different from mineral oils (from crude oil). Mineral oils are hydrocarbons, mainly alkanes. Vegetable oils have much more complex molecules with many carbon, hydrogen and oxygen atoms.

QUESTIONS

7 Briefly describe how to make lavender oil by steam distillation.

Biofuels

You will find out:

> the meaning of 'biofuels'

> why biofuels are better for the environment than fossil fuels

> using biofuels raises its own problems

Fuel from rubbish

Rubbish, buried in landfill sites, breaks down and releases methane gas. It is the same as the natural gas found with petroleum – it's a good fuel. This methane normally escapes into the atmosphere. That is a waste. So, at some rubbish dumps, it is collected, piped to power stations and burned to generate electricity.

FIGURE 1: Collecting methane gas from a landfill site.

Green fuel

Fuels from plants

Wood is a **biofuel**. Other animal or vegetable materials, burned for heating or light, are also biofuels. Nowadays, most biofuels come from plants. Some are natural, such as sunflower oil. Others are made from a plant product, such as ethanol made by fermenting sugar.

Biofuels can be burned directly or mixed with 'normal' fuels (petrol or diesel). They are often called **green fuels** because they are better for the environment than fossil fuels.

Why biofuels are 'greener' than fossil fuels

Fossil fuels take millions of years to form. We are using them faster than they form, so they will run out. They are **non-renewable fuels**. In contrast, biofuels are **renewable fuels**. Although they get burned and used up, more plants can be grown to make more fuel.

Both fossil fuels and biofuels produce carbon dioxide, a **greenhouse gas**, when they are burned. However, biofuels are **'carbon neutral'**, giving out the same amount of carbon dioxide as the plants took in while growing. Also, biofuels are sulfur-free: there is no sulfur dioxide pollution.

Did you know?

The original diesel engine ran on peanut oil, in Paris, 1900.

FIGURE 2: Biodiesel, alone or mixed with ordinary diesel, can be used in farm and other vehicles.

QUESTIONS

1 Give the name of the gas produced from rotting rubbish.

2 What are biofuels?

3 Give two reasons why biofuels are better for the environment than fossil fuels.

Biofuel issues

The need for new fuels

Fuels are essential. Petrol, diesel, kerosene (for aircraft) and fuel oil all come from crude oil. Although new oilfields are being discovered, crude oil is likely to become much scarcer within your lifetime. Alternative liquid fuels will be needed.

Problems with biofuels

Switching from 'black gold' to 'green fuels' brings its own problems.

Practical issues

> Plant oils are usually too viscous to use directly in vehicle engines. They must be chemically modified or mixed with normal fuels.

> Biodiesel and ethanol contain different chemicals from normal fuels. To burn correctly, they need different proportions of fuel and air. Engines need modifying or new ones designed to use 100% biofuels.

> Few garages sell biofuels because not many vehicles use them. Few people will buy biofuel vehicles until more places sell biofuels.

Economic and ethical issues

> Biofuels give less energy per litre than normal fuels, so more is needed.

> Growing crops uses energy for machinery, making fertilisers, transport and turning crops into biofuels. The energy used may be more than the energy released when the biofuel is burned.

> Growing plants for biofuels, instead of for food, means less food is produced. This leads to higher prices – a major problem for people in poor countries.

Environmental issues

> Producing enough biofuels to replace even half of petroleum fuels takes huge amounts of land. Large areas of natural countryside would have to become agricultural land.

> Switching land use to biofuel crops decreases **biodiversity** (the variety of plant and animal species in an area).

Did you know?

Rudolf Diesel (inventor of the diesel engine) predicted, a hundred years ago, that plant oils would replace crude oil products as engine fuels.

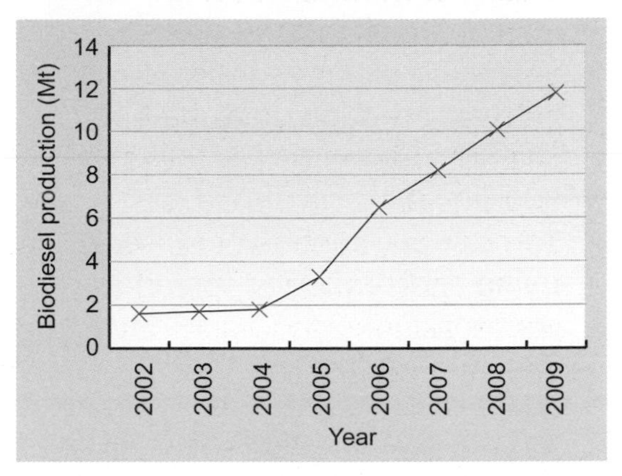

FIGURE 3: World biodiesel production is increasing. How many litres were produced in 2009? (1 Mt is 1 million tonnes, and 1 tonne is about 1100 litres)

QUESTIONS

4 Name two types of biofuel.

5 Give one disadvantage of using pure plant oil as a fuel.

6 Suggest why growing large areas of a biofuel crop decreases biodiversity.

Making biodiesel

Biodiesel can be made from plant oils, animal fats or waste vegetable oils (used cooking oil). These materials do not burn well in unmodified diesel engines, but biodiesel does.

Plant oils are reacted with an alcohol, usually methanol, to convert them into biodiesel. Compounds called esters form and the mixture of these esters is biodiesel.

The reaction also forms glycerol, which separates from biodiesel because it is denser. It is used to make soaps and cosmetics.

QUESTIONS

7 Outline the main stages in converting sunflower seeds into biodiesel.

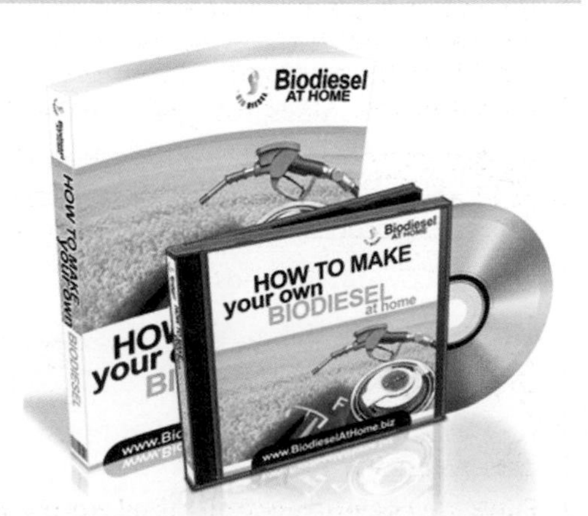

FIGURE 4: An idea for a school science project?

Oils and fats

Solid sunflower oil

Vegetable oils in your diet are better for your health than animal fats such as butter. Spreading oil on your bread is a bit messy though. Fortunately, chemists have a trick – they can turn liquid oils into solid fats. You can buy sunflower oil for frying and sunflower spread for your toast.

FIGURE 1: Chemistry turns vegetable oils into spreadable solids.

A closer look at oils and fats

Fats are a group of compounds with similar chemical structures. **Oils** are just liquid fats. Both occur naturally – usually, solid fats in animals, liquid oils in plants.

There are two types of oils and fats. You may have seen them mentioned on food labels and information about diet and health.

> As in alkene hydrocarbons, **unsaturated** oils and fats contain carbon–carbon double bonds (C=C):

> Those with two or more C=C bonds per molecule are called **polyunsaturates**.

> Those with only one are **mono-unsaturates**.

> **Saturated** fats contain no C=C bonds, only carbon–carbon single bonds.

Hard solid, soft spread or liquid?

> Animal fats, such as butter or lard, are mainly saturated fats. They are solid at room temperature.

> Unsaturated fats, such as vegetable oils, have lower melting points. They are usually liquid.

> Spreads are a mixture:

> Hard spreads, for making cakes or pastry, contain more saturated fats.

> Soft products, for spreading on bread, contain more unsaturated oils.

> 'Low-fat' spreads contain less of both and more water.

FIGURE 2: Butter is a fat, but becomes an oil when it melts.

NUTRITION INFORMATION TYPICALLY 11g PROVIDES	A	B
ENERGY	2198 kJ/534 kcal	3012 kJ/737 kcal
PROTEIN	0.3g	0.5g
CARBOHYDRATE	0.5g	Trace
of which sugars	0.5g	Trace
FAT	59.0g	81.7g
of which saturates	14.6g	54g
of which monounsaturates	28.8g	19.8g
of which polyunsaturates	12.8g	2.6g
FIBRE	Nil	Nil
SODIUM	0.63g	0.75g
Vitamin A	900 µg (113% RDA)	887 µg (111% RDA)
Vitamn D	8 µg (160% RDA)	0.76 µg (15% RDA)
RDA = Recommended Daily Allowance		

FIGURE 3: This label compares butter with a spread. Which is butter, A or B?

QUESTIONS

1 List the ways that oils and fats are similar and are different.

2 What do 'unsaturated' and 'polyunsaturated' mean?

3 Beef suet is solid. Suggest which type of fat suet contains.

Unsaturated fats

How to test for unsaturated fats

The test for an unsaturated fat is similar to the test for an alkene. It detects carbon–carbon double bonds.

1 Dissolve a small sample of fat in ethanol, to produce a solution that will mix with water.

2 Shake the solution with bromine water.

> If the orange–brown bromine goes colourless, the fat contains C=C bonds and is unsaturated.

> If the bromine stays orange–brown, no C=C bonds are present and the fat is saturated.

The quantity of bromine water that a measured amount of fat decolourises indicates how many C=C bonds the fat contains.

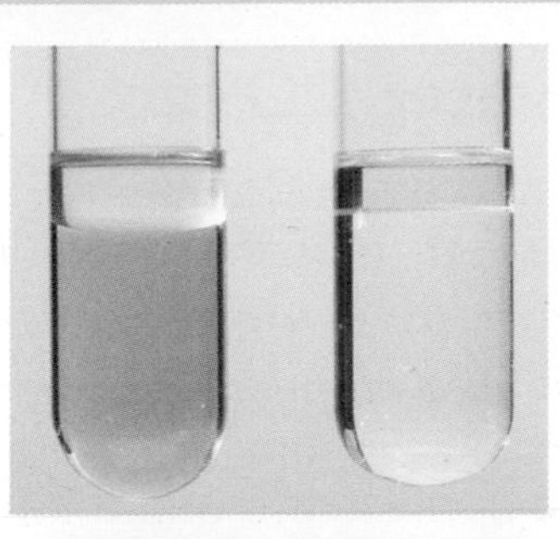

FIGURE 4: Which test tube indicates unsaturated fat?

Fats, oils and health

Everyone needs some fats in their diet, for energy and essential nutrients, but too much fat is unhealthy.

Saturated fats can increase the level of cholesterol in the blood. Tests show that polyunsaturates and, to a lesser extent, mono-unsaturates, help to keep blood cholesterol levels low. Medical advice is:

> eat foods rich in polyunsaturates, such as sunflower oil, and mono-unsaturates, such as olive oil

> avoid saturated fats, such as lard

> limit the amount of fat of all types in your food.

Oils found in fish, such as salmon and mackerel, contain omega-3 fatty acids. These compounds help to lower blood pressure, and improve mental health.

FIGURE 5: Why are oily fish beneficial to health?

QUESTIONS

4 Which part of an unsaturated fat molecule decolourises bromine?

5 Explain why a polyunsaturated fat decolourises more bromine than a mono-unsaturate.

Hardening vegetable oils (Higher tier)

Liquid vegetable oils are 'hardened' into solid spreads by converting unsaturated oils into saturated fats. The chemical reaction is called **hydrogenation**.

Hydrogen is bubbled through a mixture of the oils and nickel metal particles at about 150 °C. Hydrogen adds across the C=C double bonds in the oils. The unsaturated oil molecules 'soak up' extra hydrogen and become saturated.

The nickel is a catalyst. It makes the reaction go faster. As it is not used up, it is filtered off and used again.

$$\text{HC=CH} + \text{H–H} \rightarrow \cdots\text{CH–CH}\cdots + \text{H–H} \rightarrow \cdots\text{CH}_2\text{–CH}_2\cdots$$

unsaturated oil (liquid) → saturated fat (solid)

Note: Dotted bonds mean that the two carbon atoms are part of a longer chain in the molecule.

Did you know?

Fish live in cold water. Their fats need lower melting points to stop them solidifying. Unlike other animals, fish produce polyunsaturated oils.

QUESTIONS

6 Explain why hydrogenation is known as an addition reaction.

Emulsions

Oil and water DO mix

Normally, oil and water do not mix – but they can do under the right conditions. Oil and water mixtures are called emulsions. We use a lot of them in our everyday lives. They are in many foods, cosmetics, medical creams and the paints for walls. Most emulsions are liquid, but they do not behave like normal liquids.

You will find out:
- > what is an emulsion
- > how to make an emulsion
- > about some everyday substances that are emulsions

FIGURE 1: Milk and butter are both emulsions.

What is an emulsion?

Oil does not dissolve in water, it just floats on top. However, when shaken together, the liquids form very small droplets that do mix together. This is an **emulsion**.

Oil and vinegar salad dressing is like this. You have to shake it to mix the layers and make an emulsion. Left in the bottle, it separates out into two layers again. Some manufacturers stop salad dressing from separating out by adding a substance called an **emulsifier**.

Most emulsions contain emulsifiers to keep them together. It would be very inconvenient if emulsion paints or cosmetics kept separating while you were trying to use them.

An emulsion is made up of tiny droplets of oil suspended in water – or the other way round.

Figure 3 shows milk and butter under the microscope. Both are emulsions. Milk consists of yellow fat droplets spread out in water. Butter has water particles spread through the yellow fat.

FIGURE 2: Salad dressing. Which layer is the olive oil? Why must the bottle be shaken before use?

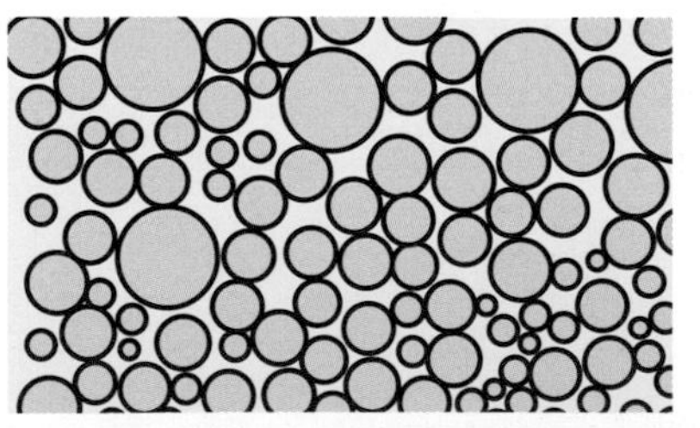

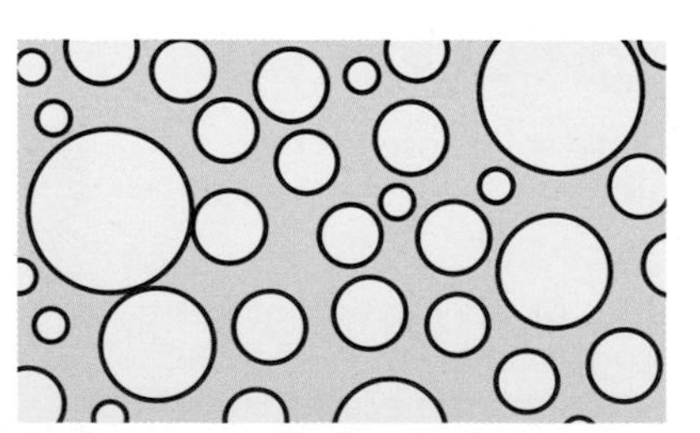

FIGURE 3: Milk and butter are both emulsions. Both emulsions contain fat and water. Which drawing is which?

QUESTIONS

1 Give three examples of emulsions that you come across in everyday life.

2 Describe the difference between the emulsions in butter and in milk.

3 What do manufacturers add to emulsions, to prevent the oil and water separating?

Did you know?

Emulsion paint is a mixture of oil paint and water. When spread on a wall, the water soaks into the plaster and then evaporates away. The oil paint reacts with oxygen to solidify.

Q ... emulsions GCSE

Useful emulsions

What makes emulsions so useful?

Any type of vegetable or mineral oil can form emulsions with water. The water-based (or **aqueous**) component is often a solution rather than pure water. By dissolving other ingredients in the oil and/or water, an unlimited range of emulsions can be made for different purposes.

Emulsions are more viscous (less runny) than the oil and less viscous than the aqueous solution. They are thin enough to spread easily, but thick enough to stay where they are spread. They can coat the surface of, for example, a lettuce leaf, a slice of bread, a wall, or your skin.

Making emulsions

Liquids that do not normally mix are **immiscible**. Emulsions are the result of forcing two immiscible liquids to mix. Emulsions are made by shaking or stirring.

> Oil-in-water emulsions (such as milk) contain droplets of oil **suspended** in an aqueous solution.

> Water-in-oil emulsions (such as butter) consist of aqueous droplets suspended in an oily liquid.

The harder you shake an emulsion, the smaller the droplets become. Shaking hard by hand gives droplets that are barely visible to the naked eye. In commercial emulsions droplet size may be much smaller – less than one thousandth of a millimetre.

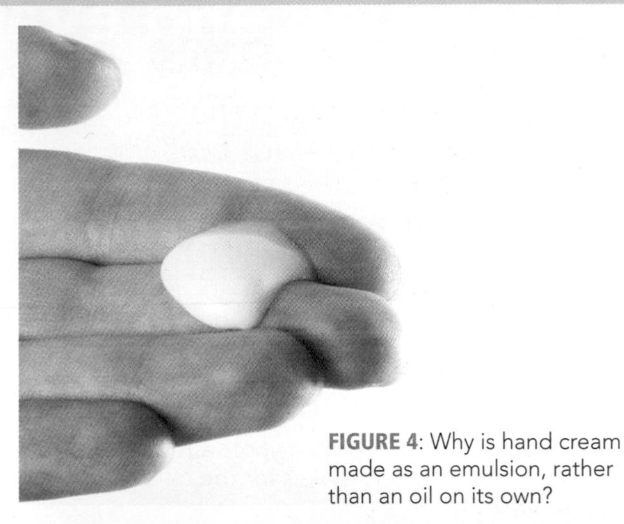

FIGURE 4: Why is hand cream made as an emulsion, rather than an oil on its own?

QUESTIONS

4 Why are oil and water described as 'immiscible liquids'?

5 Explain the difference between oil-in-water and water-in-oil emulsions.

6 Name the aqueous component of French dressing.

Emulsifiers (Higher tier)

Mayonnaise is a thick, creamy emulsion made by vigorously mixing oil, vinegar, lemon juice, and egg yolks. Unlike French dressing, mayonnaise does not separate into oil and vinegar, because the egg yolk contains an emulsifier. This makes the emulsion stable by keeping the droplets mixed.

Emulsifiers are compounds whose molecules have opposite properties at each end.

> One end is **hydrophilic** (water-loving) – it is attracted to water but not to oil.

> The other end is **hydrophobic** (water-fearing) – it gets away from the water by sticking into an oil droplet.

The oil droplets become surrounded by emulsifier molecules with their hydrophobic ends in the oil. The hydrophilic ends on the outside attract the droplets to the water. This also prevents the oil droplets joining together to form a separate oil layer again.

If the emulsion consists of water droplets in oil, the emulsifier works the other way round – with the hydrophilic ends in the droplets, and the hydrophobic ends around the outside, in the oil.

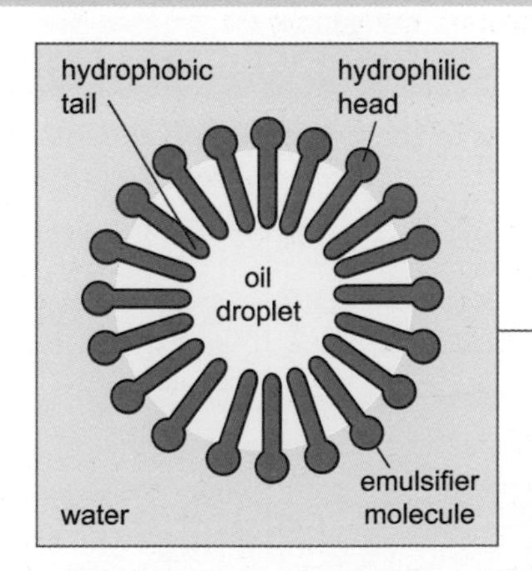

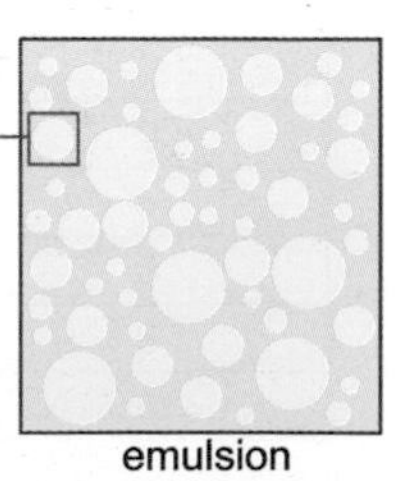

FIGURE 5: Emulsifier surrounding an oil droplet in an emulsion. Is this an oil-in-water or water-in-oil emulsion?

QUESTIONS

7 Explain, with a diagram, how an emulsifier stabilises a water-in-oil emulsion.

Preparing for assessment: Analysing and evaluating data

To achieve a good grade in science, you not only have to know and understand scientific ideas, but you need to be able to apply them to other situations and investigations. These tasks will support you in developing these skills.

Investigating emulsifiers

Hypothesis

Some students wanted to find out about the amount of emulsifier needed when making a salad dressing. Salad dressing is usually made from oil and vinegar, with mustard powder as the emulsifier.

They decided to test the hypothesis that the more mustard powder added, the longer it takes for the oil and vinegar mixture to separate.

Method

The students worked in three groups.

For each experiment, they kept constant:

- the volume of vinegar and volume of oil
- the size of the container used
- the time the mixture was shaken
- the temperature of the mixture.

They only varied the amount of mustard powder that they added.

Processing and analysing data

Here are the results of the first group of students and the graph that they plotted.

Mustard powder (g)	Time to separate (s)
0	50
2	140
4	240
6	300
8	330
10	340

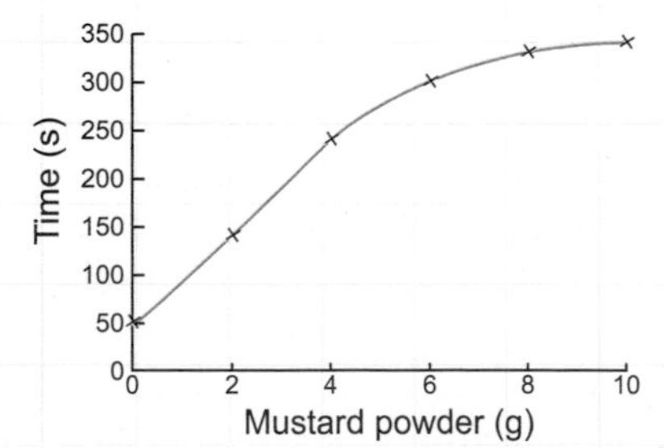

1. What conclusion can the students make from their investigation?

2. Does this conclusion support the hypothesis? Explain your answer using data from the chart or graph.

Look at both the table and the graph to make your conclusion.

Consider if your answer to question 1 is the same as the hypothesis. Look at the graph to judge the trend.

Here are the results from the second group of students.

Mustard powder (g)	Time to separate (s)
0	50
1	80
2	150
3	210
4	260
5	310

3. Plot a graph of these results.

4. Do the results from the second group of students support the hypothesis? Explain your answer.

5. How could you change the second investigation to check your answer to question 4?

The shape of your graph from question 3 is likely to be different from the first group's graph. Say how it is different and suggest a reason why the two investigations have produced different conclusions.

Look at the range of the data in both investigations and think why the pattern might be slightly different.

Evaluating data

One of the second group of students complained that it was hard to judge exactly when the oil and vinegar had separated.

6. Suggest how the students should decide when the oil and vinegar mixture had separated.

The third group decided that they needed to check the results were repeatable (that is they would get similar readings each time they carried out the investigation). They repeated each trial three times and then calculated a mean of the results:

Mustard powder (g)	Time to separate (s)			
	Trial 1	Trial 2	Trial 3	Mean
0	42	44	40	42
3	55	53	60	56
6	65	95	71	68
9	72	77	76	75
12	81	75	81	79

7. One of the results is anomalous. Identify this result by the amount of mustard added, and the trial number.

8. Look at the results from all three groups. How far does the data support or contradict the hypothesis?

9. You have been asked to advise a salad oil manufacturer about the use of an emulsifying agent such as mustard powder. Based on the results, what advice would you suggest about the maximum amount of powder to use.

Concentrate your answer on control variables. When the oil separates from the vinegar it forms two layers, you should consider some rules to use about how much the water or oil layers need to be distinct before it is judged to be separated.

Look for a result that does not match the pattern in each trial.

Compare all three sets of data against the hypothesis. Is the hypothesis right or could it be rephrased to make a better hypothesis? You should refer to all three sets of data in your answer.

Remember that manufacturers want the best performance possible whilst using the least amount of ingredients.

Connections

How Science Works

> Select and process primary and secondary data

> Analyse and interpret primary and secondary data

Science ideas

C1.6 Oils from plants (pages 164–165), Emulsions (pages 170–171)

Earth

You will find out:

> Earth is nearly a sphere and has a layered structure
> the layers are: the atmosphere, crust, mantle and core
> Earth's surface features were formed by natural forces millions of years ago

Our planet

Earth is a special place – unique in the solar system. It is the only planet where life as we know it has evolved. It supplies all that living things need: food, air, water and shelter. Humans have learned to use many more of Earth's resources. We turn them into a vast range of materials to improve our lives.

FIGURE 1: Europe from space. Only the few astronauts who went to the Moon have ever seen the whole Earth.

Planet Earth

Earth is a **planet**. This means it orbits around a star – the Sun. So do the other planets in our solar system. Earth is the largest rocky planet, but is small compared with the gaseous planets. Shaped like a ball, Earth is slightly squashed at the poles. It has a diameter of about 13 000 kilometres.

Earth is made up of layers:

> Earth's surface, called the **crust**, is rock. Water covers more than 70% of this surface.

> Below the crust is the **mantle** (molten rock).

> In the middle is the **core** (a very hot mixture of iron and nickel).

Earth has a very thin layer of air round it – the **atmosphere**.

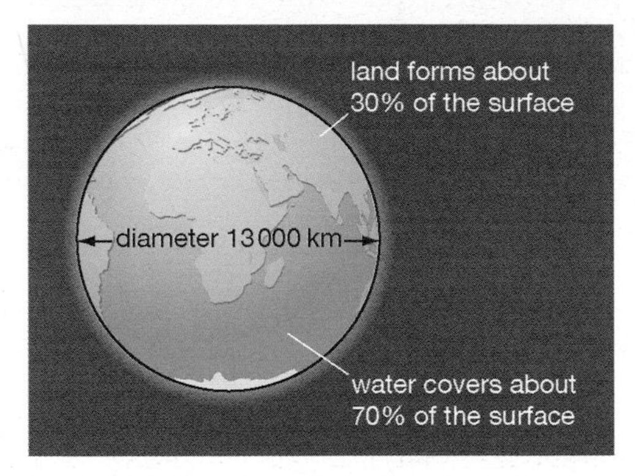

FIGURE 2: What does the pale-coloured ring, surrounding Earth, represent?

The surface of our planet

Water, mainly huge oceans, covers more than 70% of the planet. The land beside and beneath the water is formed from rocks. On land, there are mountains, valleys, hills and plains. These are found under the ocean, too. How did they form?

People once thought that:

> the centre of Earth is very hot because of heat left over from when Earth first formed

> the uneven surface was formed by the crust shrinking when Earth cooled down.

However, this is not what happened. Earth is not an unchanging ball of solid rock. It is changing all the time, but too slowly to notice.

QUESTIONS

1 Is Earth a rocky or a gaseous planet?

2 What are the four layers of Earth?

3 Estimate the percentage of Earth's surface that is covered with land.

Q ... planet Earth structure

What is Earth like inside?

Earth's three layers are very different from one another. Working from the inside outwards:

The core is at the centre. It makes up about half Earth's diameter and is very hot. The core is a mixture of iron and nickel. It has two parts:

> The inner core, in the centre, is solid iron and nickel.

> The outer core is a molten mixture of iron and nickel. Convection currents move this liquid around.

Next is the mantle. This is as thick as the core and made of very hot molten rock that flows very slowly. Convection currents make it circulate below the crust.

The crust is a thin outer layer of cold, solid rock. Its thickness is between 5 and 30 kilometres. The crust is covered in peaks (mountain chains) and hollows full of water (seas and oceans).

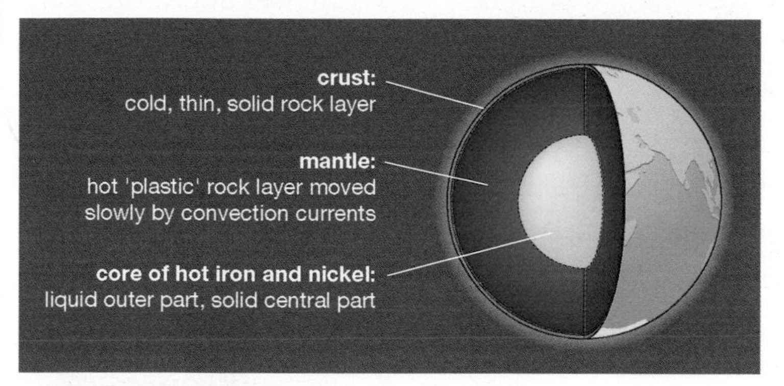

FIGURE 3: Earth's layers.

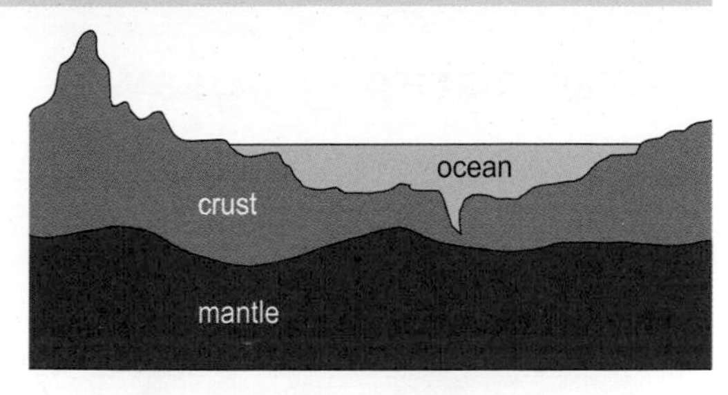

FIGURE 4: Cross-section of the crust (not to scale). The crust is thicker under the continents than under the oceans.

Did you know?

The lowest valley in the world is at the bottom of the Pacific Ocean, 11 000 metres below sea level. It is 2000 metres deeper than Mount Everest is high.

QUESTIONS

4 Which is thickest: the mantle, the inner core or the outer core?

5 Describe what makes up Earth's outer core.

How did Earth's surface features form?

When Earth formed it looked nothing like it does now. Its features have slowly changed and are still changing.

The heat inside Earth comes from radioactive materials acting like a gigantic nuclear reactor. It drives the convection currents in the outer core and mantle.

Earth's crust is broken up into separate parts called **tectonic plates.** These float on the mantle. The convection currents move the plates around, very slowly, only a few centimetres a year. Where plates push against each other, they force the land upwards, forming mountain ranges. In the same way, rocks which formed under the sea, such as limestone, were pushed up and became hills.

Over millions of years, rivers gradually wear away rocks, forming valleys and spectacular canyons. As a river approaches sea level, it slows down and deposits the soil as silt, sometimes forming a wide delta.

In some places, volcanoes on the sea bed throw out material. This builds up to form islands, such as Iceland and Hawaii.

Mountains, valleys, islands and other features have all been formed by natural forces.

FIGURE 5: Grand Canyon in Arizona is 300 kilometres long and 1.5 kilometres deep. How was it formed?

QUESTIONS

6 A new island, Surtsey, appeared in the Atlantic Ocean, near Iceland, in 1963. Explain how this could happen.

Continents on the move

You will find out:
- > continents appear to fit together like jigsaw pieces
- > about Alfred Wegener's theory of continental drift
- > how the movement of tectonic plates produced the Earth's surface features

A geography jigsaw

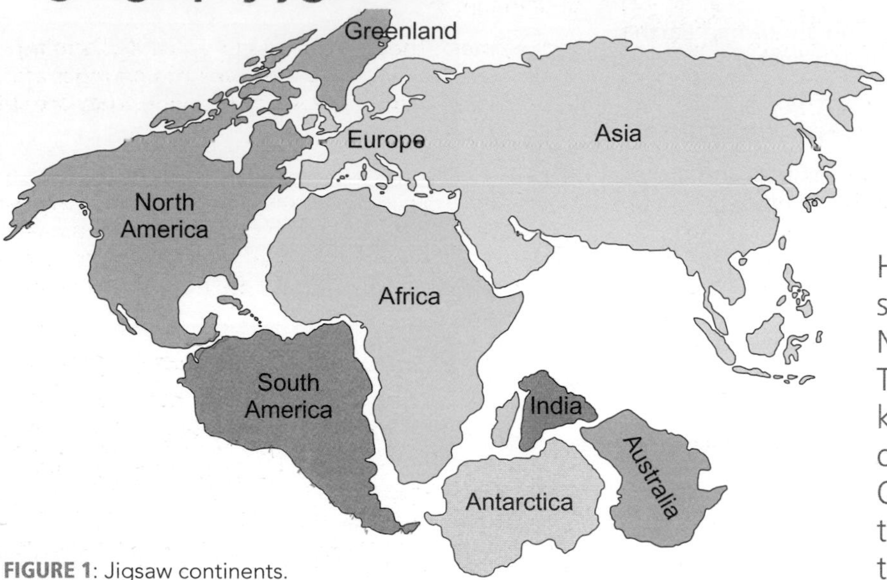

FIGURE 1: Jigsaw continents.

Have you ever noticed the shapes of Africa and of North and South America? They are thousands of kilometres apart, on opposite sides of the Atlantic Ocean. Yet, they look as if they could fit together like three jigsaw pieces.

Pangaea the supercontinent

From their shapes, Africa and America look as if they fit together. There is a bigger jigsaw if you add the other continents.

The continents fit together quite neatly. Could it be that, long ago, they were all joined?

Alfred Wegener, a German scientist, thought so. In 1915, he suggested that one huge landmass had existed. He called this supercontinent **Pangaea**. The rest of Earth would have been one vast ocean.

His theory was that Pangaea broke up about 200 million years ago. He said that the pieces drifted apart to become the present continents. This movement is called **continental drift**.

QUESTIONS

1 Suggest which parts of Africa and America fit together.

2 What did Wegener call the 'all Earth' supercontinent?

3 Explain what 'continental drift' means.

Continental drift

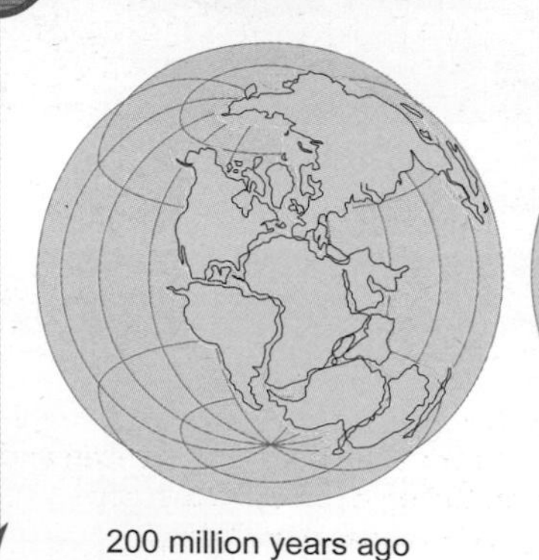

200 million years ago

100 million years ago

50 million years ago

FIGURE 2: Pangaea breaking up. Where was India, 50 million years ago?

... continental drift

Wegener's theory was correct

At the time, Wegener had no evidence to support his theory. Nor could anyone explain how huge continents could move. So, most scientists disagreed with the idea.

Later, other scientists found evidence of the same fossils occurring in Brazil and in West Africa, and matching rock layers on both sides of the Atlantic. This supported Wegener's theory and so it was accepted. However, the question remained – how can continents move?

Earth's crust and the semi-solid upper part of the mantle, make up the **lithosphere**. This is broken into a number of large pieces, called **tectonic plates**. The plates float on the liquid rock of the mantle. Heat from radioactive processes within Earth drives convection currents in the mantle. These currents carry the floating plates, and the continents that sit on top of them. This is what causes continental drift.

The tectonic plates move very slowly – a few centimetres every year – but, over millions of years, continents move thousands of kilometres.

QUESTIONS

4 Give the name for the crust and upper part of Earth's mantle?

5 Explain how continental drift occurs.

6 Which two land masses collided to form the Himalayas?

Drifting apart and coming together

While some plates are drifting apart, others are coming together, and very violently. When two tectonic plates the size of continents collide, they cause a lot of damage.

As they meet, they push the land upwards, forming mountains. The world's largest mountain ranges are formed where **continental plates** collide. They are still growing today.

When India broke away from Pangaea, it became an island in the ocean. Eventually, it crashed into the part of Asia that is now Tibet and pushed up the Himalayas. The sea bed was forced upwards, carrying fossilised sea creatures with it. These fossils are now found high in the mountains.

FIGURE 3: The Himalayas.

Did you know?

The distance between London and New York is increasing by a few centimetres a year.

What happens when continents move apart?

When tectonic plates move apart, **magma** (molten rock) escapes from the mantle, forming new crust. This is happening in the Atlantic Ocean, where the North and South American tectonic plates are moving away from the Eurasian and African plates. Magma is rising through the gap, all the way from the Arctic Ocean to the Antarctic plate, beyond the southern tip of Africa. As the magma solidifies, it forms an underwater mountain range called the **Mid-Atlantic Ridge**.

Iceland is part of the Mid-Atlantic Ridge. The underwater mountains grew so tall that they came above the ocean surface. Because the two plates are still moving apart, Iceland is getting wider (east to west) by about 2.5 centimetres every year.

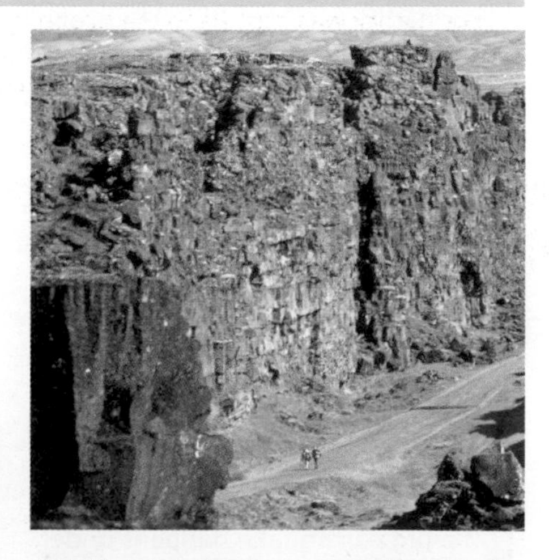

FIGURE 4: In Iceland you can walk through the gap between the North American and European tectonic plates. Is the gap getting wider or narrower? Why?

QUESTIONS

7 Figure 1 shows India off the coast of Africa. Suggest what evidence there is, that India was originally separate from Asia.

8 Explain how fossils of sea creatures finished up at the top of the Himalayas, eight kilometres above sea level.

Earthquakes and volcanoes

You will find out:

> earthquakes are linked to the movement of tectonic plates
> volcanoes form where magma bursts through the crust
> some places are much more prone to earthquakes and/or volcanic eruptions

Tsunami

On 26 December 2004, there was a huge earthquake on the floor of the Indian Ocean, off the west coast of northern Sumatra. It created a huge wave – a tsunami – which spread outwards and crashed onto the shores of Indonesia, Sri Lanka and southern India. The wave was 20 metres high, and travelled much faster than people could run away.

Did you know?

The Indian Ocean tsunami caused one of the deadliest natural disasters in modern history. It killed nearly a quarter of a million people.

FIGURE 1: One of the largest recorded tsunamis was 85 metres high.

Earthquakes

Some tectonic plates are **oceanic plates**, some are **continental plates**, and some are both. They pack together and move constantly, but very slowly. Convection currents, in the liquid mantle, carry the plates.

Plates grind past one another or collide. When two plates meet:

> pressure gradually increases

> suddenly, one plate slips over the other

> all the energy is released.

This causes an **earthquake**. It's like pushing the edges of two playing cards together – as you increase the pressure, one may buckle. Then suddenly, one slides over the other.

Most earthquakes occur along **plate boundaries**. Here, the rocks are usually thinner and weaker. The rocks are more likely to give way under pressure.

If a big earthquake happens under the sea, the sudden movement causes a huge wave. The wave spreads out in all directions. This is a **tsunami**.

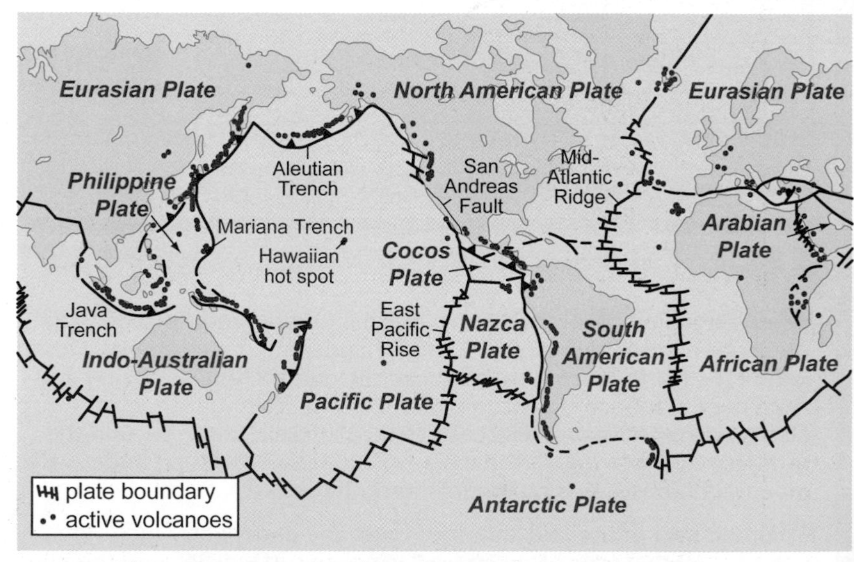

FIGURE 2: The main tectonic plate boundaries and active volcanoes.

QUESTIONS

1 Explain how a tsunami forms.

2 Explain briefly how an earthquake happens.

3 Suggest why earthquakes are more likely to happen at plate boundaries.

... earthquakes plates

Volcanoes

Plate boundaries

Like earthquakes, **volcanoes** often occur at plate boundaries. Magma, or hot molten rock, erupts through gaps between the plates. It cools, solidifies and may seal the gap, stopping the eruption. Magma stays sealed under the crust, often for hundreds of years. Eventually the pressure builds up enough for magma to burst through a **vent** – a crack or weak spot in the crust. The blast creates a crater, **lava** and gas pour out, and the familiar cone shape of volcanoes forms.

Where plates are moving apart in mid-ocean, solidifying magma forms mid-ocean ridges – underwater mountain ranges. In places, these reach the surface. This is how Iceland and New Zealand were formed.

FIGURE 3: Mount Pinatubo in the Philippines erupted in 1991. Nearly 800 people were killed and 100 000 were left homeless.

Hot spots

Volcanoes also form away from plate boundaries. Natural radioactive processes, in the core, release heat. The heat creates convection currents in the magma, causing it to circulate around the mantle. Where currents rise, the crust above gets heated. A **hot spot** occurs. This weakens the crust, allowing magma to work its way up and break through to the surface. Hawaii and the Galapagos Islands are results of hot spots.

Predicting earthquakes and volcanic eruptions

Scientists are getting better at making predictions, but there are still difficulties.

Scientists can monitor the movement of tectonic plates. They can tell when pressure is building and an earthquake is possible. However, they cannot obtain hard data about the forces involved, the friction between plates and structural weaknesses in plates. Eruptions are not the same as earthquakes, but the problems of predicting them are: the lack of sufficient valid data.

Accurate predictions are not possible. Scientists can only talk about the increasing probability of an earthquake or volcanic eruption.

QUESTIONS

4 Are the Pacific Plate and Indo–Australian Plate colliding or moving apart? How can you tell?

5 Explain what causes a volcanic hot spot.

6 Suggest why scientists cannot accurately predict earthquakes and volcanic eruptions.

Subduction zones

Tectonic plates that are 10 to 50 kilometres thick form continents. Thinner plates do not reach above sea level – they are oceanic plates. Where the two types collide, the thicker continental plate pushes the oceanic plate underneath. This is **subduction**.

The descending plate pushes magma out of the way, forcing it up to the surface. This creates a chain of volcanoes along the plate boundary.

The Andes mountain range formed this way. Dotted up the west coast of South America are extinct volcanoes which formed when an oceanic plate moved under a continental plate.

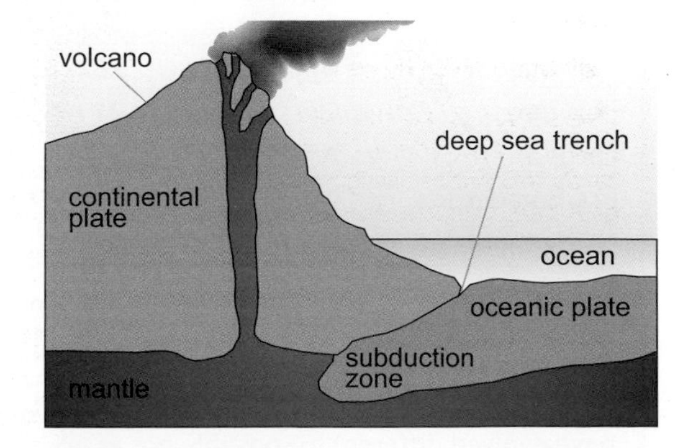

FIGURE 4: At a subduction zone, an oceanic plate is pushed down and under a continental plate.

QUESTIONS

7 Briefly describe the process of subduction.

The air we breathe

You will find out:

> air is a mixture of elements and compounds
> how the composition of the air has changed over millions of years
> the gases in air can be separated and put to use

Let there be life

When a baby is born, it must take its first breath to obtain oxygen. All animals and plants need oxygen to live. As it is an invisible, tasteless and odourless gas, we do not normally notice it. Yet, without oxygen, there would be no life on Earth.

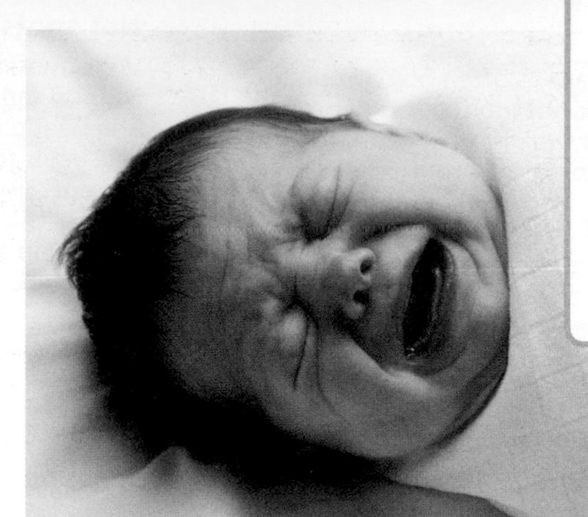

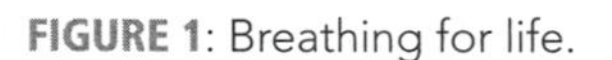

FIGURE 1: Breathing for life.

What's in air?

Air is a mixture of colourless gases – some elements, some compounds.

> Air is mainly oxygen and nitrogen, with small amounts of other gases. One of these is **argon** (a **noble gas**).

> Air contains carbon dioxide, though much less than many people think – only about 0.04%.

> Table 1 shows the composition of clean, dry air. However, air also contains between 1% and 4% water vapour. The amount varies from place to place and day to day.

> The amounts of pollutant gases, such as sulfur dioxide, also vary.

TABLE 1: Approximate percentage of the gases in air.

Gas	Formula	% of air
nitrogen	N_2	78
oxygen	O_2	21
argon	Ar	0.93
carbon dioxide	CO_2	0.04
other noble gases	He Ne Kr Xe	0.0025 in total

The composition of air has been much the same for 200 million years. Oxygen makes up 20% of the air and life has evolved to match this.

If the composition of air were different, life would have evolved differently – possibly not at all.

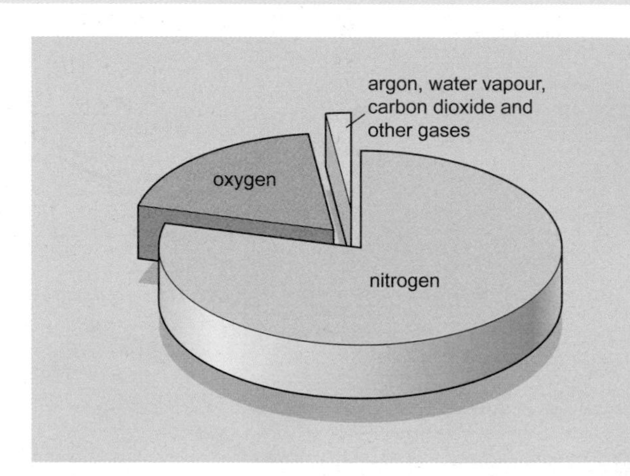

FIGURE 2: Is most of the air made up of elements or compounds?

QUESTIONS

1 Name the gas, in the air, that is needed by all living things.

2 Give the name and formula for the third most common gas in the air.

3 For how long has the composition of the atmosphere has been more or less constant?

... air composition

How did the atmosphere evolve?

The composition of air has not always been the same. Figure 3 shows the change, for three important gases, since Earth formed, 4600 million years ago.

> For the first 1000 million years, Earth had no oceans. It was covered in volcanoes. These released huge volumes of carbon dioxide. This made up most of the atmosphere, with some methane and ammonia.

> Volcanoes also released steam. As Earth cooled, this condensed to form liquid water (rain). Oceans formed and carbon dioxide dissolved in them. The proportion of carbon dioxide in the atmosphere decreased.

> About 3400 million years ago, the simple gases in the atmosphere and oceans formed the compounds needed for life to develop. Simple organisms released oxygen into the atmosphere as waste.

> The oxygen reacted with ammonia (NH_3) to produce nitrogen. Bacteria also produced nitrogen from ammonia. Nitrogen levels increased.

> The water protected the primitive life-forms in the seas from the Sun's harmful ultraviolet rays.

> Some oxygen (O_2) was converted to ozone (O_3) in the atmosphere. This protected Earth's land from ultraviolet rays, allowing land plants to appear about 400 million years ago. The plants photosynthesised, removing more carbon dioxide and increasing the amount of oxygen in the atmosphere.

> Eventually, the effects of living things and volcanoes began to balance out. The composition of the atmosphere became steady. It has stayed much the same for 200 million years.

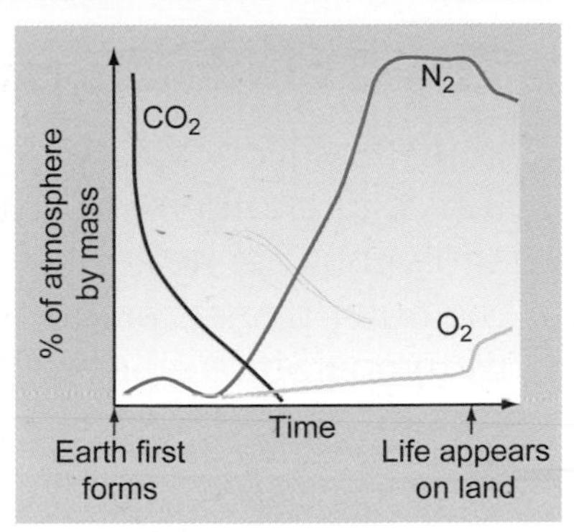

FIGURE 3: Percentage of different gases in the air, since Earth formed.

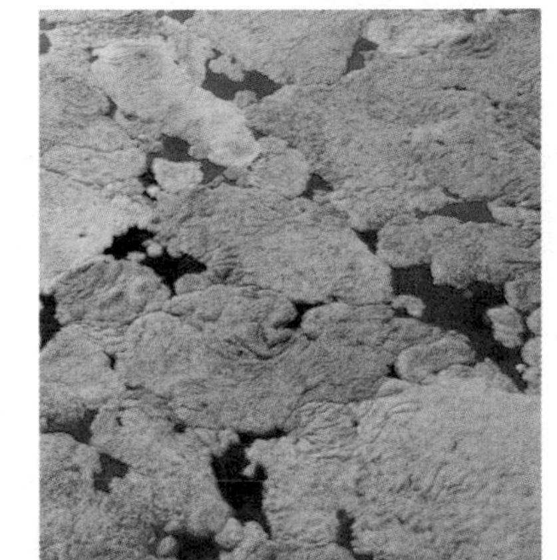

FIGURE 4: How did organisms like these algae help life to spread onto the land?

QUESTIONS

4 Air had its present composition for how long? Give your answer as a percentage of Earth's age.

5 Originally there was no oxygen or nitrogen in the atmosphere. Explain how each of these gases was produced.

Separating gases from the air (Higher tier)

There are many uses for the gases that make up air. For instance:

> liquid nitrogen – fast-freezing foods and biological specimens

> oxygen – converting iron into steel; helping patients with breathing

> argon and other noble gases – filling light bulbs; xenon headlights; 'neon' lights.

Fractional distillation separates the gases from the air. This is the same process used to separate the substances in crude oil.

Air is cooled to about -200 °C. Since this is below the boiling points of nitrogen (-196 °C) and oxygen (-183 °C), the air liquefies. As the liquid air warms, the nitrogen boils off first, leaving liquid oxygen behind.

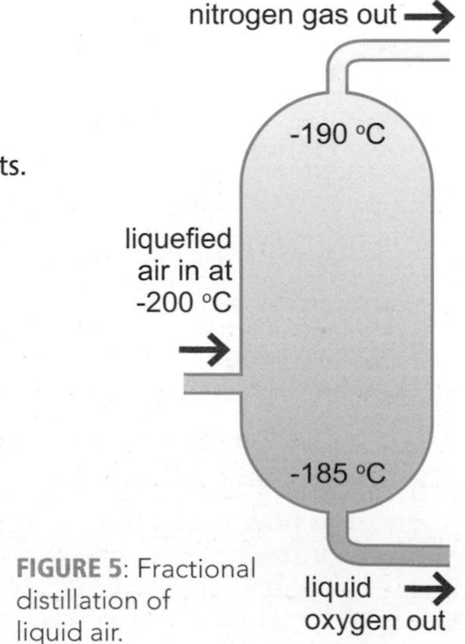

FIGURE 5: Fractional distillation of liquid air.

QUESTIONS

6 Find out how air is cooled enough to make it liquefy.

7 Explain why fractional distillation is needed to separate mixtures such as liquid air or crude oil.

The atmosphere and life

You will find out:

- why we need plants
- more about photosynthesis
- one theory of how life developed on Earth

The Goldilocks planet

In the Goldilocks story, one bowl of porridge was too hot, one was too cold, another was just right. Living things are like this. Mercury and Venus are nearer the Sun – too hot for life to exist. Mars and the outer planets are further away – too cold. Earth is in between – its temperature is just right. So Earth is the only planet in the solar system where there is life.

Venus Mercury Earth Mars Jupiter Saturn Uranus Neptune Pluto

500 400 100 0 -100 -200

temperature in °C

planets not shown to scale

FIGURE 1: Average daytime temperatures of planets.

Right for life

What's 'just right' for living things?

There are many theories about how life developed on Earth, but nobody really knows. We do know that all living things need the right conditions:

- oxygen – from the atmosphere
- liquid water – taken in through roots for example, or by drinking
- temperatures between 0 and 50 °C.

Only Earth has all these conditions. The atmospheres of Venus and Mars are mainly carbon dioxide. Venus is covered in clouds of sulfuric acid. Mars is dry.

Animals and plants take in oxygen and give out carbon dioxide. This is respiration.

Why are plants important to life?

Plants make their own food by **photosynthesis**. Leaves absorb energy from sunlight to combine carbon dioxide with water and make glucose and oxygen. The oxygen is released into the environment.

The equation for photosynthesis:

carbon dioxide + water + energy → glucose + oxygen

$6CO_2 + 6H_2O + \text{energy} \rightarrow C_6H_{12}O_6 + 6O_2$

Plants are essential because they:

- remove carbon dioxide produced by respiration and burning fuels
- produce the oxygen needed for life.

This is how the oxygen entered the atmosphere millions of years ago. It is how the concentration of oxygen in the air nowadays stays constant.

FIGURE 2: The Mars Rover. Why is there no life on the surface of Mars?

QUESTIONS

1 Give two reasons why animals could not exist on Earth without plants.

2 In which part of plants does photosynthesis take place?

How did life on Earth begin? (Higher tier)

Earth is about 4600 million years old. Primitive forms of life appeared about 3400 million years ago. Modern humans appeared only 200 000 years ago.

There is uncertainty about how life began because there is no evidence. The first primitive life-forms did not form fossils. Also, no one has succeeded in creating life from non-living materials.

Here is one theory.

> For the first billion years, Earth's atmosphere was mainly carbon dioxide, with some methane, ammonia, hydrogen and water vapour. The water vapour eventually condensed to form oceans.

> Lightning provided energy to break chemical bonds and split molecules. The fragments recombined in different ways, forming new compounds.

> These new compounds included **amino acids** (from which all proteins are built up), **sugars** and other carbon compounds needed to make **DNA**. These compounds are the basis of life. DNA allows cells to duplicate and organisms to reproduce.

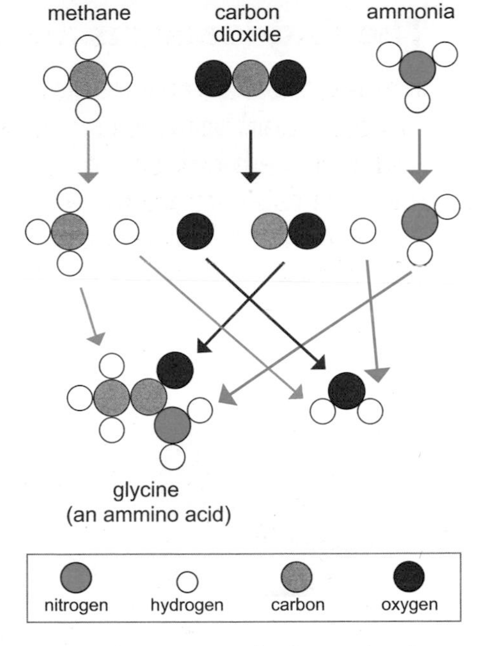

FIGURE 3: Rearranging broken molecules to make an amino acid. What else forms?

QUESTIONS

3 Name the main compound in the Earth's early atmosphere.

4 Suggest how lightning was involved in creating life.

5 What type of compound is needed to make proteins?

The Miller-Urey experiment (Higher tier)

In 1952, two American scientists, Miller and Urey, devised an experiment to test the theory of how life began. They mixed methane, ammonia and hydrogen in a sterile glass bulb. This simulated the early atmosphere.

A flask of water represented the ocean. They circulated water vapour from the 'ocean' through the 'atmosphere', and made electric sparks to simulate lightning.

After a week they analysed the mixture, known as 'primordial soup'. They found many different **organic compounds** – carbon compounds that normally come from living organisms. These included glycine and several other amino acids, and sugars, including ribose. This was significant because D in DNA stands for deoxyribose, a closely related sugar.

Miller and Urey did not create life. They did show that molecules essential for living things could be made by a natural process from Earth's early atmosphere.

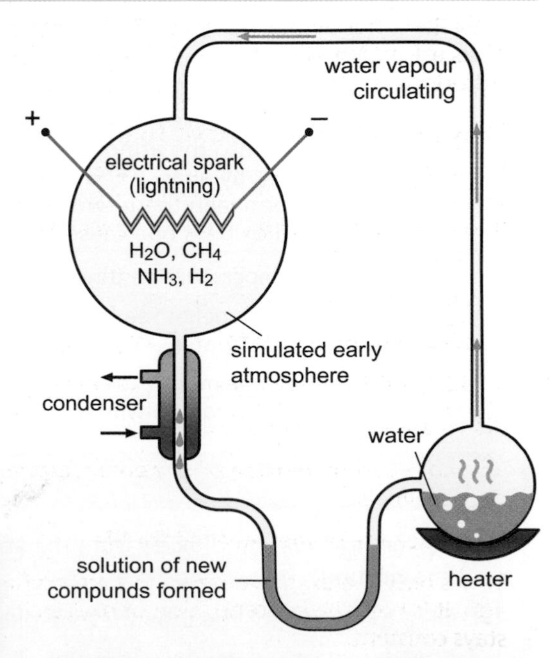

FIGURE 4: A simplified diagram of Miller and Urey's apparatus.

QUESTIONS

6 Find out what 'primordial' means and then explain why the mixture was called 'primordial soup'.

Carbon dioxide levels

You will find out:

> about links between carbon dioxide, oceans, rocks and fossil fuels

> how carbon dioxide levels in the atmosphere are changing

> these changes affect the whole Earth

Melting icebergs

Ice sheets cover massive areas of the world around both poles. At the edges, lumps of ice break off, forming icebergs. Icebergs can be enormous. They melt as they drift into warmer water. Ice is now melting into the oceans faster than snowfall can replace it. This is seen as a worrying sign of global warming.

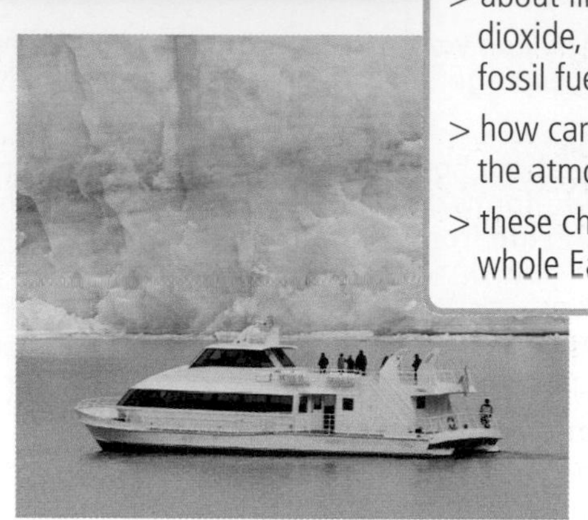

FIGURE 1: If too much of the ice sheet melts, sea level will rise, flooding low-lying land.

Carbon dioxide in the atmosphere

Where did the carbon dioxide go?

Carbon dioxide made up most of Earth's early atmosphere. Now it is only about 0.04%. The decrease was because carbon dioxide:

> dissolved in newly forming oceans – It was absorbed by primitive organisms in the oceans.

> was removed by plants – It was used by plants that could photosynthesise.

> was converted by marine organisms into calcium carbonate for shells and skeletons.

Over millions of years, organisms removed carbon dioxide from the air. The organisms became the carbon compounds found in fossil fuels. The shells and skeletons of dead marine organisms formed limestone and chalk (rocks made of calcium carbonate). These fossil fuels and rocks are called **carbon sinks**.

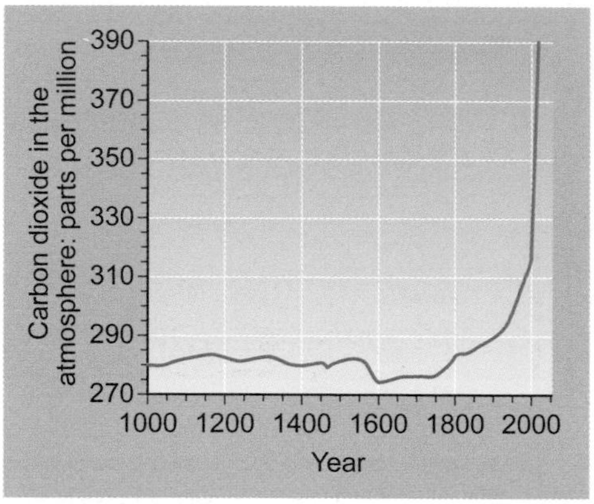

FIGURE 2: Why have carbon dioxide levels risen rapidly since the Industrial Revolution (around 1750) and even faster since 1900?

Global warming

For 250 years, people have used more and more machines for industry and transport. These machines burn fossil fuels and put more carbon dioxide into the atmosphere. Earth's average temperature has risen by almost 1 °C since 1890. Most scientists agree that there is a link between these two facts. They think that the increase in carbon dioxide is one cause of **global warming**.

If Earth's average temperature continues to increase:

> polar ice sheets could melt – Sea levels rise and parts of the world would disappear under water.

> weather could become more extreme – It becomes stormier and hotter in summer, and colder in winter.

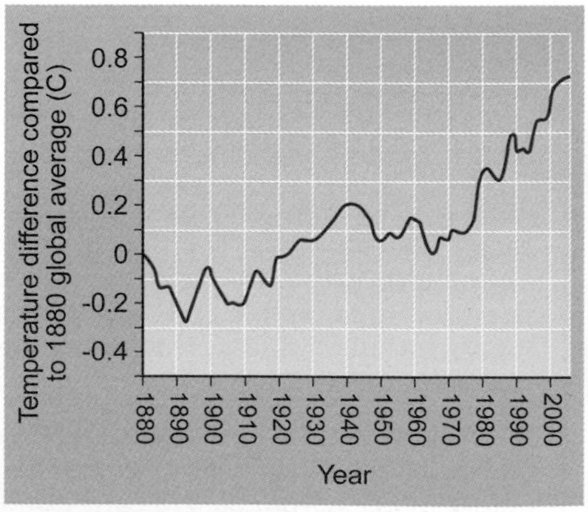

FIGURE 3: Average rise in temperature of the atmosphere since 1880.

QUESTIONS

1 Explain how carbon dioxide from the atmosphere is trapped in the ground.

2 Give the name for the idea that the planet is getting warmer.

3 Suggest what may happen if sea levels rise.

Carbon recycling

Carbon dioxide dissolves in rain and in the oceans, removing it from the atmosphere.

Both animals and plants take in oxygen and give out carbon dioxide – the **respiration** process. In **photosynthesis**, green plants take in carbon dioxide and give out oxygen.

Dead plants and animals decay. Oxygen from the air or water converts them back into carbon dioxide, with the help of bacteria, fungi, and other organisms.

Without oxygen, for instance in mud underwater, dead material is not oxidised to carbon dioxide. It decays to carbon and hydrocarbons found in fossil fuels. This takes millions of years.

Burning fossil fuels and biofuels, such as wood, also returns carbon dioxide to the atmosphere.

Biofuels are said to be **carbon-neutral** because the carbon dioxide released, when they burn, is only the same amount as that removed by photosynthesis.

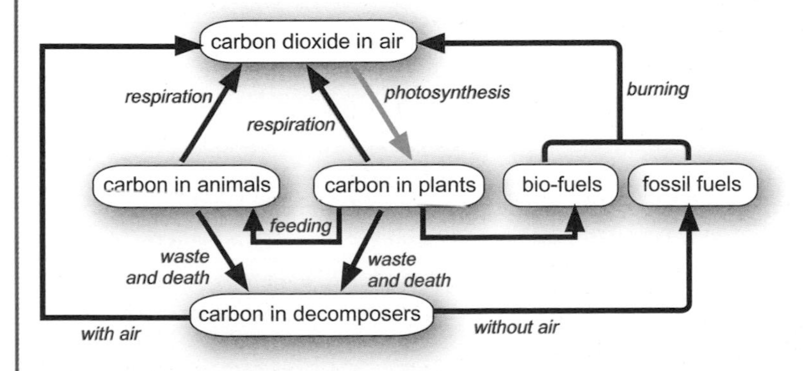

FIGURE 4: The carbon cycle.

Did you know?

Average sea level rose 20 centimetres during the 20th century. A further rise of nearly one metre is predicted for this century.

QUESTIONS

4 Describe the 'carbon cycle'.

5 Give two ways in which carbon dioxide is (a) removed from the atmosphere (b) put back.

6 Explain why biofuels are carbon-neutral.

So what's the problem?

Modern societies burn enormous amounts of fuels for manufacturing, generating electricity and transport. Carbon dioxide is being put into the atmosphere much faster than natural processes can take it out. In the past 250 years, since the Industrial Revolution, the concentration of carbon dioxide in the atmosphere has increased from 275 to 390 parts per million.

Carbon dioxide is a greenhouse gas. It traps energy from the Sun. Increasing amounts of carbon dioxide are likely to make Earth hotter in future. However, Nature is complex and there are many other factors that affect global temperature.

A further problem is that carbon dioxide dissolves in water to form carbonic acid. Higher concentrations of carbon dioxide in the atmosphere mean that more dissolves in the seas and oceans. Thus they become slightly more acidic (lower pH). This upsets ecosystems, killing coral for instance.

QUESTIONS

7 Carbon compounds in fossil fuels were originally formed from carbon dioxide in the atmosphere. Explain why fossil fuels are not considered to be carbon-neutral like biofuels.

Checklist C1.5–C1.7

To achieve your forecast grade in the exam you will need to revise

Use this checklist to see what you can do now. Refer back to the relevant topic in this book if you are not sure. Look across the three columns to see how you can progress. **Bold** text means Higher tier only.

Remember that you will need to be able to use these ideas in various ways, such as:

> interpreting pictures, diagrams and graphs
> applying ideas to new situations
> explaining ethical implications
> suggesting some benefits and risks to society
> drawing conclusions from evidence you are given.

Look at pages 276–297 for more information about exams and how you will be assessed.

To aim for a grade E	To aim for a grade C	To aim for a grade A
Recall that hydrocarbons are broken down (cracked) to produce alkanes and unsaturated hydrocarbons called alkenes. Recall that cracking produces fuels.	Understand that the vapours are either passed over a hot catalyst or mixed with steam at a very high temperature so that they thermally decompose. Recall that alkenes have the general formula C_nH_{2n}. Know that alkenes have a double bond (=). Know that bromine water turns from orange to colourless if alkenes are added.	
Recall that alkenes can be used to make polymers, such as poly(ethene) and poly(propene). Know that many small molecules (monomers) join together to form very large molecules (polymers).	Describe the useful applications of polymers that are being developed, such as new packaging materials, waterproof coatings for fabrics, dental polymers, wound dressings, hydrogels, smart materials (including shape memory polymers).	
Know that many polymers are not biodegradable, so they are not broken down by microbes and this can lead to problems with waste disposal.	Explain why plastic bags made from polymers and cornstarch can break down more easily.	
Recall that ethanol can be produced by reacting ethene with steam. Understand that ethanol can be produced by fermentation with yeast, using renewable resources.	Recall that fermentation can be represented by: sugar → carbon dioxide + ethanol	
Explain why vegetable oils are important foods and fuels.	Describe how plant material can be processed to produce plant oils.	**Explain why foods cooked in oils have different flavours from foods cooked in water.**

To aim for a grade E	To aim for a grade C	To aim for a grade A
Understand that oils do not dissolve in water. Know that oils can be used to produce emulsions.	Know that emulsions are thicker than oil or water. Know that emulsions provide better texture, coating ability and appearance, such as in salad dressings, ice creams, cosmetics and paints.	**Describe hydrophilic and hydrophobic properties of molecules.**
Recall that vegetable oils that are unsaturated contain double carbon=carbon bonds.	Know that bromine water can be used to test for unsaturated vegetable oils.	**Describe how vegetable oils can be hardened, and why this makes them more useful.**
Recognise that Earth's crust, the atmosphere and the oceans are the only source of minerals and other resources that humans need.	Explain and evaluate the effects of human activities on the atmosphere.	Explain and evaluate theories of the changes that have occurred and are occurring in Earth's atmosphere.
Describe that Earth has a core, mantle and crust, and atmosphere. Know that the crust and the upper part of the mantle are cracked into a number of large pieces.	Describe the processes that cause the tectonic plates to move at relative speeds of a few centimetres per year.	Explain that the mantle is mostly solid, but that it is able to move slowly.
Recall that earthquakes and/or volcanic eruptions occur at the boundaries between tectonic plates.	Explain why scientists cannot accurately predict when earthquakes and volcanic eruptions will occur.	
Know that during the first billion years of Earth's existence there was intense volcanic activity.	Explain how volcanic activity released the gases that formed the early atmosphere and water vapour that condensed to form the oceans.	
Know that there are several theories about how the atmosphere was formed.	Describe a theory that suggests the composition of Earth's early atmosphere.	Explain the ideas behind the theory about the composition of Earth's early atmosphere.
Know that there are many theories as to how life was formed billions of years ago.	Recall that plants and algae produced the oxygen that is now in the atmosphere.	Describe a theory on the origin of life. Explain why we do not know how life was first formed.
Describe how most of the carbon from the carbon dioxide in the air gradually became locked up in sedimentary rocks, as carbonates and fossil fuels. Explain why the release of carbon dioxide by burning fossil fuels increases the level of carbon dioxide in the atmosphere.	Explain how the oceans act as a reservoir for carbon dioxide. Know that the increased amount of carbon dioxide absorbed by the oceans has an impact on the marine environment.	**Explain how air can be fractionally distilled to provide a source of raw materials used in a variety of industrial processes.**

Exam-style questions: Foundation C1.5–1.7

1. The table shows the temperature and the percentage composition of the atmospheres of Earth and Mars today.

Gas	Percentage composition (%)	
	Earth today	Mars today
nitrogen	78	3
oxygen	20	0.13
argon	0.97	1.6
water vapour	0.4	0.03
carbon dioxide	0.04	95

Average surface temperature: Earth 20°C Mars -60°C

(a) Use information from the table to help you to answer these questions.

AO1 **(i)** Name the main gas in Earth's atmosphere today. [1]

AO2 **(ii)** Name the main gas that was in Earth's atmosphere billions of years ago. [1]

AO3 **(iii)** Earth's surface is mainly covered with water. There is no liquid water on the surface of Mars. Suggest why. [2]

AO1 **(b)** The diagram shows part of Earth and ways that carbon dioxide can be removed from Earth's atmosphere.

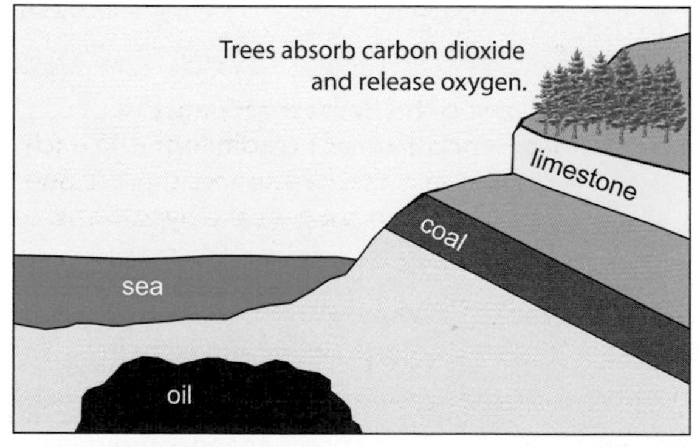

Give three ways that carbon dioxide can be removed from Earth's atmosphere. [3]

2. The British Heart foundation is keen to prevent coronary heart disease, which kills many people. They know that eating lots of saturated fat increases the risk of developing the disease.

Type of oil	Saturated fat (g per 100 g of oil)	Polyunsaturated fat (g per 100 g of oil)
palm	45	8
soya	16	57
olive	14	11
maize	16	49

(a) Use the information from the table to answer each question.

AO2 **(i)** Which type of oil appears to be best for your heart? Explain your answer. [1]

AO3 **(ii)** Countries that use only olive oil in food preparation have less heart disease than other countries that use other types of oil. Suggest a possible reason for this. [1]

(b) Unsaturated fats react with bromine water, changing it from orange–yellow to colourless. This reaction can be used to compare the unsaturation of low spread fats.

Scientists added bromine water from a burette to equal amounts of three low fat spreads until it was no longer decolourised and the orange–yellow colour was permanent.

Low fat spread	Volume of bromine water added (cm^3)
olive spread	8.9
sunflower spread	32.3
palm oil spread	7.4

AO3 **(i)** What would you see when the first few drops of bromine water are added to each spread? [1]

AO3 **(ii)** What do these results tell you about sunflower spread compared with olive spread and palm spread? [2]

3. Blue plastic rope is made from poly(propene).

Poly(propene) is a polymer made from propene.

Propene is made by cracking saturated hydrocarbons from crude oil.

AO1 **(a)** Poly(propene) molecules are made from propene molecules by a polymerisation reaction. Describe what happens in a polymerisation reaction. [2]

AO1 **(b)** Poly(propene) ropes are not biodegradable. Suggest why this is an advantage in the rope's use, but a disadvantage when it is discarded. [3]

AO1 recall the science AO2 apply your knowledge AO3 evaluate and analyse the evidence

WORKED EXAMPLE – Foundation tier

Alfred Wegener in 1915 suggested that the continents had once been joined together. He knew that South America and Africa looked as if they could join together like a jigsaw.

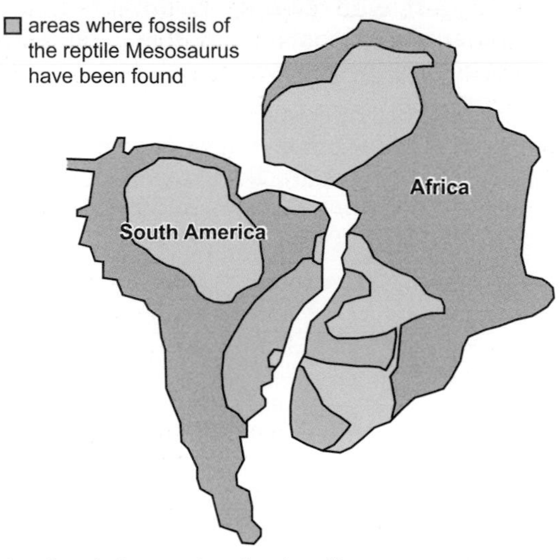

(a) Using information in the diagram, suggest two pieces of evidence that confirmed his hypothesis. [2]

The shapes of the two continents look to fit together. There are similar rocks found on the coasts of S America and Africa. The same fossils are found on both continents.

(b) Despite this evidence, Wegener's hypothesis was rejected because there was no theory about how Earth's crust could move. Scientists later discovered that Earth's tectonic plates do move a few centimetres each year.

Explain the processes that lead to the tectonic plates moving. [2]

Radioactive processes inside the earth generate heat that warms the mantle, causing convection currents in the mantle. The mantle is like a jelly, and can only move very slowly, carrying the plates with it.

(c) Earthquakes happen when plate movement is stopped temporarily by hard rocks in Earth's crust.

Explain why scientists cannot accurately predict when earthquakes will happen. [2]

Scientists still do not know enough information about Earth's crust to know where the hard bits of rock are, and how much force is needed to crumble them so they cannot accurately predict earthquakes.

How to raise your grade!

Take note of these comments – they will help you to raise your grade.

The first part of the answer repeats the information and so is not creditworthy. In each of the second and third sentences there is one piece of evidence to support the hypothesis. The answer would receive the two marks.

This well-constructed answer explains the convection effect, including the source of the heat, and why the plate movement is so slow.

In this answer, the candidate has given two reasons, showing that they understand why earthquakes cannot be predicted accurately. Two marks would be given.

Exam-style questions: Higher C1.5–1.7

1. Ethanol is a used as a substitute for petrol in many countries. There are two methods of producing ethanol.

Method 1: Ethene obtained from crude oil is reacted with steam and a catalyst.

Method 2: Sugar from crops is fermented with yeast.

AO2 **(a)** Which of the two methods of production involves using renewable sources? Explain your answer. [2]

AO2 **(b)** Explain why a car powered by ethanol produced by Method 1 produces pollution that counts towards carbon dioxide emissions and global warming, whilst a car powered by ethanol from Method 2 does not. [2]

AO2 **(c)** Here is the unbalanced symbol equation for Method 2.

$$C_6H_{12}O_6 \rightarrow ....C_2H_5OH + ...CO_2$$

Balance the equation. [1]

AO2 **(d)** Vegetable oils can be converted into biodiesel to fuel vehicles.

Converting food crops such as sugar and vegetable oils into fuels for vehicles has consequences for the environment and also for people.

Suggest, with reasons, one advantage and one disadvantage of using food crops as fuels rather than food. [4]

2. Seeds and nuts that are rich in oils are important sources of food and fuel. The plant material is crushed, and the oil removed by pressing or steam distillation.

Here is some information about several oils that are used in foods.

Oil	olive oil	rapeseed oil	sunflower oil
Boiling point (°C)	300	381	326
Saturated fats (%)	13.1	7	10
Unsaturated fats (%)	74.2	84	66

Scientists believe that oils are healthier if they are low in saturated fats and high in polyunsaturated fats.

AO1 **(a)** Explain the difference between a saturated fat and an unsaturated fat. [2]

AO3 **(b)** Which of the three oils is likely to be most healthy? Explain your answer. [2]

AO3 **(c)** Describe how you would test each oil to compare their unsaturated fat content. [3]

AO2 **(d)** Suggest, with reasons, one advantage and one disadvantage of cooking potatoes in oil rather than boiling them in water. [4]

3. Air is a mixture of many gases. Many of them need to be separated for use in industry and medicine. They are separated by fractional distillation. Here is a flow diagram of what happens, and the melting and boiling points of the gases involved.

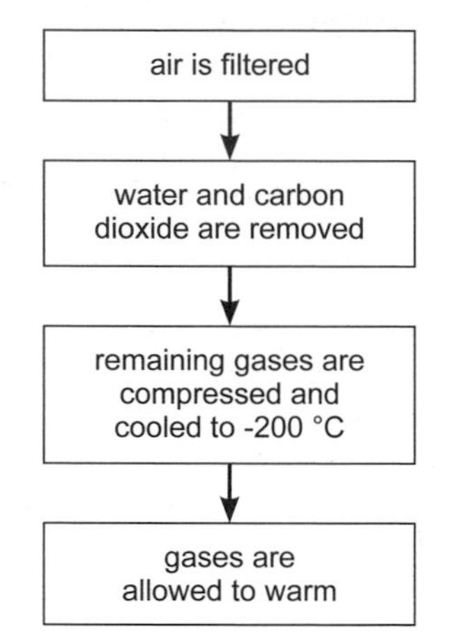

Gas	Melting point (°C)	Boiling point (°C)
argon	-189.0	-185.9
carbon dioxide	-78.5	-78.5
helium	-272.2	-269.0
nitrogen	-210.0	-195.9
oxygen	-219.0	-183.0
water vapour	0.0	100.0

AO2 **(a)** Explain why the water vapour and carbon dioxide are removed before cooling. Use information from the table in your answer. [2]

(b) After compression and cooling to -200 °C the helium is removed.

AO2 **(i)** Explain why the helium is removed now. [2]

AO2 **(ii)** The remaining gases are allowed to warm up. Which gas will vaporise first? [1]

AO1 recall the science | AO2 apply your knowledge | AO3 evaluate and analyse the evidence

WORKED EXAMPLE – Higher tier

There are many theories about how Earth's atmosphere has developed and how life began on Earth billions of years ago. Today, Earth's atmosphere is a mixture of 80% nitrogen, 20% oxygen and small amounts of other gases.

(a) Describe how the early atmosphere was different to today's, and how it was formed. [3]

There was little nitrogen, no oxygen, but lots of carbon dioxide, methane, and ammonia. These gases had been produced by volcanoes, and the water vapour produced condensed to form the oceans.

It was in the oceans with this early atmosphere that the earliest forms of life began.

(b) Explain why there are many theories about how life began. [2]

We do not know the conditions on Earth when life began, and so we have not been able to copy them. We also have no evidence of the earliest microscopic forms of life.

Miller and Urey carried out an experiment into the origins of life. They used this apparatus for their investigation. It contained water and the gases present in the early atmosphere.

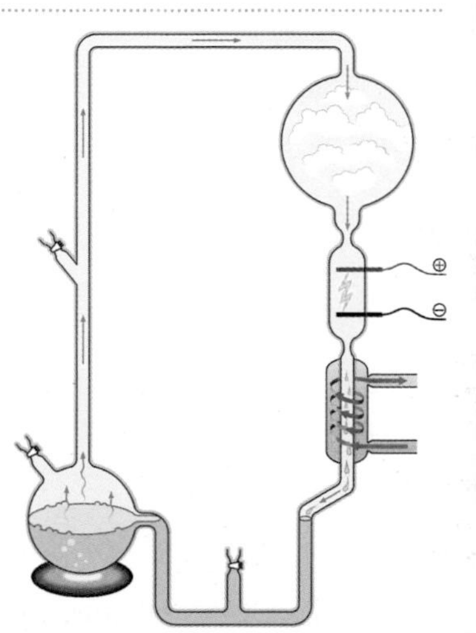

(c) Explain what the sparks model, and why they were used. [2]

The sparks model lightning and it was thought that the early atmosphere was subject to frequent lightning storms that made little molecules join together to make bigger ones.

(d) What hypothesis do you think the scientists were testing? [2]

That simple molecules present in the early Earth can be converted into some complex molecules present in life.

(e) After a week, the scientists found more complex molecules in the water. Name the type of molecules they found, and explain why this supported the hypothesis. [3]

They found amino acids which are the molecules that proteins are made from, proving that life was formed.

How to raise your grade!

Take note of these comments – they will help you to raise your grade.

The answer follows the question by answering each part in turn. This makes sure that all parts of the question are covered in the answer.

The answer shows some understanding of the reason why there are many theories but fails to make it clear that insufficient knowledge has led to many theories developing.

The candidate clearly realises that the sparks are to simulate lightning. More detail is needed about the high temperatures providing energy to allow the small molecules to become larger ones.

The answer clearly relates the key features of the investigation into a workable hypothesis.

The candidate correctly names the type of molecules found, asserts that these are needed to make proteins, but fails to make the link to the hypothesis that is required.

Physics P1.1–1.3

What you should know

Energy and its transfer

Objects get hotter when they gain energy by heating processes and cooler when they lose energy.

We use thermometers to measure the temperature of something. The units of temperature are degrees Celsius.

Energy can be transferred from place to place and different names can be used to describe energy in different places.

Name the three different ways in which heating or cooling can transfer energy.

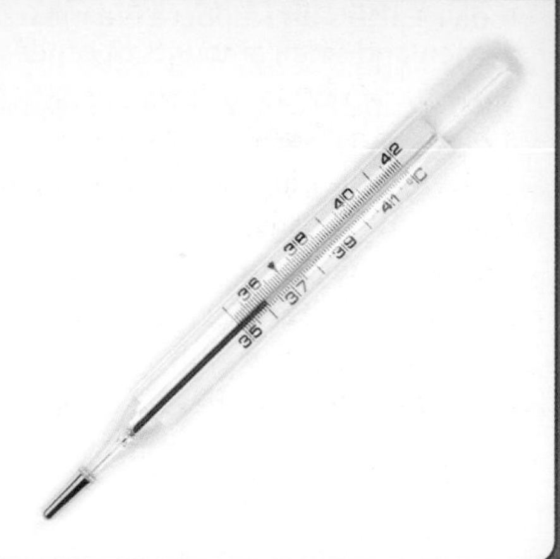

Electric circuits

Symbols are used to draw circuit diagrams.

Electric current is measured in amps, using an ammeter.

The potential difference between two points of a circuit is measured in volts, using a voltmeter.

There are two types of circuits, called series and parallel.

Explain why mains electricity is more dangerous than electricity from batteries.

Electrical appliances

Electrical devices make use of a variety of effects caused by electric currents.

Turning off electrical appliances when they are not in use is one way to 'save energy'.

Describe some good and some bad effects of heating caused by an electric current.

You will find out

The transfer of energy by heating processes and the factors that affect the rate at which that energy is transferred

> Energy can be transferred from place to place by work or by heating processes.

> Different heating processes are important in different situations.

> There are ways to evaluate appliances and materials that transfer energy by heating.

> The effectiveness of building insulation depends on the material's properties and its cost.

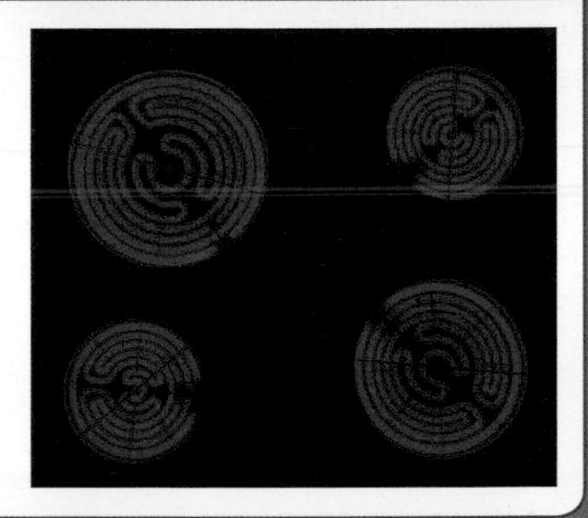

Energy and efficiency

> Appliances that transfer energy rarely transfer all the energy to the place where it is needed.

> The efficiency of appliances allows us to evaluate how cost effective they are and to try to improve them.

> By calculating the efficiency of appliances, they can be compared.

> There are methods to reduce the amount of energy that appliances waste.

The usefulness of electrical appliances

> Electrical appliances are often a quick and convenient way to transfer energy.

> Calculations show how much energy an appliance transfers.

> Calculations can show how much different appliances cost to run.

Energy

Energy and entertainment

Have you ever seen your favourite group performing at a live gig? It's exciting. Lots of light, sound, movement. And massive electricity bills later! All these are connected with energy, but how can a scientist describe what is going on?

FIGURE 1: There is a lot of energy on stage. How would you describe it?

You will find out:
- > all objects have energy
- > energy moves around, but it cannot appear or disappear

What is energy?

You may have learned about different 'forms' of **energy** – light, sound, heat, movement, chemical and so on. This is not quite true however, because energy is not physical 'stuff'. It cannot exist in different forms. Light, sound and so on are ways of helping you to see where energy was stored and if it was being **transferred** (moved from place to place).

A simple way to describe what happens when a candle burns is to say that chemical energy, in the wax, changes to heat energy and light energy. Now you know that energy is just energy – there are not different forms. A more accurate description is to say that, when the candle is burned, energy stored in the candle is transferred to its surroundings by **infrared radiation** and light.

FIGURE 2: How can a burning candle be described in terms of energy?

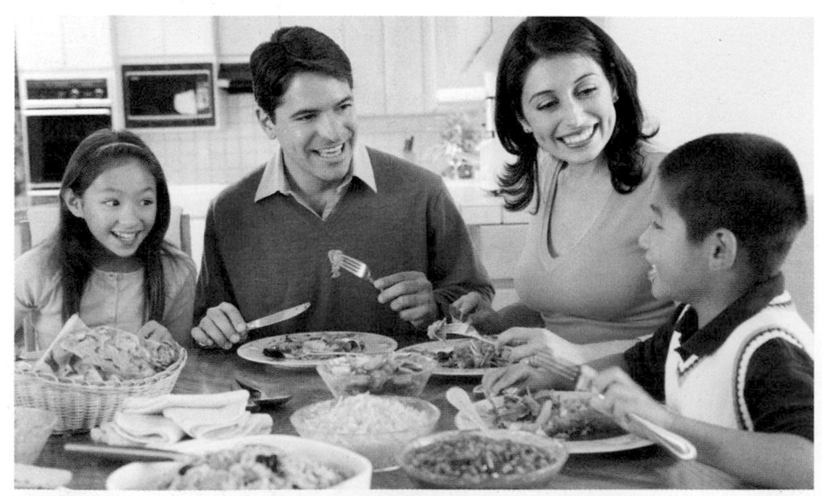

FIGURE 3: How is energy being transferred here?

Did you know?

Absolutely every object there is has energy, even when it is not very easy to detect or to see what it is doing.

QUESTIONS

1 Explain why it is not quite true to talk about different 'types of energy'.

2 Describe what happens when a battery-powered torch is switched on. Use the terms 'energy stores' and 'energy transfers'.

Storing and transferring energy

More about energy

Everything stores energy. Although energy cannot be seen or picked up, you can tell whether something has lots of energy or little energy.

Energy enables things to happen. Something with lots of energy will be able to do lots: it may move, make a sound or, if it burns, heat the surroundings.

Something with less energy cannot do as much. A large piece of explosive has more energy than a small piece: it will move more earth and make a bigger hole when it explodes.

Transferring energy

Mug heaters that plug into the USB port on a computer keep your coffee warm. Most of the heaters are claimed to keep the coffee at about 40 °C. Actually they just stop your coffee going cold quite so quickly.

You could say that electrical energy, from the computer, changes to heat energy, in the coffee. However, it is more accurate to describe how the energy is being transferred (moved from one place to another):

> The **electric current**, in the mains supply, transfers energy from where it is generated to your home.

> Electric current transfers energy through your computer to the mug heater, which gets warmer.

> Energy is transferred through the mug to the coffee.

> Energy is also transferred away from the hot coffee into the surroundings (that's why hot coffee cools down).

> When the electric current from the computer transfers energy to the coffee as fast as the coffee transfers energy to the surroundings, the temperature of the coffee stops changing.

FIGURE 4: Knowing about energy can help you decide whether or not to buy one of these.

QUESTIONS

3 Is electricity an energy store or a way of transferring energy?

4 Suggest how you could compare two different mug heaters. What would you do to make a valid and reliable comparison?

5 Choose another example of an 'energy change'. Using energy transfers, describe what is really happening.

Energy and chemistry

Chemical reactions show that energy is not any kind of physical substance. In a chemical reaction, the total mass of chemicals before the reaction is the same as the total mass of chemicals formed in the reaction. Yet, the reaction will have taken in or given out energy. Burning fuel and heating the surroundings is an example.

QUESTIONS

6 The kilocalorie (kcal) figure, in the Nutritional Information section on food packaging, tells you how much energy different foods can transfer to you when you eat them. Find out how scientists compare energy in different foods.

Infrared radiation

Chilly satellites

It's cold in space. Satellites have to be made from materials that will not break when they get very cold. Satellites can also get very hot, when the Sun shines on them. So, scientists have invented a paint that changes colour when the temperature changes. It absorbs energy when the satellite is cold and stops absorbing energy when the satellite is in danger of getting too hot.

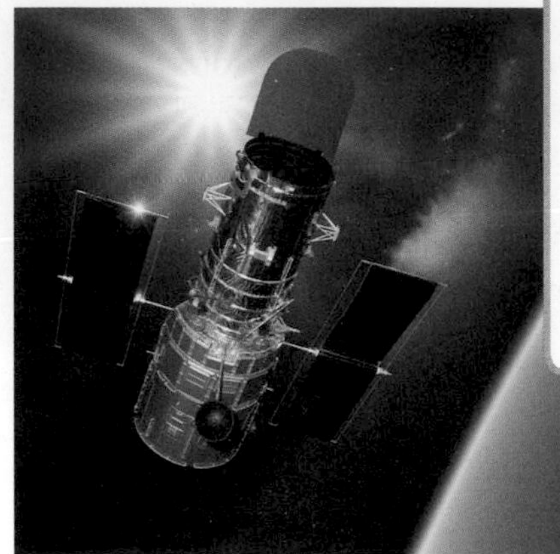

FIGURE 1: It's very hot in the sunshine and very cold in Earth's shadow.

You will find out:

- > all objects emit and absorb infrared radiation
- > hot objects emit more infrared radiation in a given time than cooler ones
- > an object's surface affects how well it emits and absorbs infrared radiation

Infrared radiation

Emission of infrared radiation

Hold your hand next to, but not touching, a warm radiator. You can feel **infrared radiation**. The hot radiator is emitting infrared radiation – and that is why it is called a radiator.

All objects **emit** infrared radiation. In a fixed time, a hot object emits more infrared radiation than a cooler one. The bigger the difference in temperature between an object and its surroundings, the faster energy is transferred.

Measuring infrared radiation

Scientists can measure infrared radiation with a thermopile. An 'ear thermometer' that a doctor uses, to measure your body temperature, is a thermopile. It measures the infrared radiation from your eardrum.

Leslie's cube is a hollow cube with four different surfaces: shiny black, dull black, shiny white and dull white. If the cube in Figure 3 is full of boiling water, the results show that, in the same length of time:

- > black surfaces give out more infrared radiation than white surfaces
- > dull surfaces give out more infrared radiation than shiny surfaces.

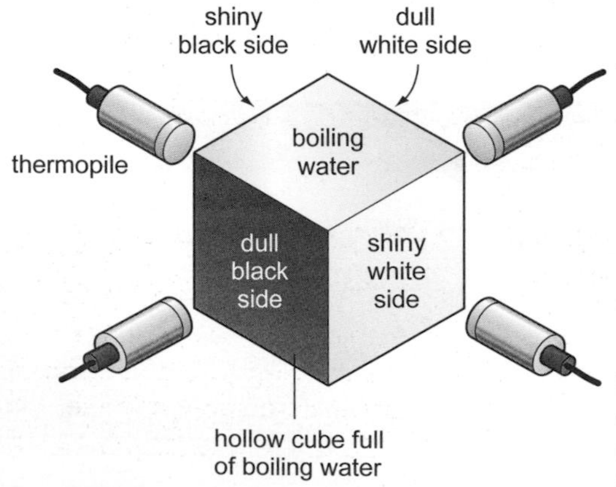

FIGURE 3: Using a Leslie's cube to compare infrared radiation from different surfaces.

Did you know?

The temperature outside the International Space Station ranges from -157 °C to 121 °C. It takes very clever engineering to keep the inside temperature constant.

FIGURE 2: A bonfire might emit so much infrared radiation that you cannot stand very close.

QUESTIONS

1 Explain what 'emitting infrared radiation' means.

2 Which instrument do scientists use to measure infrared radiation?

3 List the four surfaces of the Leslie's cube in the order that they emit infrared radiation, starting with the greatest rate.

... infrared radiation for students

Emission, absorption and uses

Emission and absorption

Figure 4 shows another way to compare the infrared radiation emitted by different surfaces. The thermometers show that the temperature of the water in the dull black beaker decreases faster than the water in the shiny silver beaker.

The dull black beaker is transferring energy by emitting infrared radiation at a greater rate than the silver beaker.

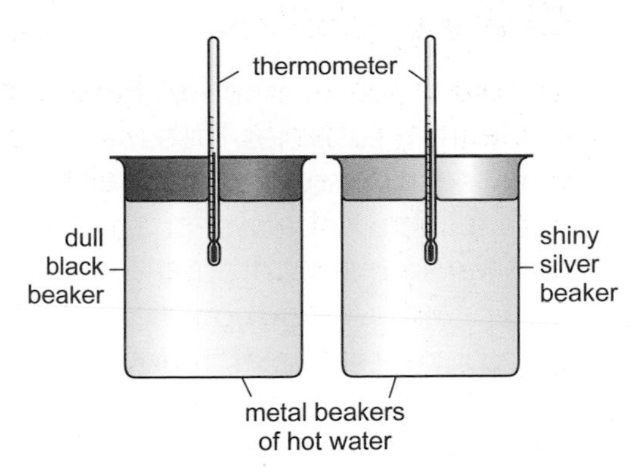

FIGURE 4: Explain how you could use this apparatus to show: the absorption of infrared radiation by dull black surfaces is at a greater rate than by shiny silver surfaces.

Using infrared radiation

People use colour to affect the **emission** and **absorption** of infrared radiation. There are examples all around us:

> Walkers or climbers suffering from hypothermia are wrapped in shiny silver 'space blankets', to stop them cooling down too quickly.

> Buildings in hot countries are usually white or light coloured so that they **absorb** less infrared radiation and are cooler inside.

> Fins on the backs of refrigerators are black, to increase the rate at which they emit infrared radiation and so cool quickly.

Vacuum flasks often have stainless steel insides and outsides. Some have glass with a silvery coating. For either type, there is a vacuum between the two layers.

> The shiny surface inside reflects infrared radiation back into the flask – to keep hot drinks hot.

> The shiny surface outside reflects infrared radiation from the surroundings away from the flask – to keep cold drinks cold.

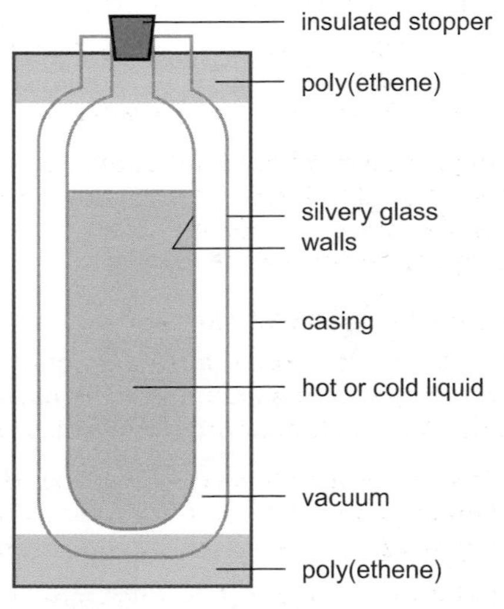

FIGURE 5: A vacuum flask with silvery glass surfaces to reflect infrared radiation.

QUESTIONS

4 Explain how you would make sure that the results from an infrared radiation experiment are valid and reliable.

5 Suggest a colour for a fire fighter's uniform. Explain your choice.

6 Discuss how you might design a solar cooker, to heat water using infrared radiation from the Sun.

Infrared radiation and global warming

Some scientists fear that, as global warming makes ice at the North and South poles melt, global warming will increase. The dark-coloured seawater absorbs infrared radiation faster from the Sun than the light-coloured ice does. No one knows if they are right yet, because there are many other things that affect global warming too.

QUESTIONS

7 Explain why seawater absorbs infrared radiation faster than snow and ice.

Kinetic theory

You will find out:

- > how diagrams can be used to model states of matter
- > particles of solid, liquids and gases have different amounts of energy

Dry ice

'Dry ice' is solid carbon dioxide. It does not melt. It sublimes. This means it changes from solid to gas and does not form a liquid in between. You can make artificial smoke by dropping dry ice into water. Very cold carbon dioxide gas forms and condenses water vapour from the air; you see a mist of tiny water droplets.

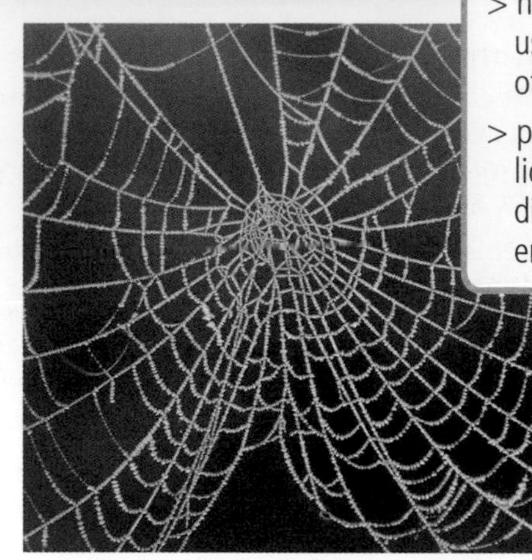

FIGURE 1: Spider webs catch water vapour from the air.

Particle energy

States of matter and the particle model

There are three **states of matter**: solid, liquid and gas. All consist of **particles** but are arranged in different ways. The differences in properties between solid, liquid and gas states are due to the arrangement of the particles.

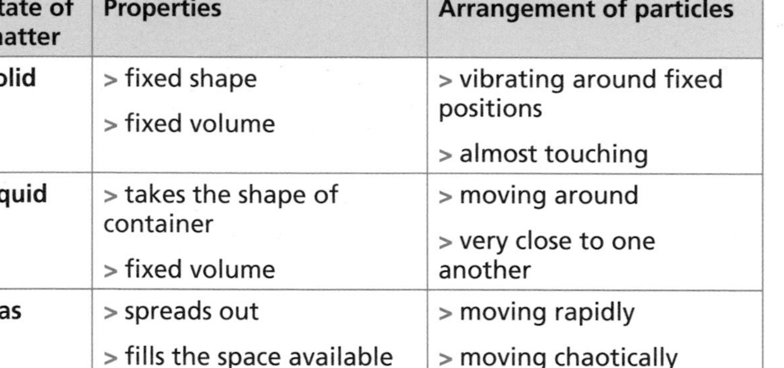

State of matter	Properties	Arrangement of particles
solid	> fixed shape > fixed volume	> vibrating around fixed positions > almost touching
liquid	> takes the shape of container > fixed volume	> moving around > very close to one another
gas	> spreads out > fills the space available	> moving rapidly > moving chaotically

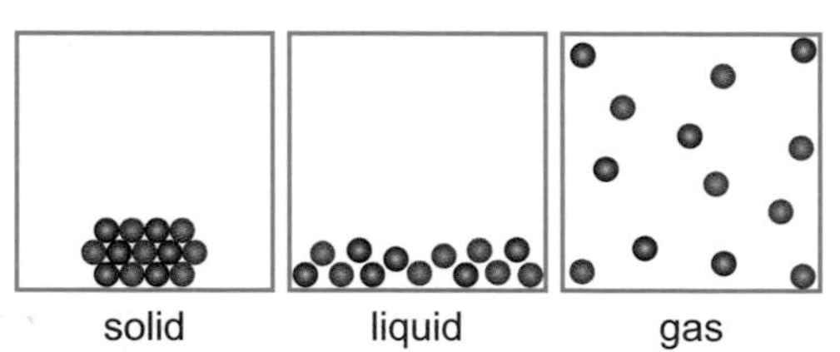

FIGURE 2: Arrangement of particles in each state of matter.

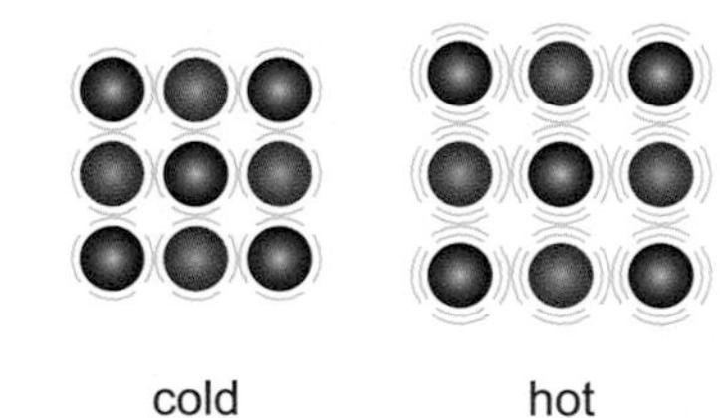

FIGURE 3: When a solid is heated, its particles vibrate more. Why do solids expand slightly, when heated?

Particles, energy and movement

In solids, particles vibrate (jiggle around) about their fixed positions. In liquids and gases, particles do not have fixed positions. They move around, but more freely in gases than in liquids.

Moving particles have **energy**. They have different amounts in solids, liquids and gases:

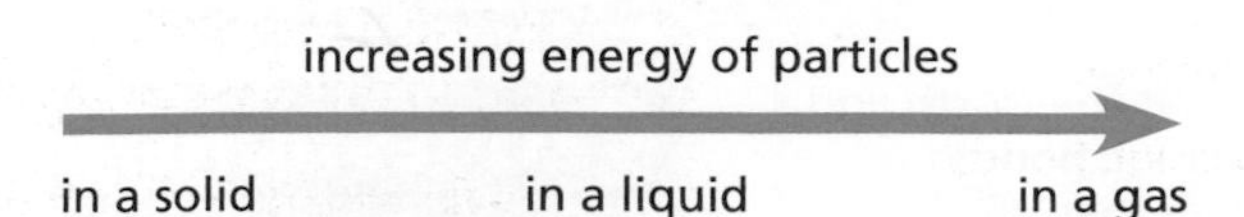

The greater the energy that particles in a material have, the more likely it is to be a liquid or even a gas. **Kinetic theory** describes the energy of particles and the way that they move.

QUESTIONS

1 Why do solids have fixed shapes and volumes?

2 In which state of matter do particles have the greatest energy?

3 Describe how the arrangement of particles in a liquid is different from those in a gas.

... particles in solids liquids gases

Bonds between particles

In all states of matter, there are **bonds** between particles. These bonds are strongest in solids and weakest in gases.

Melting is a change of state from solid to liquid. When energy is transferred to a solid by heating, its particles vibrate more vigorously. However, they remain in their fixed positions because of bonds between them. If they gain sufficient energy, they vibrate enough to break these bonds; they become free to move around. The solid melts and a liquid forms. The traditional explanation is that the bonds are still there, but break and reform easily, which is why a liquid keeps a set volume, but not a set shape.

Boiling is a change of state from liquid to gas. When energy is transferred to a liquid by heating, its particles move around faster. If they gain sufficient energy, they move fast enough to break the weak bonds that hold them close to one another. They escape from the liquid and move around randomly and chaotically. The liquid boils and a gas forms.

The reverse of boiling is **condensing** and the reverse of melting is **freezing**. In each case, particles transfer energy to their surroundings. There comes a point at which bonds form again and gas becomes liquid (condensation) and liquid becomes solid (solidification).

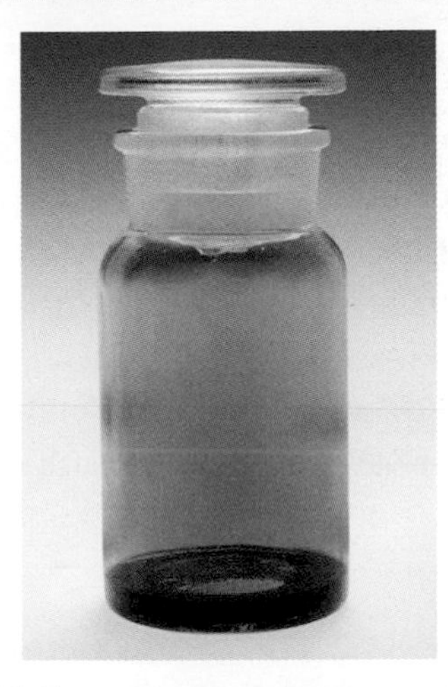

FIGURE 4: Even at room temperature, some particles in liquid bromine have enough energy to become a gas. Bromine liquid and gas are both reddish brown.

QUESTIONS

4 Suggest what happens to the particles in a solid, when it is heated.

5 Give the names for these changes of state: (a) liquid to gas (b) liquid to solid (c) gas to liquid.

FIGURE 5: A boiling kettle of water produces steam. How could you change steam (water in its gas state) into ice?

What's a particle?

The word 'particle' is used a lot in science. In simple kinetic theory, particles are imagined to be tiny solid spheres. However, in more sophisticated kinetic theory, particles vary in size, shape and mass. This variety affects the amount of energy they have at a given temperature, how fast they move and the strength of bonds between them. More sophisticated kinetic theory describes particles as:

> in a solid – held by strong bonds and vibrating.

The strength of the bonds varies from one solid to another.

> in a liquid – moving around one another, but with weak bonds preventing them from escaping.

The strength of the weak bonds varies from one liquid to another.

> in a gas – moving rapidly and chaotically, colliding with one another and bouncing off.

There are only very weak bonds between the particles.

Did you know?

Frost forms directly from water vapour in the air by the reverse of sublimation – it's called deposition.

QUESTIONS

6 Explain why (a) solids have different melting points to one another (b) liquids have different boiling points to one another.

Conduction and convection

You will find out:

> energy may be transferred by conduction and convection
> how the particle model of matter and kinetic theory explains the transfer of energy
> why metals are particularly good conductors

Lava lamps

The movement you see in lava lamps is due to convection currents. Solid wax, at the bottom of the lamp, melts when it is warmed by a light bulb. It becomes less dense than the liquid, which is oil, and moves upwards. As it rises, it cools, becomes denser and sinks back down again.

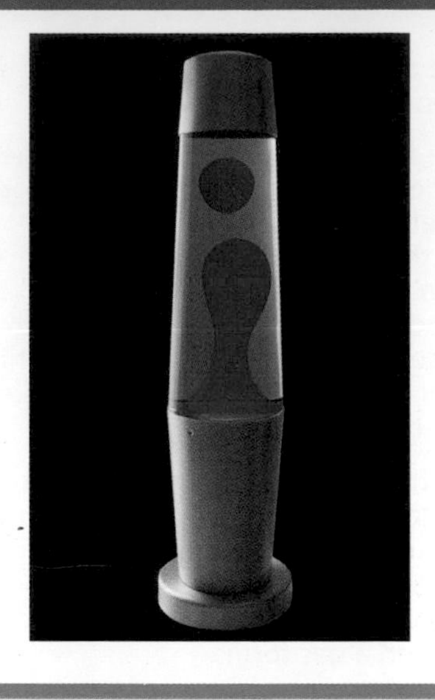

FIGURE 1: A decorative effect of convection currents.

Transferring by heating

Conduction

Conduction is how energy is transferred through solids when they are heated. Energy can also be transferred from one solid to another by conduction.

Some materials transfer energy more easily than others. **Conductors** transfer energy easily: for example, metals. **Insulators** do not transfer energy easily: for example, glass, ceramics and plastics.

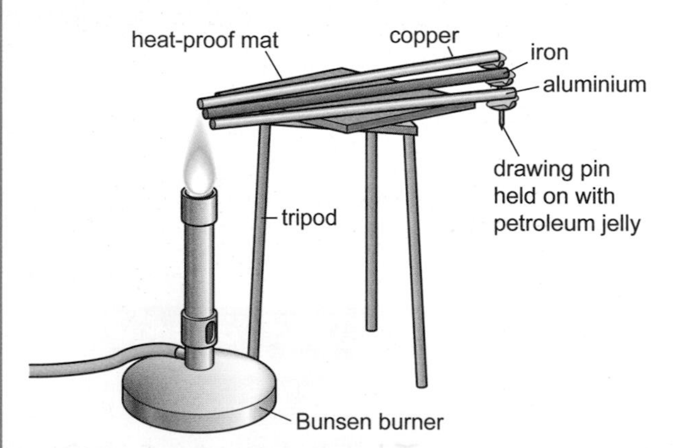

FIGURE 2: When the petroleum jelly melts, the drawing pin drops off. How would you decide which of the three metals is the best conductor?

Convection

Convection is the main way that energy is transferred in liquids and gases. Warm liquids and warm gases rise upwards because they are less dense than cooler liquids or gases. This is how hot air balloons work.

Figure 3 shows a **convection current** in a beaker of water. As the water warms, it rises up the side of the beaker, carrying the coloured dye with it. You can see the coloured dye falling down the other side of the beaker, as the water cools and becomes denser again.

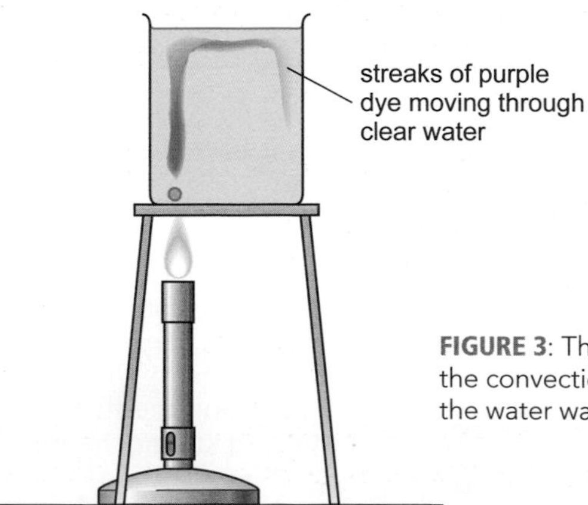

FIGURE 3: The dye traces the convection current as the water warms.

QUESTIONS

1 What is a conductor?

2 Suggest three materials that are insulators.

3 Explain what happens to liquids and gases when they are warmed.

4 Describe a convection current.

Transferring the energy

Energetic particles

In all materials, particles vibrate – they jiggle around. In solids, particles vibrate about fixed positions. In gases and liquids, particles are free to move.

If one end of a solid is heated, particles at that end gain energy and vibrate more. They shake their neighbouring particles. The vibrations are passed from particle to particle. Energy is transferred and spreads out through the material. This is conduction.

Particles in liquids and gases have more energy than particles in a solid. When heated, particles gain energy and move more rapidly. They move further away from one another. Therefore, the liquid or gas expands and becomes less dense. It rises into the cold areas and denser, cold liquid or gas falls into the warm areas. Convection currents are created which transfer energy from place to place.

Liquids and gases are poor conductors. Their particles are not in a regular array, touching one another, and so they cannot pass energy easily, from particle to particle, through vibrations.

Using conduction and convection

Choosing the right materials is important.

> Most saucepans are made from metal, with plastic handles. The metal conducts energy from the cooker to the food. Plastic handles do not conduct the energy to your hand.

> Double glazed windows have air or a gas such as **argon** between the panes of glass. Gases are poor conductors. A narrow gap between the panes reduces the chance of convection currents being set up.

> Computers, and other devices, use forced convection. A fan blows cool air across hot components. The air is heated and rises. It has the same effect as ordinary convection, but faster because the components always have cool air next to them.

FIGURE 4: Processing speeds might be limited by cooling speeds in future computers.

Did you know?

Vultures use convection currents in the air to gain height – one vulture was recorded at a height of eleven kilometres.

QUESTIONS

5 Explain why solids transfer energy mainly by conduction.

6 Describe how a convection heater warms a room.

7 Using conduction, suggest why serving dishes are usually made from glass or china.

Why are metals good conductors?

In metals, like all solids, some energy is transferred by vibrating particles – but they have another way to transfer energy.

In a metal lattice, outer electrons are freed from their atoms. 'Losing' these electrons means that the metal atoms become positively charged ions. Heating the metal causes these ions to vibrate more vigorously. They have more energy. This energy is transferred from hotter to cooler parts of the metal by the free electrons; they collide with the vibrating ions and gain energy. This energy is transferred to ions in cooler parts of the metal.

QUESTIONS

8 Devise an experiment to find out whether or not the rate of energy transfer varies along a strip of copper. Explain how you could tell if any change is linear.

Q ... metals as conductors

Evaporation and condensation

> **You will find out:**
> - > energy is transferred in evaporation and condensation
> - > how the particle model of matter and kinetic theory explain evaporation

Cooling sun sprays

Harmful ultraviolet radiation from the Sun can cause sunburn and skin damage. After-sun sprays can soothe skin that has been exposed to the Sun. Adverts for some say they have a cooling effect. Spray them on and one of the ingredients evaporates quickly, making your skin feel cool.

FIGURE 1: Spray to cool.

Evaporation and condensation

You have seen that energy can be transferred by infrared radiation, conduction and convection. Here are two more ways: **evaporation** and **condensation**. Both are changes of state:

> evaporation – liquid changes to gas

> condensation – gas changes to liquid.

Evaporation is not the same as boiling. Boiling happens only when a liquid is at its boiling point. Evaporation happens at temperatures below the boiling point.

Condensation is the name used when a gas cools and changes back to a liquid, either at the boiling point or at a lower temperature.

Evaporation and energy

> Particles in a liquid have a range of energies but there is an average value. Some particles have more energy than the average. Some have less.

> Particles in a liquid have less energy than particles in a gas.

> Bonds between particles hold the particles in the liquid.

To evaporate, particles must have sufficient energy to break away from the other particles. Particles with high energy escape, at the surface, and become gas particles. The average energy of particles left behind decreases, so the liquid cools.

Because the liquid is cooler, it absorbs energy from its surroundings, so its surroundings cool down as well. Energy is transferred from the surroundings to the liquid.

FIGURE 2: The solvent in nail varnish evaporates quickly, leaving the coloured pigment behind.

Did you know?

Raindrops form by condensation. A typical raindrop falls about five metres in a second.

QUESTIONS

1 Describe the difference between evaporation and boiling.

2 Water at 60 °C is changing from a liquid to a gas. What is this change of state called?

3 The bonds between particles in liquid A are stronger than those in liquid B. Suggest which liquid will evaporate most easily and say why.

Evaporation and environment

Factors affecting evaporation

Some days, puddles dry up faster than on other days. When they 'dry up', the water evaporates.

Evaporation takes place at the surface of a liquid. Liquids evaporate most quickly when:

> It is warm.

The average energy of the particles in the liquid is higher than when the temperature is lower. More particles have enough energy to break free from the bonds that hold them together as a liquid.

> The liquid is spread out.

The liquid has a larger surface area and, therefore, more particles at the surface.

> It is windy.

The air above a liquid can only hold a certain proportion of liquid particles before it is saturated (unable to hold any more). If particles leave the liquid and blow away, there is room for more to evaporate.

Remember

Vapour is another word for gas. This is where the term 'evaporation' comes from.

The water cycle

Water evaporates from Earth's surface. Water vapour (water in a gas state) is carried upwards, by convection currents in the air, and gradually cools.

As the water vapour cools, it transfers energy to the surrounding air. Some of the particles no longer have enough energy to remain as gas particles. Bonds form between them and they condense into a liquid.

Drops of water form and, when they become big enough, they fall as raindrops.

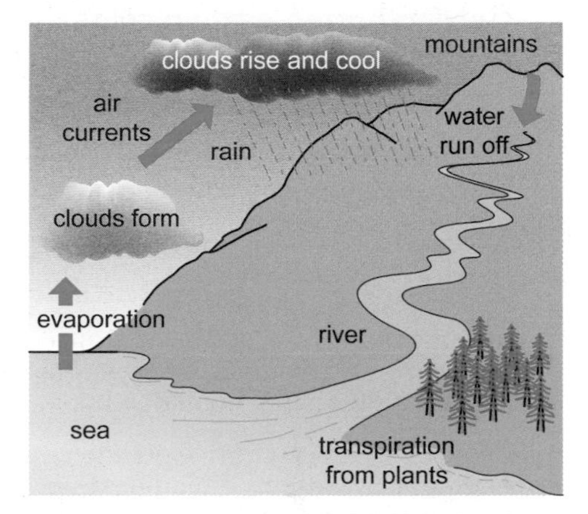

FIGURE 3: It is often wet in hilly areas near the sea.

QUESTIONS

4 Explain why you can dry your hair faster using a hair drier than using a towel.

5 During the day, you notice that the temperature is going down. Is it more likely or less likely to rain? Explain your answer.

6 A gas condenses to a liquid. Will make its surroundings warmer or cooler? Explain your answer.

Evaporation and pressure

Climbing high mountains can make your mouth and lips feel dry. Moisture evaporates very quickly from your skin. The climate in high mountains can be almost as dry as desert areas. Why is it so dry?

At high altitudes, wind speeds are usually higher and air pressure is lower. The force of the air pushing down on the surface of any liquid is less. Therefore, liquid particles escape more easily and objects 'dry' more quickly.

QUESTIONS

7 Using your knowledge about particles and energy, explain why lower air pressures make liquids evaporate more quickly.

Rate of energy transfer

You will find out:

- > which factors affect the rate of energy transfer
- > how the rate of energy transfer can be controlled

Good enough to eat

Baked Alaska pudding is amazing! The outside is meringue, straight from a hot oven. The inside is frozen ice cream. So why doesn't the ice cream melt? The meringue is so poor at transferring energy that the ice cream in the centre stays cold – even when the meringue gets hot enough to cook.

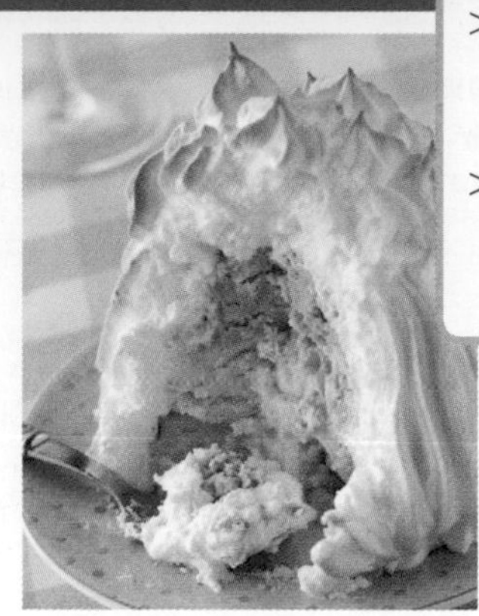

FIGURE 1: Different rates of energy transfer in one pudding.

Energy transfer

Getting hotter, getting colder

A hot cup of tea goes cold if you do not drink it quickly. Energy is transferred from the hot cup of tea into the cooler room. The tea cools down and the room warms up – but not enough to notice.

When a child eats an ice lolly slowly, it nearly always melts. This is because energy is transferred from the warm surroundings into the cold ice lolly. The air cools down a tiny bit, but it is negligible.

How quickly will something heat up or cool down? It all depends on the **rate of energy transfer**.

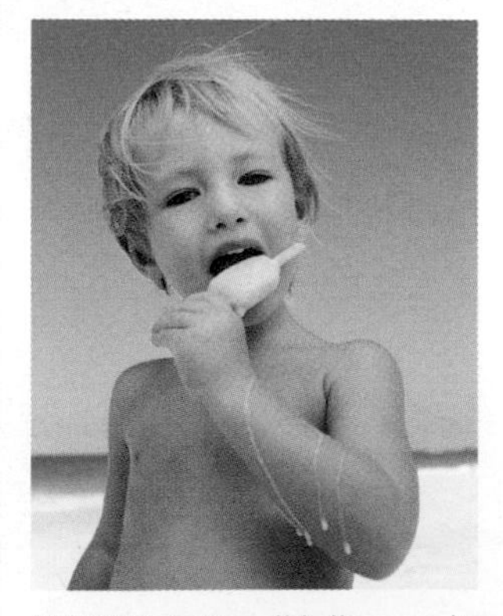

FIGURE 2: Do all lollies melt as quickly as one another?

Did you know?

A polar bear's fur is so poor at transferring energy that the bear will not show up on an infrared camera.

Transferring energy

An object transfers energy to and from its surroundings at its surface. How quickly an object transfers energy depends on:

> the material it is made from.

A good conductor transfers energy at a faster rate, to and from its surface, than a poor conductor.

> its size and shape.

A large **surface area** transfers energy at a faster rate, to or from the surroundings, than a smaller surface area.

> the type of surface it has.

Dull, dark surfaces transfer energy at a faster rate, to or from the surroundings, than shiny, light-coloured surfaces.

> the temperature difference between the object and its surroundings.

The bigger the temperature difference, the faster the rate that energy is transferred to or from the object.

FIGURE 3: A heat sink is made of metal and has lots of fins. Suggest why.

QUESTIONS

1 Where does the energy come from to melt a cold ice lolly?

2 Which factors affect the rate at which an object transfers energy?

Conduction, infrared radiation and convection

Energy is transferred though a material by conduction (the main way in solids) and convection (the main way in gases and liquids). When a hot object cools in air, energy is transferred to its surface by conduction. At the surface, energy is transferred to the air by infrared radiation. The energy then spreads out by convection.

Surface area

Cut a strip of copper in half. Cut one half into lots of smaller pieces. Heat all the pieces to the same temperature. Drop the half piece into a beaker of water and the small pieces into another beaker containing the same amount of water. What happens to the water temperature?

In both cases the temperature rises, but it rises quicker in the water with the small pieces of copper. This is because, together, the small pieces have a greater surface area (it includes all the cut edges).

In both cases, the higher the temperature of the copper, the greater the rate at which energy is transferred to the water.

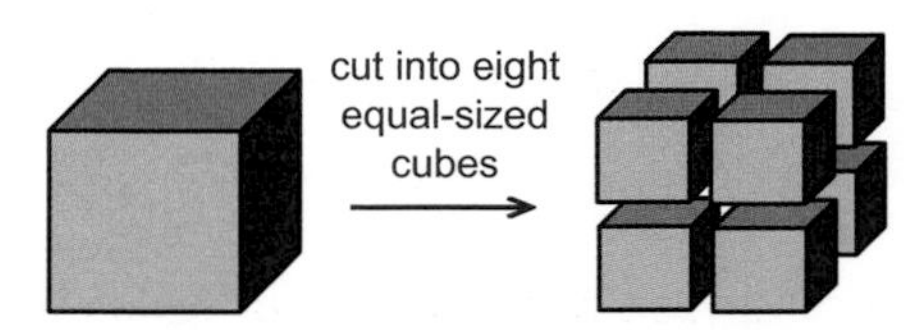

2 cm x 2 cm x 2 cm cube
volume = 8 cm^3
surface area = 24 cm^2

1 cm x 1 cm x 1 cm each cube
volume = 1 cm^3
surface area = 6 cm^2
total volume of the eight cubes = 8 cm^3
total surface area of the eight cubes = 48 cm^2

FIGURE 4: What is the ratio of surface area to volume for (a) the single cube and (b) the eight smaller cubes?

Animal adaptations

Animals that live where the temperature is:

> very cold, tend to grow bigger.

Big animals have a lower ratio of surface area to volume than small animals. This decreases the rate they transfer energy to their surroundings.

> very hot, often have long legs or large ears.

A high ratio of surface area to volume increases the rate of energy transfer to the surroundings. This helps to keep them cool.

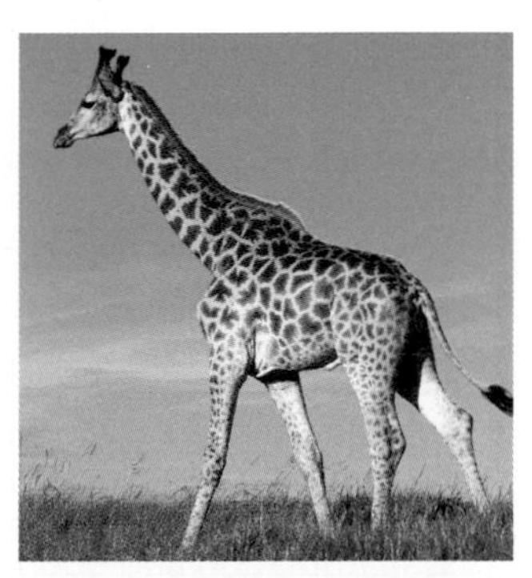

FIGURE 5: Why is a giraffe better suited to live in hot climates than in cold climates?

QUESTIONS

3 Explain why a potato cooks more quickly, if it is cut into small pieces, rather than left whole.

4 Which has the lowest ratio of surface area to volume, an elephant or a mouse?

Cooling curves

Heat water in a beaker until it is just boiling. As the water cools, measure its temperature. Record the room temperature as well. If you record the results on a graph, you may see a cooling curve.

QUESTIONS

5 Explain how the data in Figure 6 provides evidence for this statement: "The bigger the temperature difference between an object and its surroundings, the greater the rate of energy transfer between them."

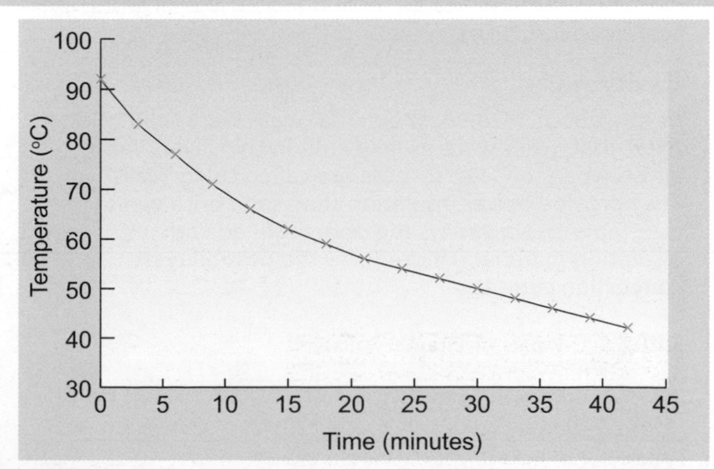

FIGURE 6: A typical cooling curve for water. The room temperature was 20 °C.

Insulating buildings

You will find out:

> U-values are a measure of the effectiveness of insulation

> the payback time of installing insulation varies

Better buildings?

Years ago, before people thought about using energy efficiently, no one worried about how much energy they used to heat their homes. Nowadays, most people are concerned about using energy carefully. All new buildings must waste as little energy as possible.

FIGURE 1: We cannot all live in houses like this one, but we can improve the homes that we do live in.

Home insulation

U-values

Using less energy for heating saves money. Also, less **fossil fuels** are burned, and this helps to protect the **environment**. Buildings are **insulated** to reduce the transfer of energy from them to their surroundings. Double glazing, cavity wall insulation and good loft insulation all reduce the amount of energy needed to keep homes warm.

U-values show how well a material acts as a **thermal insulator**. A low U-value means it is hard for energy to flow through a material, so the material is a good insulator. To insulate a loft, a material with a low U-value is needed.

In many insulators, the material traps bubbles of air (or gas). This combination makes them good insulators.

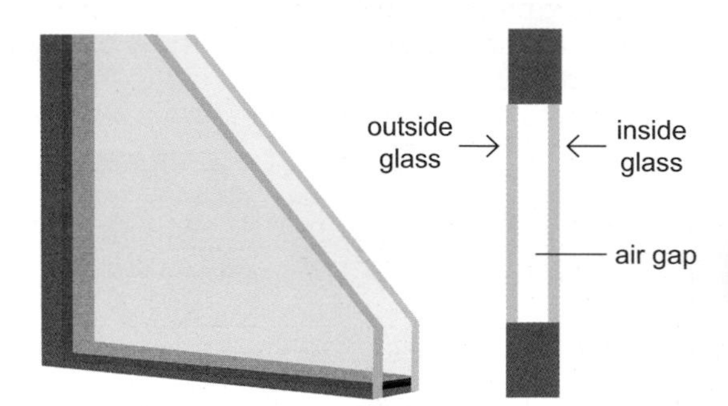

FIGURE 2: Glass and air together offer effective insulation.

Double glazing

Glass is a reasonable insulator. A single glazed window (single pane of glass) has a U-value of about 5.0 W/m^{2}K. A double glazed window, with air or argon trapped between the two panes of glass, has a U-value of about 3.0 W/m^{2}K. This is because air and argon, like all gases, are very good insulators.

Cavity walls

Before about 1920, most external walls were solid brick. After that, walls were usually built in two layers, with air between the layers. These are called cavity walls. They provide better insulation than solid brick walls (see Table 1). Nowadays the cavity is filled with an insulating material. This reduces the possibility of **convection currents.**

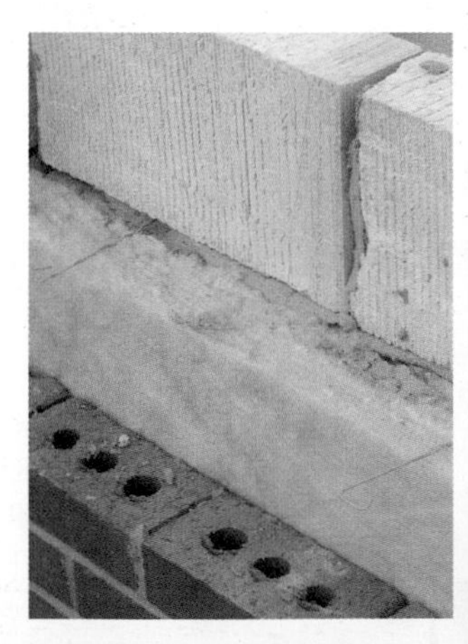

FIGURE 3: Insulation fills the cavity in this wall.

TABLE 1: U-values of external walls.

Material	Approximate U-value
solid wall	2.2 W/m^{2}K
cavity wall (unfilled)	1.0 W/m^{2}K
cavity wall (insulated)	0.6 W/m^{2}K

QUESTIONS

1 Explain why it is a good idea to use less energy.

2 Explain why cavity walls and double glazed windows have lower U-values than solid walls and single glazed windows.

Q ... home insulation ... energy efficiency

How much can we save?

By installing insulation, people in an average UK home could reduce their heating costs. They would reduce energy wasted through roofs, walls and windows. Unfortunately, the work and materials actually cost money – so is insulation worthwhile?

Payback time is the time it takes to save as much money as the insulation cost to install.

TABLE 2: Payback time for different insulation methods

Method of insulation	Installation cost (£)	Annual saving (£)	Payback time (years)
loft insulation	240	60	4
cavity wall insulation	360	60	6
draught proofing doors and windows	45	15	3
double glazing	2400	30	80

Draught proofing doors and windows pay back the quickest (Table 2). Double glazing takes a lifetime to pay back, yet installing it is still very popular. Rather than to save on fuel bills, people often install double glazing to reduce noise, from outside, or to save money on repainting wooden window frames.

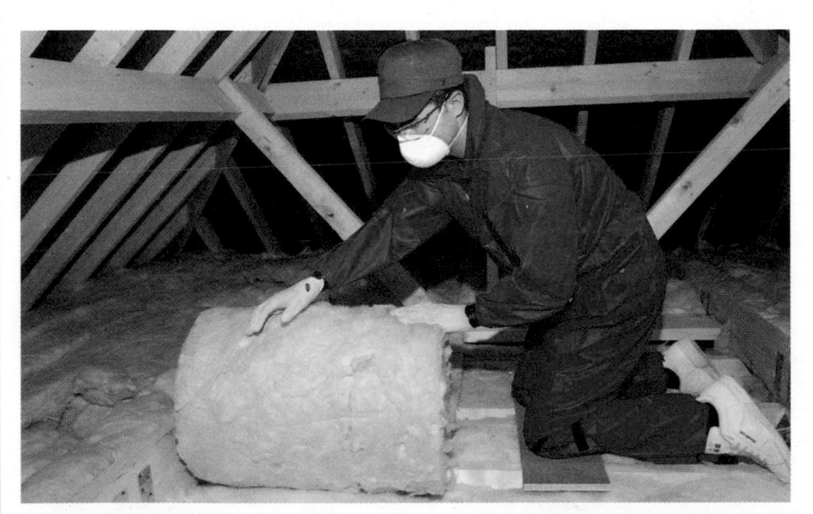

FIGURE 4: Would you recommend that people install loft insulation?

Did you know?

Roofs covered in soil and grass are an effective and 'environmentally friendly' way to insulate a building.

QUESTIONS

3 Which type of insulation is cheapest to install?

4 Describe what 'payback time' means.

5 If draught proofing lasts 5 years before it needs replacing, and loft insulation lasts 10 years, suggest which you would install and why.

The wider picture

When governments and local authorities give grants that help people to pay for insulation, they must also consider the energy used in manufacturing the insulation. For example, cavity wall insulation saves energy being used in homes, but factories use energy to make the foam.

6 Imagine you are on a local council. You are deciding whether or not to give grants for installing home insulation. Discuss all the factors you would consider and what other information you would need before making a decision.

Preparing for assessment: Applying your knowledge

To achieve a good grade in science, you not only have to know and understand scientific ideas, but you need to be able to apply them to other situations and investigations. These tasks will support you in developing these skills.

Steam power

In the 18th century, Britain was on the verge of an industrial revolution that would bring about huge changes. People in their thousands would move from living in the country and working on the land to working in factories or mines, often in rapidly expanding towns and cities. New technologies would enable the large-scale production of textiles, building materials and iron. It was the development of the steam engine that made large factories and deep mines possible on a scale never previously seen.

Two of the people behind the development of steam power were Matthew Boulton and James Watt. Engines of their design powered machines, pumped water and lifted people and materials from the bottom of mineshafts.

A Boulton & Watt engine needs a supply of steam from a boiler. The boiler burns a fuel, such as coal, to heat water and turn it into steam. The steam goes into the engine and pushes a piston along a cylinder (imagine blowing into a syringe and making the plunger move). The piston is linked to a wheel, which then turns (rather like when your leg pushes down on a bicycle pedal, making the wheels turn). This rotating movement can be linked to, for example, a water pump, or a cable down a mineshaft.

These engines did not move very quickly (they turned far slower than a car engine), but they were large and powerful. They were much more efficient than the crude engines in use before, but not very efficient by today's standards. Compared with the energy content of the fuel, only about 5% reached the machines being driven by the engine. In comparison, a gas turbine in a modern power station will run at about 60% efficiency.

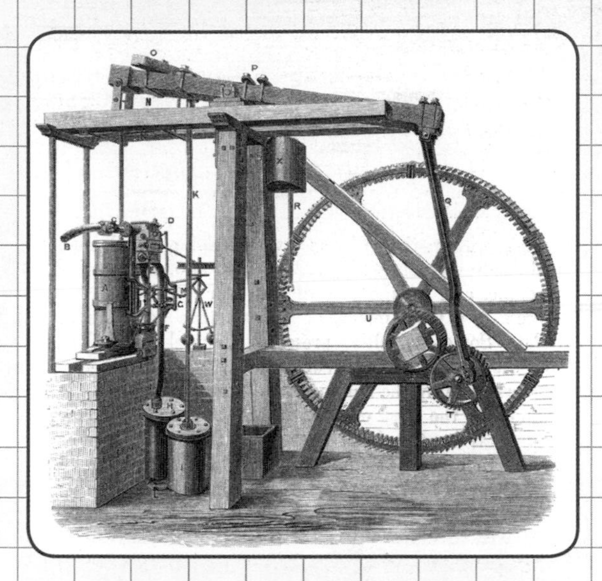

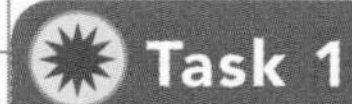

Task 1

What do you think it would be like to be near a Boulton & Watt engine? Think about the way in which the engine works. Describe the sight, sounds and smells.

Task 2

Tell the story of the transfer of energy, starting with the fuel in the boiler and finishing with the machine being driven by the steam engine.

Task 3

Most of the energy ends up being transferred to the environment. How does this happen? Use the equation for efficiency as part of your answer.

$$\text{efficiency} = \frac{\text{useful energy out}}{\text{total energy in}}$$

Task 4

Why did factory and mine owners use Boulton & Watt engines, even though the engines were so inefficient?

Maximise your grade

	Answer includes showing that you…
	know that energy cannot be created or destroyed.
	know that energy can be stored and transferred in various ways.
E	know that particles of solids, liquids and gases have different amounts of energy.
	know that when energy is transferred, only part of it may be usefully transferred – the rest is 'wasted'.
	understand how energy is transferred when there is a change of state.
C	understand that wasted energy is eventually transferred to the surroundings as heat, spreading out and becoming less useful.
A	understand the idea of input energy, useful energy output and the efficiency of energy transfer.
	can evaluate the efficiency of machines.

Specific heat capacity

> **You will find out:**
> - > the meaning of specific heat capacity
> - > how solar panels work
> - > how to calculate the amount of energy absorbed by materials

Hot pies

Many people like pies, but why does the filling seem hotter than the pastry around it? It's because the pastry transfers energy to the surroundings (and, therefore, cools down) more quickly than the filling. It's all to do with specific heat capacity. The pastry is relatively dry, but the filling has a high water content.

FIGURE 1: Is the filling hotter than the pastry?

Absorbing energy

Water is often used as a coolant in machinery. This is because it can absorb a lot of energy, but only get a little bit warmer. Other materials are not as good at absorbing energy – their temperatures increases more. For example, 1000 **joules** (J) of energy will raise the temperature of 100 g of water by 2.4 °C and 100 g of iron by 27.7 °C.

The energy needed to give the same rise in temperature, in identical masses of different materials, varies. **Specific heat capacity** is a measure of this.

Specific heat capacity is the energy needed to raise the temperature of one kilogram of a material by one degree Celsius. Its unit is J/kg °C.

Metals have lower specific heat capacities than materials such as rocks and liquids.

TABLE 1: Some specific heat capacities.

Material	Specific heat capacity (J/kg °C)
water	4200
limestone	910
glass	840
iron	460
copper	390

Solar panels

Solar panels use infrared radiation from the Sun to heat water. They are usually fitted to the roofs of buildings. Water flows through pipes in the panel (Figure 3) and absorbs infrared radiation from the Sun. The water gains energy and becomes warmer. The water is usually not hot enough to use by itself, but it can be 'topped up' with energy from the electricity supply or a boiler.

256 g of copper 110 g of limestone 24 g of water

FIGURE 2: 1000 joules of energy will raise the temperature of these masses of materials by 10 °C.

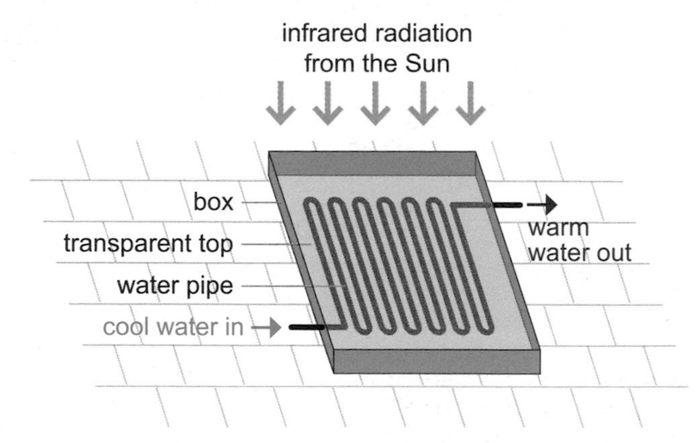

FIGURE 3: A solar panel uses the Sun's energy to heat water. Why are the pipes black?

QUESTIONS

1 What happens to the temperature of a material when it absorbs energy?

2 Put these in order of their specific heat capacity: glass, water, copper.

3 Solar panels can store energy only for a short time. Use ideas of energy transfer to explain why.

Q ... specific heat capacity GCSE ... solar panels for students

Calculating absorbed energy

The energy transferred to a material as it warms up (or transferred away as it cools down) can be calculated using the equation:

$E = m \times \theta \times c$

where

E is energy transferred in joules, (J)

m is mass in kilograms, (kg)

θ is temperature change in degrees Celsius, (°C)

c is specific heat capacity in J/kg °C

Worked example

A piece of iron of mass 50 g is heated until its temperature rises from 21 °C to 38 °C. How much energy has it absorbed? The specific heat capacity of iron is 460 J/kg (°C).

Change the mass to kilograms: m = 0.05 kg

Calculate the temperature rise: θ = 38 – 21 = 17 °C

c = 460 J/kg °C

Substitute these values in the equation $E = m \times \theta \times c$

$E = 0.05 \times 17 \times 460 = 391$ J

Remember

A solar panel is not the same as a solar cell. A solar cell uses the Sun's energy to generate electricity.

Did you know?

Place the same masses of cooking oil and of water in two identical saucepans and heat them on identical hobs. By the time the water boils (at 100 °C) the cooking oil will be 130 °C

QUESTIONS

4 100 g of glass is heated until its temperature rises from 19 °C to 33 °C. How much energy has it absorbed? The specific heat capacity of glass is 840 J/kg °C.

5 250 g of the brick inside a storage heater (specific heat capacity = 910 J/kg °C) absorbs 5000 joules of energy. How much will its temperature rise?

Oil-filled radiators

Oil-filled radiators are one option for heating small rooms. They contain oil that has a specific heat capacity of 2000 J/kg °C. A small oil-filled radiator contains 5 kg of oil.

FIGURE 4: The oil in this type of radiator is heated by electricity.

QUESTIONS

6 Calculate the energy needed to heat a small oil-filled radiator, from 20 °C to 60 °C. Compare this with the energy required if the radiator were filled with water.

7 In practice, the energy needed to heat a small radiator is slightly more than the calculated value. Explain why, using your knowledge of energy transfer.

Energy transfer and waste

Is this car a good idea?

Many car manufacturers now make electric cars, which can be recharged from the mains. They seem like a good idea as they produce less pollution than petrol cars. Are they really a good idea? Energy diagrams are one of the tools that scientists use to compare different devices and help them to decide which is more suitable.

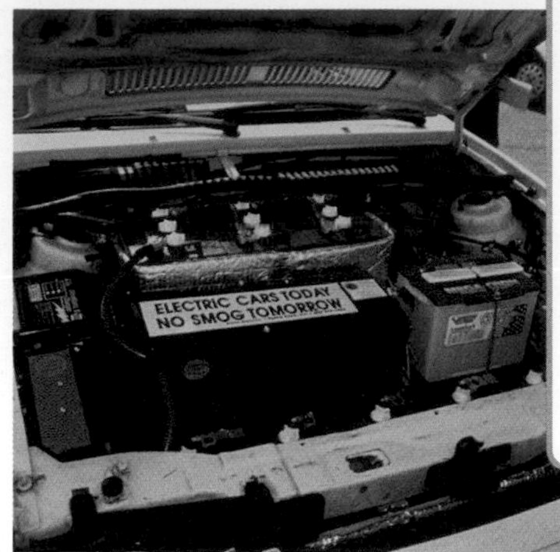

You will find out:

- > energy can be transferred usefully, stored or dissipated, but cannot be created or destroyed
- > how to draw and interpret Sankey diagrams
- > when energy is transferred, only part is usefully transferred; the rest is 'wasted'

FIGURE 1: Are electric cars a good idea?

Sankey diagrams

The purpose of an electric light bulb, unsurprisingly, is to produce light. However, the old filament light bulb became very hot. Energy that does not produce light is **wasted energy**.

Energy can be transferred usefully, stored or dissipated – but cannot be created or destroyed. This is the Conservation of Energy law.

A **Sankey diagram** shows energy transfers that happen in a device. They show the energy:

> used by a device (**energy input**)

> given out to do different things (**energy output**).

Some is usefully transferred (does what the device was designed for) and the rest is 'wasted'.

The widths of the arrows are drawn to scale. They show the relative proportions of input and output energy. The values are given on the diagram.

The total amount of energy stays the same (the law of Conservation of Energy).

Figure 2 shows:

energy input for light bulb = 100 J

energy output = useful energy + wasted energy = 10 J + 90 J = 100 J

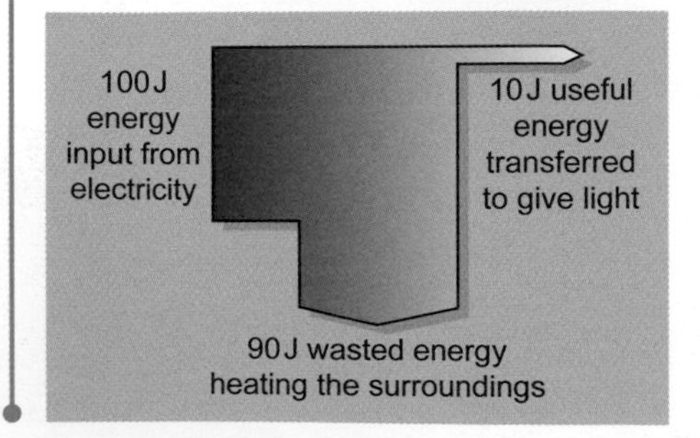

FIGURE 2: Sankey diagram of energy inputs and outputs per second.

Did you know?

The European Union (EU) banned the import of filament light bulbs into EU countries from 1 September 2009.

QUESTIONS

1 What does the law of Conservation of Energy say?

2 Explain what the arrow width in a Sankey diagram shows.

3 A Bunsen burner heats water in a beaker. 1000 J of energy are transferred into 400 J of useful energy, to warm the water, and 600 J of wasted energy to heat the room. Draw a Sankey diagram, to scale, showing this.

Q ... Sankey diagram AND law of conservation of energy

Wasted energy

Missing values

The Law of Conservation of Energy may be used to work out missing values in an energy transfer.

In Figure 3, a Sankey diagram for a car engine, the total energy output must be 1000 J because the input energy is 1000 J.

useful energy to move the car + wasted energy that heats the surroundings = 1000 J

300 J + wasted energy heating the surroundings = 1000 J

wasted energy heating the surroundings = 1000 J – 300 J = 700 J

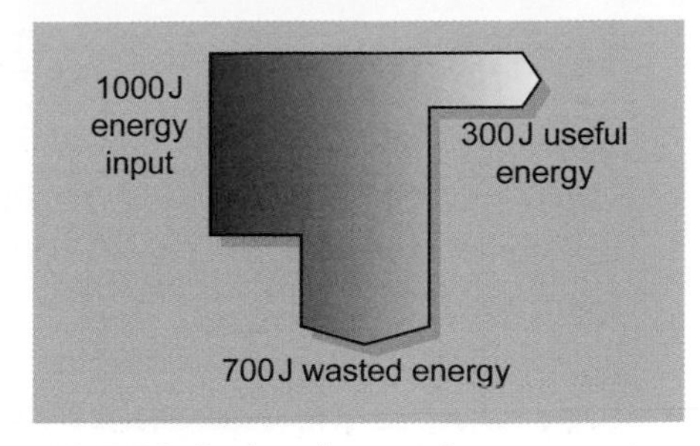

FIGURE 3: Sankey diagram for a car engine.

More wasted energy

Most wasted energy from a device heats up the surroundings. However, it may spread out in more than one way. In an electric fan, for example, some wasted energy heats the surrounding air and some makes sound (see Figure 4).

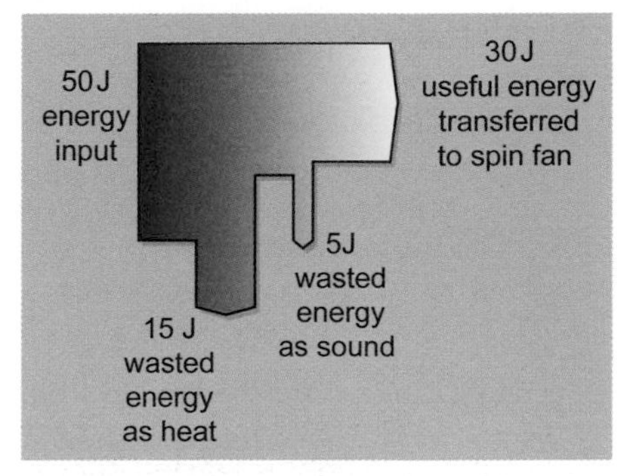

FIGURE 4: Sankey diagram for an electric fan.

QUESTIONS

4 An athlete gains 40 000 J of energy from her food. She uses 800 J to train. How much is wasted?

5 The input to an electric motor is 5000 J. 1500 J of the output heats the engine and the surrounding air. How much energy is usefully transferred?

6 How much energy does the electric fan, described in Figure 4, waste?

Comparing devices

There are lots of ways to produce light. How do they compare?

The Sankey diagrams in Figure 5 show:

> It takes 30 candles to provide as much light as one filament light bulb.
> The filament bulb heats the surroundings far more than the fluorescent bulb.
> The fluorescent bulb will cost much less money to run than the filament bulb.

Remember
Wasted energy becomes increasingly spread out and less useful.

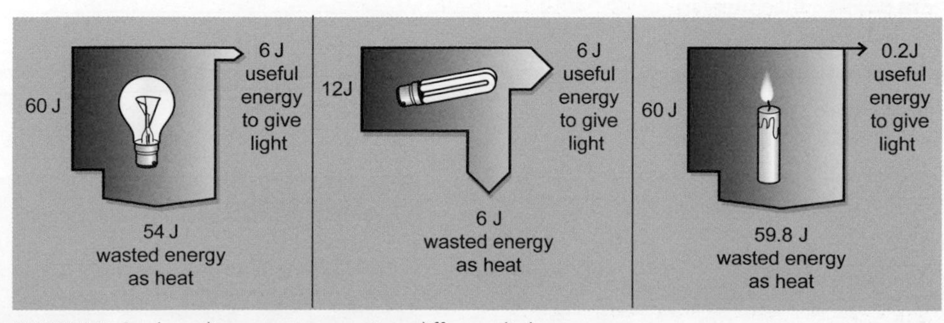

FIGURE 5: Sankey diagrams comparing different light sources.

QUESTIONS

7 Discuss the validity of the conclusions given for Figure 5.

Efficiency

You will find out:
- > the meaning of efficiency in energy transfers
- > how to calculate efficiency

Dying for efficiency?

Explorers walking to the North Pole or the South Pole pull their food with them on sledges. If they miscalculate their food needs, they will starve – as Sir Ranulph Fiennes nearly did in 1993. It's important to know how efficiently they can use the energy in the food they take.

FIGURE 1: The Arctic and the Antarctic are lonely and dangerous places.

Being efficient

What is efficiency?

An efficient person is someone who gets the work done without fuss or wasting energy. An efficient device is similar. It uses as little energy as possible to do what it is designed for. For example, electrical devices provide light or heating, they move objects and they make sound.

Scientists describe **efficiency** as:

$$\text{efficiency} = \frac{\text{useful energy out}}{\text{total energy in}}$$

'Total energy in' is the energy transferred to the device. 'Useful energy out' is the work the device does.

Power is the rate of transfer of energy. Efficiency can also be described as:

$$\text{efficiency} = \frac{\text{useful power out}}{\text{total power in}}$$

FIGURE 2: Cranes use energy to do useful work. Some are more efficient than others.

Comparing efficiency

Knowing about the efficiency of devices can help people to decide which one is the best buy – efficient devices cost less to run.

Cranes lift and move things around – that is, the work they do. Suppose:

Crane X uses 2000 J energy to do 800 J useful work.

Crane Y uses 3000 J energy to do 900 J useful work.

It would be better to buy Crane X, because, for every 1000 J of energy that each crane uses:

- > Crane X can do 400 J of useful work
- > Crane Y can do 300 J of useful work

Crane X is more efficient than Crane Y.

QUESTIONS

1 Write down the equation to work out efficiency.

2 Explain why an efficient machine costs less to run than an inefficient machine.

3 How much energy would Crane X use, to do 600 J of useful work?

Efficiency and 'wasted' energy

An efficient device does not waste much energy. The device does what it was designed for, using as much as possible of the energy put in.

You can use energy diagrams to:

- > calculate the efficiency of devices
- > compare the efficiency of different devices.

Did you know?

The only device that could be 100% efficient is an electric heater, such as an immersion heater.

Q ... efficiency equation

Figure 3 shows a Sankey diagram for a petrol engine. To calculate the efficiency of the petrol engine:

$$\text{efficiency} = \frac{\text{useful energy out}}{\text{total energy in}}$$

total energy input from combustion of petrol = 1000 J

useful energy output to move the car = 300 J

The efficiency of the petrol engine = $\frac{300\text{ J}}{1000\text{ J}} = 0.3$

To convert an efficiency of 0.3 to a percentage, multiply by 100:

0.3 × 100 gives an efficiency of 30%

You can use decimals or percentages to describe efficiency.

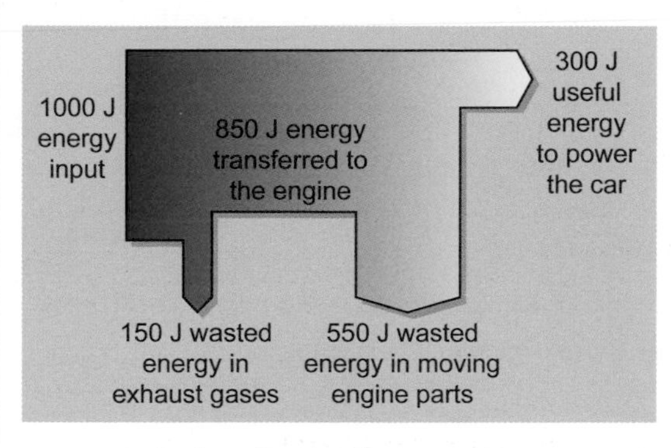

FIGURE 3: Sankey diagram for a petrol engine.

Using Sankey diagrams to show efficiency

Sankey diagrams can be used to compare different devices. In Figure 4, you can see that the diesel engine is more efficient than the petrol engine.

In a steam–diesel hybrid car, wasted energy is used to heat water and, eventually, produce useful energy to drive the car at low speeds. In this way, one-third of the wasted energy is recycled to do useful work. Figure 4 shows that the efficiency of the steam-diesel hybrid is:

$$\text{efficiency} = \frac{(40\text{ J} + 20\text{ J})}{100\text{ J}} = 0.6 \text{ or } 60\%$$

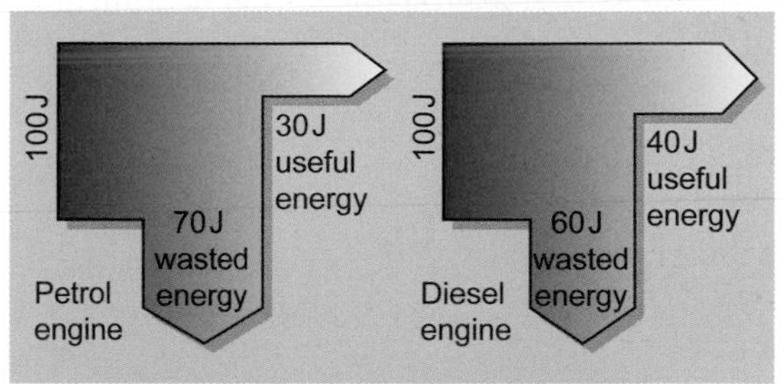

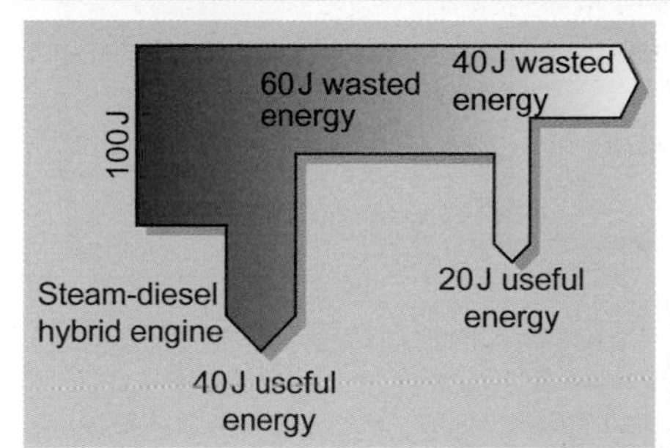

FIGURE 4: Sankey diagrams to compare different types of engine.

QUESTIONS

4 Draw an energy diagram for a device that is 25% efficient.

5 Describe why lubricants improve the efficiency of engines.

Remember

Most wasted energy heats the surroundings.

Perpetual motion machines

Perpetual motion machines would be 100% efficient – if they existed. Early scientists spent many years trying to design perpetual motion machines. Occasionally inventors still try to make them today.

FIGURE 5: This executive toy appears to be a perpetual motion machine. It swings continuously without 'running down', but it has a battery and an electromagnet in its base.

QUESTIONS

6 Discuss why a true perpetual motion machine is impossible.

Preparing for assessment: Analysing and evaluating data

To achieve a good grade in science, you not only have to know and understand scientific ideas, but you need to be able to apply them to other situations and investigations. These tasks will support you in developing these skills.

Investigating energy loss and insulation

Jo was researching energy loss during heating and cooling. He particularly wanted to find out about the effectiveness of insulation to reduce energy loss from an object.

Plan

Jo decided to measure the energy loss from a container of water as it cooled.

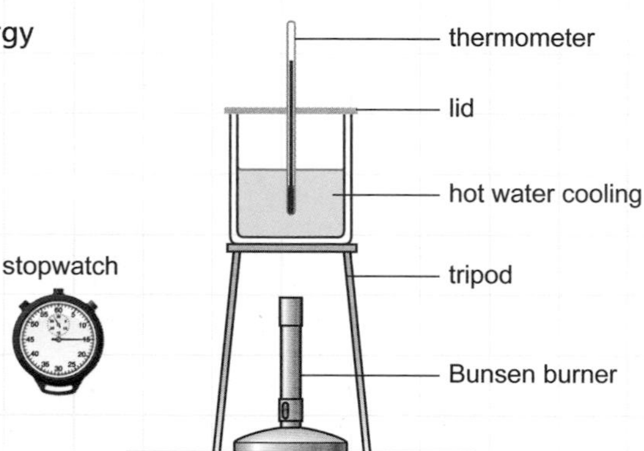

Method

Jo used a Bunsen burner to heat water, in a beaker, to boiling point. He used a thermometer to measure the temperature of the water. Using a stopwatch, he recorded the temperature every minute, for eight minutes after turning off the burner.

He then wrapped different thicknesses of insulation around the container and repeated the procedure.

Analysing data

Jo undertook four tests, using 0, 1, 2 and 3 layers of insulation. Here are his results.

Time (mins)	Temperature (°C)			
	Layers of insulation			
	0	1	2	3
0	100	100	100	100
1	75	80	84	88
2	55	65	73	80
3	40	52	64	75
4	30	44	61	72
5	24	40	54	70
6	21	37	51	69
7	21	35	49	68
8	21	34	48	68

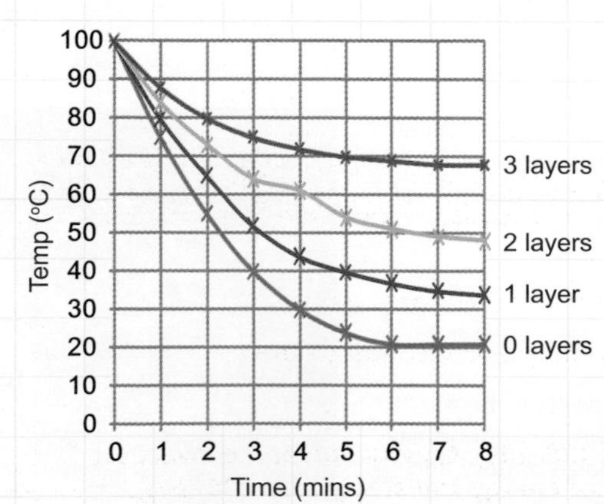

1. For each set of readings, Jo has connected the points on the graph with a curve. Explain why he has used (a) a smoothed curve, rather than (b) a single straight line of best fit or (c) joined each of the points with straight lines.

He noticed that, for three layers of insulation, the temperature did not drop at all for the last minute. He assumed that this was because the insulation was so good that there was not one complete degree drop in the water's temperature. He could not explain why there was no temperature change for the last two minutes when he had used no insulation.

2. Can you explain this?

3. Jo also saw that one result did not seem quite to fit the pattern. Which result was that? He was not sure whether he should conclude that this result was (a) right (b) possibly wrong (c) definitely wrong.

What could he do to check whether (a) (b) or (c) was the correct conclusion?

For 0 layers of insulation, plot the results and then draw options (b) and (c). How well do these show what is likely to be happening?

Think about the temperature of the water before he started, and how this might vary.

Interpreting data

Finally, Jo used a test tube containing wax. He heated it in the container of hot water until the wax had completely melted. He put the thermometer in the liquid wax. He allowed the wax to cool, measuring temperature and time. Jo then drew the cooling curve.

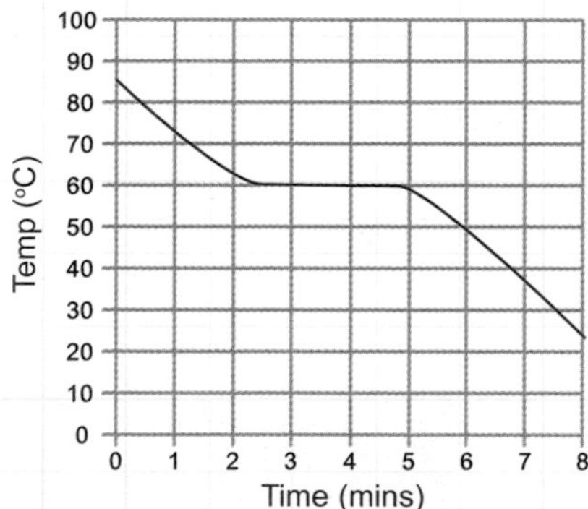

4. How could Jo describe the 'story of the graph'? What pattern or patterns does it have?

It might help to think of the graph as having three parts to its story: 0–2 minutes, 2–5 minutes and 5–8 minutes.

Connections

How Science Works

> Select and process primary and secondary data

> Analyse and interpret primary and secondary data

Science ideas

P1.1 Kinetic theory (pages 198–199), Conduction and convection (pages 200–201) and Rate of energy transfer (pages 204–205).

Electrical appliances

You will find out:

- > examples of appliances that transfer energy from electricity
- > some of the alternatives to mains electricity and some of their limitations
- > alternatives to using electricity for tasks that need energy

People power

The radio in Figure 1 does not need mains electricity or batteries. Winding a handle turns a small electricity generator that makes electricity to charge up a built-in rechargeable battery. The 'clockwork radio' was invented in the 1990s by Trevor Baylis. Since then, he has also invented clockwork torches, clockwork mobile phones, clockwork laptop computers and even a clockwork laser for eye surgery.

FIGURE 1: What advantages might 'clockwork' electricity bring to remote communities?

Using electricity

Mains electricity

Mains electricity is convenient, clean, safe and reliable. You just flick a switch and there is the energy when you need it. People do not usually want the electricity for itself. They use a wide range of **appliances** to transfer the energy supplied by electricity into something useful:

> Hairdryers provide hot air.

> Televisions provide light and sound.

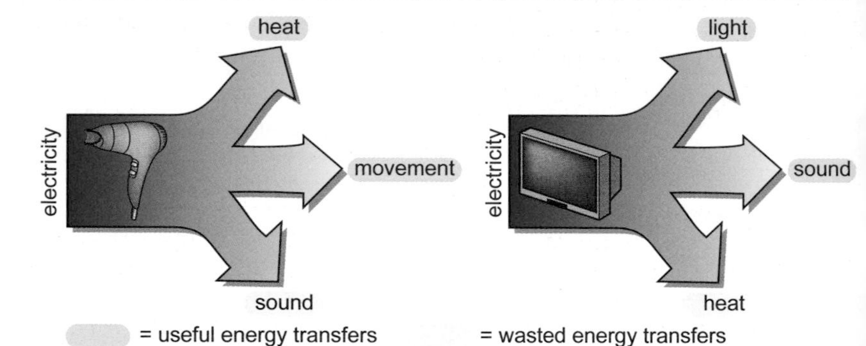

FIGURE 2: Energy transfer in useful appliances.

Batteries

Batteries provide energy for electricity in places where there is no mains electricity. Batteries do not contain electricity. They store energy in chemicals. The energy is transferred by an electrical current when the battery is connected in a circuit.

Heaters rarely use batteries – they need a lot of energy and batteries only store a small quantity (compared with the mains). A heater that ran on batteries would not be very hot and the batteries would go flat very quickly!

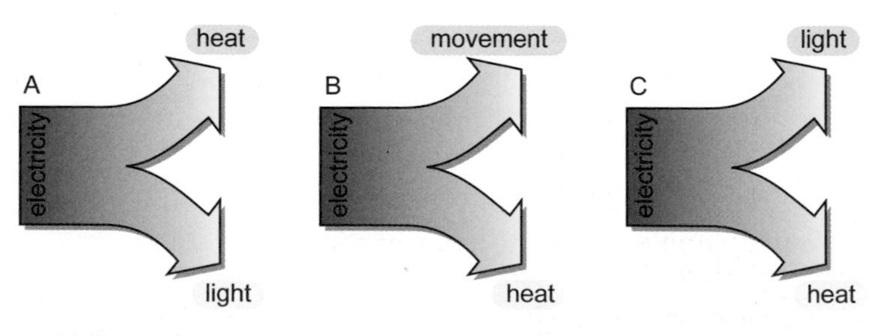

= useful energy transfers = wasted energy transfers

FIGURE 3: Match these diagrams to these appliances: light bulb, electric fire, motor.

Did you know?

About one-third of the world's population do not have mains electricity – that's about two billion people.

QUESTIONS

1 Draw a diagram to show the energy transfers in a radio run (a) on mains electricity (b) by batteries.

2 Suggest two places where it would be best to run a radio using (a) mains electricity (b) batteries.

Q ... mains electricity GCSE ... batteries GCSE

Alternatives to electricity

People can cook using energy stored in chemicals as gas or as **biomass**. Before mains electricity was available, there were candles, gas lamps or oil lamps to provide light. Human, animals or steam can provide power to move things. In many places, these methods are still used.

All of these examples transfer the energy stored in chemicals, either in food or fuel, into useful output. Many of them are smelly, dirty or hard to use, and so many people prefer to use electricity. However, whilst using mains electricity may appear **pollution**-free, generating electricity in a **power station** is not.

Where there is no alternative

Sometimes it seems as though there is no alternative to electricity. It is hard to imagine using interactive television, computer games or the internet without electricity. However, scientists are gradually developing ways other than mains electricity, to supply the energy needed.

Solar power and the energy stored in clockwork springs are two ways of recharging batteries. This will allow people without mains electricity, such as in the developing world, to use telephones, computers and many other appliances.

FIGURE 4: Trains usually use energy transferred through diesel or electric motors, and sometimes by steam, from coal. Suggest the disadvantages of each for a railway.

QUESTIONS

3 Give three examples of appliances that transfer energy supplied by electricity. For each example, give an alternative way of supplying the energy.

4 Suggest disadvantages, including dangers, of heating water by (a) electricity (b) gas (c) an open fire (d) sunlight.

LEDs

Light-emitting diodes (LEDs) do not have a filament inside them, unlike most light bulbs.

All atoms have electrons orbiting around the nucleus in energy levels. Electrons furthest from the nucleus have more energy. When a voltage is connected across an LED, it makes some electrons jump to a higher level.

They then fall back again, giving out a little bit of light. The light is always the same colour because the electrons always give out the same quantity of energy as they fall from the higher to the lower level. Different materials give different colours because they have different electron energy levels.

FIGURE 5: LEDs look like tiny, coloured light bulbs, but actually they are quite different.

QUESTIONS

5 What does the section 'LEDs' tell you about the colour of light and its energy?

6 LEDs last longer than ordinary light bulbs. Suggest a reason why.

Energy and appliances

You will find out:

- > what the power rating of an appliance means
- > how to calculate the energy that an appliance transfers from the mains

Green lawns

Mowing grass is a job for a fine day, but it still takes effort. A good idea might be to use energy from the sunshine to provide the power, instead. There are now robot lawn mowers that do just that – they have photovoltaic panels to collect energy radiated by the Sun.

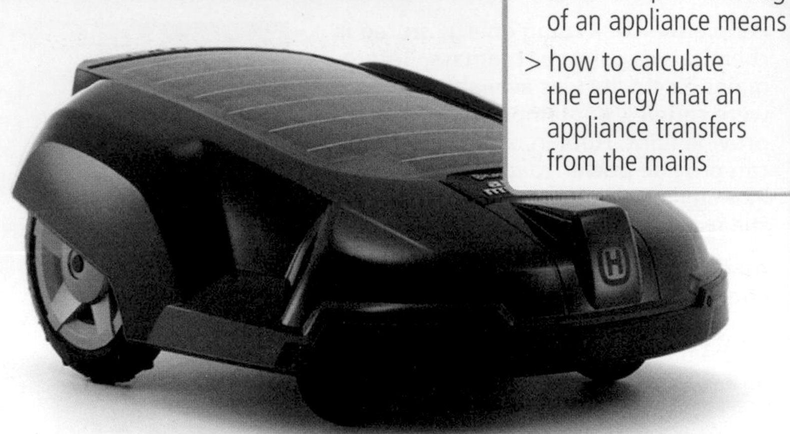

FIGURE 1: This mower has a power rating of 32 W.

Appliance power

Energy and electricity

Energy transfers from a power station to homes, schools and industry as an electrical **current**. **Electrons**, in the metal cables, transfer **charge** down the line. The rate at which this happens is measured in **amperes** or amps (A).

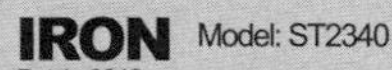

IRON Model: ST2340
Batch: 0812
AC230 - 240V ~ 50Hz 1840 -2000W
Produced for IRON STORES LTD, Yeovil, BA20 BRD
CAUTION: DO NOT IMMERSE IN WATER
HOUSE HOLD USE ONLY
UNPLUG FROM THE MAIN SUPPLY WHEN NOT IN USE
CE
327/165

FIGURE 2: An appliance's power rating label.

Power

Mixing the ingredients to make a cake is quite hard work. You transfer energy to move the spoon, and waste some in heat and noise. If you had to make lots of cakes, you might use an electric mixer to provide beating power.

Power is the rate at which something can do work. In an **appliance**:

- > Power is the rate of energy transfer.
- > The higher the power, the more current it needs.

A high power electric mixer will draw a high current. Low-powered appliances, such as radios, only draw a low current.

Power is measured in **watts** (W).

Power rating

All electrical appliances have a **power rating**, given in watts or kilowatts (kW). An appliance with a power rating of 1 W transfers 1 joule (J) of energy each second.

Electrical current in one 60 W light bulb is transferred at 60 J of energy each second, giving light and infrared radiation.

Did you know?

One joule of energy will take an electric car four times further than one joule in a petrol-driven car.

TABLE 1: Power ratings of some household appliances.

Appliance	Power rating
laptop computer	100 W
electric kettle	2.4 kW
hairdryer	1.5 kW
radio	4 W
television	50 W
microwave oven	1.2 kW
electric fire	2.0 kW
lawn mower	1.3 kW

QUESTIONS

1 Explain what the power rating of an appliance means.

2 Which appliance, from the table, transfers energy at the greatest rate?

3 A desktop computer has a power rating of 1.2 kW. This is equivalent to how many laptops?

Q ... power energy ... amps watts joules

How much energy is used?

To find the total energy an appliance uses, you need to know the quantity of:

> joules transferred each second

> seconds the appliance is working.

The number of joules transferred each second is the power rating. You can calculate the energy transferred using:

total electrical energy transferred by an appliance (in joules, J) = power rating (in watts, W) × time (in seconds, s)

The equation is written as

$E = P \times t$

Worked example

A 1.2 kW microwave oven works for 5 minutes.

To find out the total energy used:

> change the values into the correct units

1.2 kW = 1200 W and 5 minutes = 5 × 60 = 300 s

> then use the equation $E = P \times t$

Total energy transferred in the microwave oven = 1200 × 300 = 360 000 J

Remember

1 kilowatt is 1000 watts. 1 watt is the transfer of 1 joule of energy per second.

FIGURE 3: How could you adjust the power of the microwaves in this oven?

QUESTIONS

4 Calculate the energy transferred by a laptop computer, used for 15 minutes.

5 A television transfers 1000 J of energy. For how many seconds was it switched on?

Heat and current

Wires for electric fire filaments are very thin. It is harder for the current to flow through the wire and more of the energy carried by the current transfers into infrared radiation. It is what they are designed to do. They have greater electrical resistance than thicker wires. Thin wires become hotter than thick wires of the same metal.

Most electricity wires are copper, even though copper is more expensive than, for example, steel. Comparing copper and steel of the same thickness, current flows through the copper more easily and less energy is wasted heating up the surroundings.

Large currents carry more energy than smaller currents. This means that appliances need the right thickness of wire for the current they use, so that the wires do not overheat.

FIGURE 4: Cable thickness should suit the current it carries.

QUESTIONS

6 Explain why electricity cables are thicker for a cooker than a telephone, and why this is important.

The cost of electricity

You will find out:

> how to calculate the energy an appliance transfers from the mains
> how to calculate the cost of the electricity used
> what you pay for when you buy electricity

The FIFA World Cup

World Cup football matches need electricity for floodlighting, for public address systems and for many other things. An arena for a major match needs about five million watts of electricity. That's as much as a town of 5000 people use. It's a huge electricity bill!

FIGURE 1: After reading this page, try to estimate the bill for a match that lasts 90 minutes.

Energy from electricity

Electric current does not get used up – the same size of current, flows out of a house as flowed into it. The electricity bill is for the electrical energy used in the house. Energy was transferred to electrical appliances by the current and transferred in the appliances to useful output.

Some appliances use much more energy than others:

> Electric cookers transfer thousands of joules of energy every second, to heat food.

> Radios transfer only about four joules of energy into sound each second.

Joules and kilowatt-hours

Electricity supply companies measure the energy used by consumers in **kilowatt-hours** (kWh). It is more convenient than joules – an average electricity bill would be millions of joules.

Using a two kilowatt oven for an hour:

$$2000\ \text{W} \times 60\ \text{s} \times 60\ \text{s} = 7\ 200\ 000\ \text{J}$$

The kilowatt-hour is the same type of unit as a joule, but it is used for larger quantities of energy. The equation for calculating the energy is the same:

$$E = P \times t$$

but with different units:

total energy transferred by an appliance (in kilowatt-hours, kWh) = power rating (in kilowatts, kW) × time (in hours, h)

Using a two kilowatt oven for one hour:

$$2\ \text{kW} \times 1\ \text{h} = 2\ \text{kWh}$$

FIGURE 2: Listening to music uses energy. Who pays?

Did you know?

Playing a computer game for an hour uses the same amount of energy as boiling the water for one cup of coffee.

QUESTIONS

1 If a 1 kW electric fire is left on for 1 hour, how much energy does it use in (a) kWh, (b) J?

2 Calculate the kWh that 10 laptops, rated at 100 W each, use in 1 hour.

3 A plasma screen television has a power rating of 200 W. How much energy does it use, in kWh, if left on for 2 hours?

How much does it cost?

Electricity companies charge for the number of kilowatt-hours of energy used. They call the energy used 'Units'. 1 Unit = 1 kWh.

To find the cost of using an appliance:

total cost = number of kilowatt-hours used × cost per kilowatt-hour

Heating a ready-meal, in a 1.2 kW microwave oven for 5 minutes, uses:

$1.2 \times {}^{5}/_{60} = 0.1$ kWh of energy = 0.1 Units on an electricity bill

If each kilowatt hour of energy costs 12p, the cost of heating the ready-meal will be:

total cost = 0.1 × 12 = 1.2p

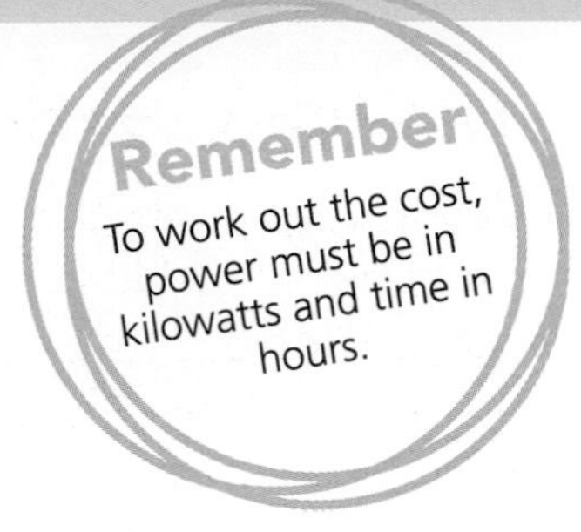

LANCASHIRE
ELECTRIC

Mr Pritchard
74 Green Avenue
Preston
LANCASHIRE

Account number
6645/3526/936B

Date
30 June 2010

Electricity bill

	£	£
Previous bill 30 March 2010	224.67	
Payment received 7 April 2010	224.67 CR	
Balance brought forward		0.00
Present electricity charges		197.04
VAT @ 5% on present charges		9.85
	TOTAL	206.89

Payment Due
by 25 July 2010
£206.89

How your bill is calculated

Meter number	METER READINGS Present	Previous	Units used	Rate	Unit cost	Charge £
06039	45698	44096	1602	Normal	11.52p	184.55
	Standing charge 30 Mar 10 – 30 Jun 10					12.49
					Present electricity charges	**197.04**

FIGURE 3: On this electricity bill, how is the number of 'Units used' calculated?

QUESTIONS

4 If electricity costs 12p per kilowatt hour, how much would it cost to cut a large lawn that takes 2 hours to mow, with a 1.0 kW mower?

5 To mix some concrete, one mixer, rated at 600 W, takes 1 hour. A different mixer, rated at 400 W, takes 2 hours. Explain which will be the cheapest to use.

Switching to standby

The Energy Saving Trust is concerned that, as people switch to digital television, there will be a large increase in the emission of greenhouse gases from power stations.

People tend to leave set-top boxes that receive the digital TV on standby, when they are not using them. Set-top boxes use about 7 W of power when they are on standby. This compares with an analogue video recorder which uses about 3 W of power on standby. The problem is that so many manufacturers are competing to sell set-top boxes that some are using cheap, inefficient components to keep the prices low.

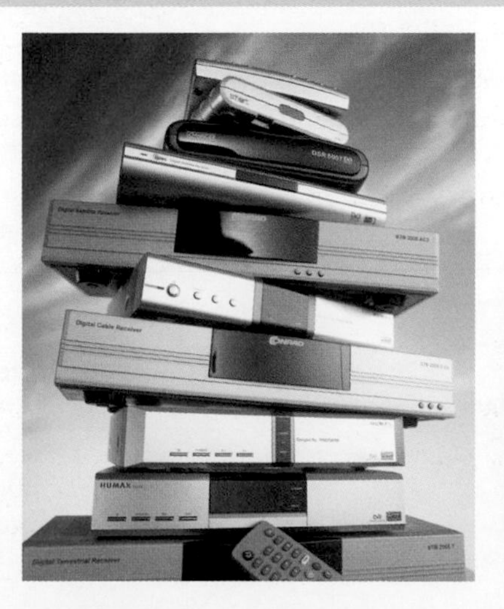

FIGURE 4: How much money could you save by switching off a set-top box, instead of leaving on standby?

QUESTIONS

6 Suggest how it will affect energy consumption, when all radio and TV appliances are changed from analogue to digital?

Checklist P1.1–1.3

To achieve your forecast grade in the exam you will need to revise

Use this checklist to see what you can do now. Refer back to the relevant topic in this book if you are not sure. Look across the three columns to see how you can progress.

Remember that you will need to be able to use these ideas in various ways, such as:

> interpreting pictures, diagrams and graphs
> applying ideas to new situations
> explaining ethical implications
> suggesting some benefits and risks to society
> drawing conclusions from evidence that you are given.

Look at pages 276–297 for more information about exams and how you will be assessed.

To aim for a grade E	To aim for a grade C	To aim for a grade A
Recall that all objects have energy that can move around, but cannot appear or disappear.	Understand that all objects have energy that moves around and that it cannot appear or disappear.	
Recall that all objects emit and absorb infrared radiation.	Understand that all objects emit and absorb infrared radiation and that hot objects radiate at a faster rate than cool ones.	
Describe how dark, matt surfaces and light, shiny surfaces absorb, emit and reflect infrared radiation.	Describe where absorption, emission and reflection of infrared radiation is used in practice.	Explain how the absorption, emission and reflection of infrared radiation is used in practice.
Understand that there are three states of matter (solid, liquid and gas) and recognise diagrams that model these. Recall that particles of solids, liquids and gases have different amounts of energy.	Use kinetic theory and the strength of bonds between particles when explaining the different states of matter. Understand energy transfer in boiling, melting, condensing and freezing.	Explain the concept of particles in the different states of matter. Explain how energy is transferred when there is a change of state.
Describe conduction, convection, evaporation and condensation as energy transfers that involve particles. Recall that, in convection, particles move apart, making fluid less dense.	Explain conduction, convection, evaporation and condensation as energy transfers involving particles. Understand how, in convection, particles move apart, making fluid less dense. Describe the use of conduction and convection in devices.	Explain that a material may be a good conductor or an insulator depending on the arrangement and movement of particles and that this explains why metals are particularly good conductors.
Describe evaporation and condensation as changes of state and know that evaporation has a cooling effect because of energy transfer.	Explain that the cooling effect of evaporation is due to the loss of high kinetic energy particles from a liquid. Explain how factors in the environment affect the rate of evaporation and condensation and describe the effect, such as in the water cycle.	

To aim for a grade E	To aim for a grade C	To aim for a grade A
Understand that the rate of energy transfers in heating and cooling depends on surface area, volume, material and type of surface. Recall that the bigger the temperature difference, the faster the rate at which energy is transferred by heating.	Describe the factors that affect the rate of energy transfer in heating and cooling. Understand that animals in cold and warm climates are adapted to decrease or increase the rate of energy transfer. Describe devices that make use of energy transfer.	Explain methods of measuring the rate of energy transfer by cooling.
Recall that U-values measure how effective a material is as an insulator and that the lower its U-value, the better it insulates.	Understand that different types of insulation pay for themselves over different lengths of time and make calculations to show this.	Evaluate different types of insulation.
Understand that solar panels use radiation from the Sun to heat water.	Explain how solar panels make use of infrared radiation from the Sun to heat buildings or water.	
Describe specific heat capacity as the amount of energy required to change the temperature of one kilogram of the substance by one degree Celsius.	Carry out calculations using the equation $E = m \times c \times \theta$. Understand that different materials have different specific heat capacities.	Evaluate materials according to their specific heat capacities.
Understand that energy cannot be created or destroyed. Recall that, when energy is transferred, only part of it may be usefully transferred – the rest is 'wasted'.	Understand that wasted energy is eventually transferred to the surroundings as heat, spreading out and becoming less useful.	
Draw Sankey diagrams to show energy transfer and waste. Calculate the efficiency of an appliance using a Sankey diagram.	Understand how to compare the efficiency of a range of appliances using Sankey diagrams.	Evaluate a range of appliances using Sankey diagrams.
Describe examples of appliances that transfer energy from electricity.	Explain how appliances transfer energy from electricity. Describe some of the alternatives to mains electricity and some alternatives to using electricity for tasks that need energy.	
Understand that electrons transfer a charge, to give an electrical current. Explain what the power rating of an appliance means.	Calculate and compare the amount of energy that an appliance transfers in a given time.	Explain how current, energy, power and time are related.
Calculate the amount of energy transferred from the mains using kilowatt-hours.	Calculate the cost of the electricity used by individual appliances. Interpret electricity meter readings and calculate the cost of the electricity used.	

Exam-style questions: Foundation P1.1–1.3

In the examination, equations will be given on a separate equation sheet.
Write down the equation that you will use. Show clearly how you work out your answer.

1. This question is about the different ways in which heat can be transferred.

AO1 **(a)** Name **three** ways that energy is transferred from one place to another by heating. [3]

AO1 **(b)** Which is the only type of energy transfer by heating that can travel in a vacuum? [1]

2. Here is a diagram of a vacuum flask, which is used to keep tea or coffee hot (or to keep ice cubes cold).

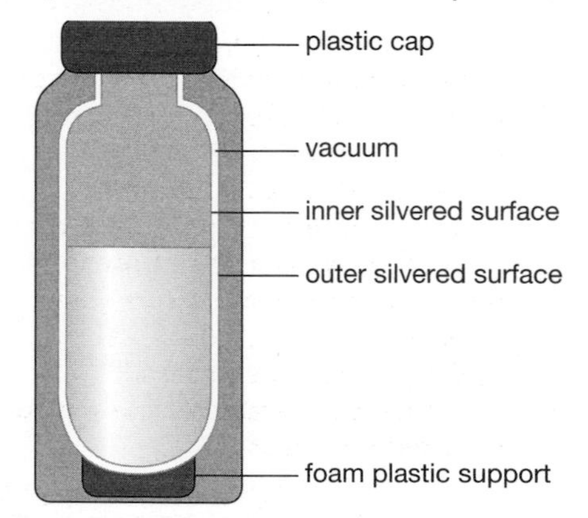

Explain how each of the following helps to ensure that the vacuum flask works well.

AO2 **(i)** plastic cap [1]

AO2 **(ii)** vacuum [1]

AO2 **(iii)** inner silvered surface [1]

AO2 **(iv)** outer silvered surface [1]

AO2 **(v)** foam plastic support [1]

3. A student does an experiment, using a copper rod, and a glass rod, each 20 cm long, with a drawing pin fixed to the end of each rod by petroleum jelly.

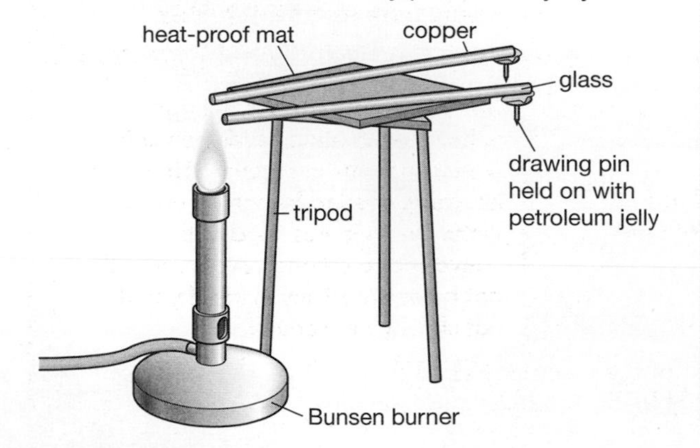

He heats the other ends of the rods with a Bunsen burner, and writes down how long it takes for each drawing pin to fall off. Here are the student's results.

Type of rod	Time for drawing pin to fall (seconds)
glass	62
copper	73

AO3 **(a)** Is this what you might expect? Explain whether or not you expected these results. [3]

AO3 **(b)** Explain **why** you think this. [2]

4. Three similar houses have different heating systems. One has an oil-fuelled boiler, the second has a wood-burning stove, and the third has solar panels fitted to the roof.

AO2 **(a)** Give **one** advantage and **one** disadvantage for each heating method. [6]

AO2 **(b)** Which heating method should have least effect on global warming? [1]

AO2 **(c)** Explain your reasons for your answer to part (b). [3]

5. Energy transferred for an appliance depends on its power and the time it is switched on. The equation is:

$$E = P \times t$$

(a) How much energy is transferred when:

AO2 **(i)** a 60 watt bulb is switched on for 8 seconds? [1]

AO2 **(ii)** a 10 watt bulb is switched on for 2 minutes? [1]

AO2 **(b)** A 3 kW kettle is used for a total of 4.5 hours in one week. Electricity costs 12 p per kWh.

Calculate the cost of using the the kettle during this week. [2]

AO1 recall the science AO2 apply your knowledge AO3 evaluate and analyse the evidence

WORKED EXAMPLE – Foundation tier

This question is about energy transfer and efficiency. Two equations you will need are:

$E = P \times t$

$$\%\text{ efficiency of a device} = \frac{\text{useful energy output}}{\text{total energy input}} \times 100$$

(a) A portable electric drill uses energy from the chemicals in a battery to rotate a drill bit. For every 150 J of energy input, 10 J is wasted as sound, 35 J is wasted as heat and 105 J is output as kinetic energy.

(i) Draw a Sankey diagram to show this. [1]

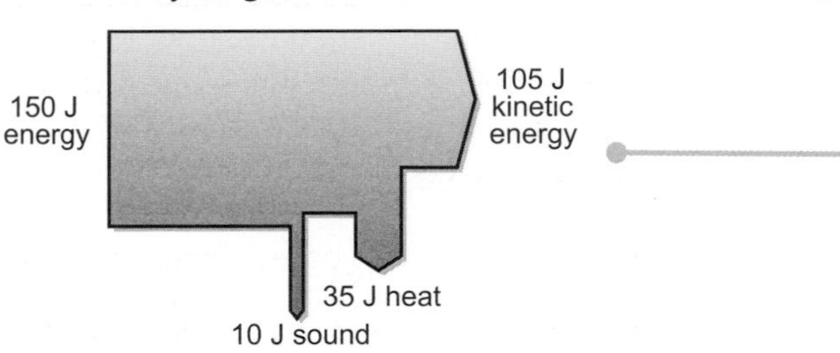

(ii) Calculate the % efficiency of the electric drill. [1]

% efficiency = (105 ÷ 150) × 100% = 70%

(b) A light is powered by an electric motor. Here is the Sankey diagram.

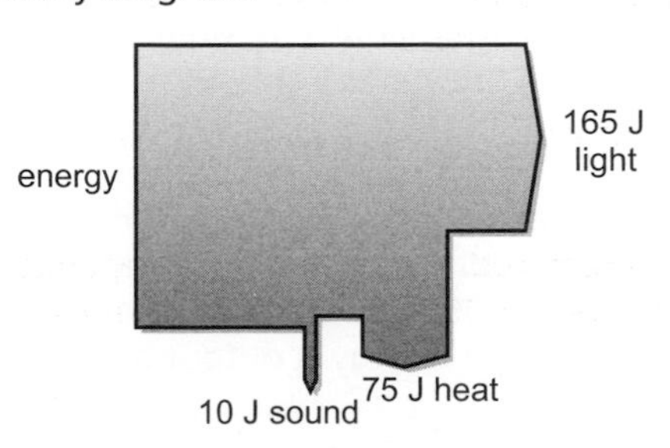

(i) Calculate the % efficiency of the light. [2]

Total energy input = 10 + 75 + 165 J = 250 J

% efficiency = (165 ÷ 250) × 100% = 66%

(ii) If this was used as a heater, what would the % efficiency be? [1]

% efficiency = (75 ÷ 250) × 100% = 30%

(c) Which of these two devices involves the transfer of more energy: a 2 kW kettle running for 30 seconds or a 50 W light bulb running for 15 minutes? [2]

kettle: E = P × t   2 × 1000 × 30 = 60 000 J

light: E = P × t   50 × 15 × 60 = 45 000 J

The kettle transfers more energy than the light does.

How to raise your grade!

Take note of these comments – they will help you to raise your grade.

You will have to select the appropriate equation from the sheet and use the correct units.

In a Sankey diagram, the thickness of the line represents the amount of energy – the thicker the line, the more energy involved – which makes the diagram easier to understand. Use a ruler to draw the diagram. However, do not spend time in an exam making sure your Sankey diagram is perfect – as long as the general differences in thicknesses are accurate, you will be awarded the marks.

As no energy can be created or destroyed, if there is a total energy output equal to 10 + 75 + 165 J = 250 J, there must be the same amount of energy input.

Although it is the same device, this time the useful energy is transferred by heating, not light, so the equation is (heating efficiency = energy ÷ total energy) not (light = energy ÷ total energy).

Make sure that you use the right units. The equation needs power in watts, and time in seconds. Therefore, kilowatts need to be converted to watts, and minutes need to be converted to seconds. Also, you will not be awarded any marks if you do not show your working.

Exam-style questions: Higher P1.1–1.3

In the examination, equations will be given on a separate equation sheet.
Write down the equation that you will use. Show clearly how you work out your answer.

1. Normal matter can exist in three states: as a solid, or a liquid or a gas.

The particles in solids, liquids and gases can be drawn as small circles, inside a rectangle that represents a container. The circles are drawn so as to best represent the basic properties of a solid or liquid or gas. Look at this diagram.

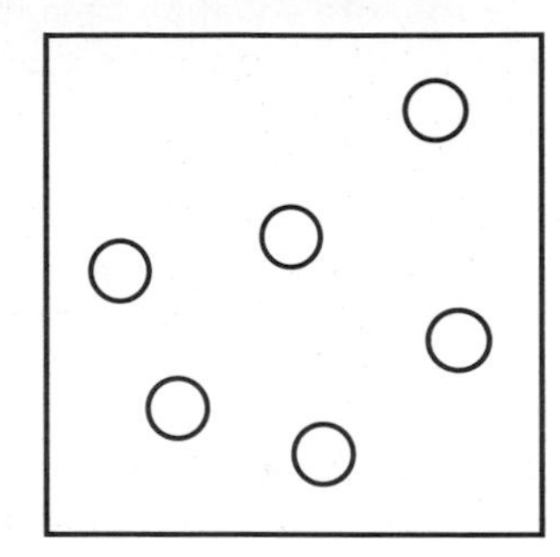

AO1 **(a)** Does it represent a solid, a liquid or a gas? [1]

AO1 **(b)** Give **two** reasons for your decision. [2]

AO1 **(c)** Draw a box diagram for the other two states of matter, labelling which is which. [4]

AO1 **2 (a)** Which part of the electromagnetic spectrum is usually associated with transferring energy by heating? [1]

(b) An amateur astronomer builds a dome for a telescope, using metal sheets that are dark matt in colour. During the afternoon of a sunny summer's day, she happens to touch the inside surface of the metal dome.

That evening, she paints the outside of the dome with white, shiny gloss paint. The following afternoon, with the weather very similar to the previous day, she again touches the inside surface of the dome.

AO2 **(i)** What difference would she find between touching the dome on the first day and on the second day? [1]

AO2 **(ii)** Explain the differences she finds between the two days. [2]

AO2 **(iii)** Explain why her neighbours might ask her to paint her dome using white matt paint instead. [2]

AO3 **3.** *In this question you will be assessed on using good English, organising information clearly and using specialist terms where appropriate.*

On a very cold winter's day, a postman is making deliveries. He holds onto a metal hand rail, but soon lets go, as it feels so cold.

He then picks up a box protected by polystyrene, and carries it with ease, as it does not seem to be particularly cold.

He wonders why this is so, as both the metal and the polystyrene were outside for long enough to have the same temperature.

Explain fully why the postman finds these differences. [6]

AO1 **4. (a)** If a cup of tea is too hot to drink, blowing on it helps to cool it down. Explain why this works. [3]

AO2 **(b)** Some people cool down a hot drink by pouring some into the saucer, and after a time, tipping it back into the cup. Explain why this works. [3]

AO3 **5. (a)** A torch wastes 17% of its energy output as heat. The rest is output as light.

Calculate the efficiency of the torch.

AO2 **(b)** A battery provides an electric motor with 600 J of energy. 30 J is wasted as sound, 60 J is wasted as heat, 150 J is wasted in other ways, and the remaining energy is output as useful kinetic energy.

What is the percentage efficiency of the motor? [2]

6. A person uses a 2 kW kettle for 15 minutes; he uses a 5 kW electric oven for 90 minutes; a 800 W microwave for 30 minutes, and he has a 100 W light on for 2 hours.

AO2 **(a)** Using the equation:

$$E = P \times t$$

calculate the energy used. [2]

AO2 **(b)** If the price of electricity is 15 p per kilowatt-hour, how much does this energy use cost him? [1]

AO1 recall the science | AO2 apply your knowledge | AO3 evaluate and analyse the evidence

WORKED EXAMPLE – Higher tier

(a) A student has been finding out about U-values. He learns that the equation for this is:

rate of heat loss (W) = U-value × area (m^2) × temperature difference (°C)

He finds out that a tiled roof with no insulation has a U-value of 2 W/m^2 °C, but the same tiled roof with insulation has a U-value of 0.4 W/m^2 °C.

The student's tiled roof at home has an area of 80 m^2 and the internal temperature of his house is 23 °C.

If the outside temperature is 8 °C, how would the rate of energy loss per second change if the roof had been insulated. [3]

Tiled roof, no insulation: rate of heat loss = 2 × 80 × (23–8) = 2400 W (joules per second).

Tiled roof, with insulation: rate of heat loss = 0.4 × 80 × (23–8) = 480 W (joules per second).

The difference between them is 1920 W (joules per second).

(b) The student decides to find out how to reduce the amount of energy wasted. He gets quotations for installation costs and savings for each insulation method.

He puts all the information into a table.

Type of insulation	Total cost of installation (£)	Annual saving (£)
cavity wall filling	500	75
carpets	400	50
double glazing	4000	60
draught-proof doors	80	20
loft insulation	250	100

(i) To get best value for money, by calculating the payback time, which method would it be best to install first? Which method would it be best to leave until all the rest have been installed? Explain your reasons. [3]

Payback time (time taken for each method to pay for itself) = total cost ÷ annual saving

Cavity wall filling 500 ÷ 75 = 6.7 years
Carpets 400 ÷ 50 = 8 years
Double glazing 4000 ÷ 60 = 66.7 years
Draught-proof doors 80 ÷ 20 = 4 years
Loft insulation 250 ÷ 100 = 2.5 years

Install loft insulation first (smallest payback time); and install double glazing last (largest payback time).

(ii) Draught-proofing the doors saves the least money. Explain why it should not be left until last. [2]

Draft-proofing the doors only takes 4 years to pay for itself (£80 cost; saving £20 per year). Other methods take longer to save the amount of money they cost.

How to raise your grade!

Take note of these comments – they will help you to raise your grade.

When a question asks you to compare two things, usually you (a) find or calculate the first quantity; (b) find or calculate the second quantity;
(c) calculate the difference (and explain it if you are asked to in the question). In this case, you need to use the equation and write down your working and the answer for a non-insulated tiled roof and then do the same for an insulated roof with tiles. Then write down the difference.

Even though you only need to find the best and the worst, you need to calculate the payback time for all of them.

It is important to understand exactly what the question is asking. This question refers to the best value for money, which means the payback times need to be calculated and compared. The question requires both the savings and the installation costs to be considered.

Physics P1.4–1.5

What you should know

Electricity generation

Electricity can come from the mains or from batteries. Mains electricity is generated in power stations. Electricity from batteries comes from the chemicals that they contain.

Fossil fuels are non-renewable energy resources.

Renewable energy resources are those are always available or can be replaced as we use them.

List five different types of renewable energy resource.

Waves

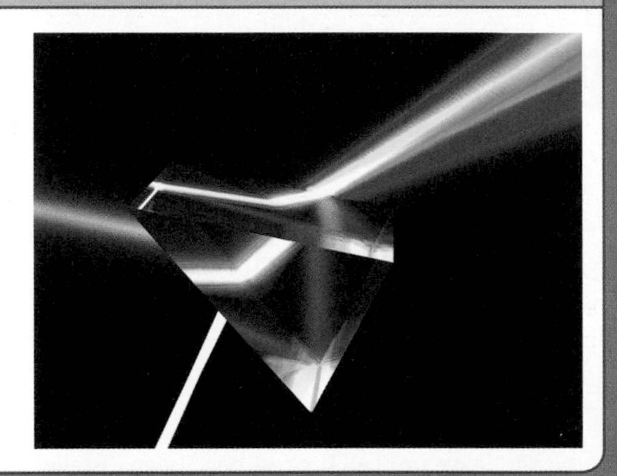

Light and sound transfer energy by waves.

Light can be split using a prism to make a spectrum.

List as many sources of light and sound as you can.

Earth and space

Earth spins on its axis and orbits the Sun.

The solar system consists of the Sun, Earth and other planets.

Artificial satellites orbit Earth. They are used for communications, weather forecasting, military purposes and research.

List as many of the bodies in the solar system as you can.

You will find out

Methods we use to generate electricity

> Various energy sources can be used to generate the electricity we need.

> Using each energy source has advantages and disadvantages, which can be evaluated to decide which energy source is best to use in a particular situation.

> Electricity is distributed via the National Grid.

The use of waves for communication

> Electromagnetic radiations travel as waves, transfer energy and can travel through a vacuum.

> Sound and some mechanical waves are longitudinal, and cannot travel through a vacuum.

> The electromagnetic spectrum is electromagnetic radiation covering a continuous range of wavelengths.

> Hazards and uses of different parts of the electromagnetic spectrum depend on wavelength and frequency.

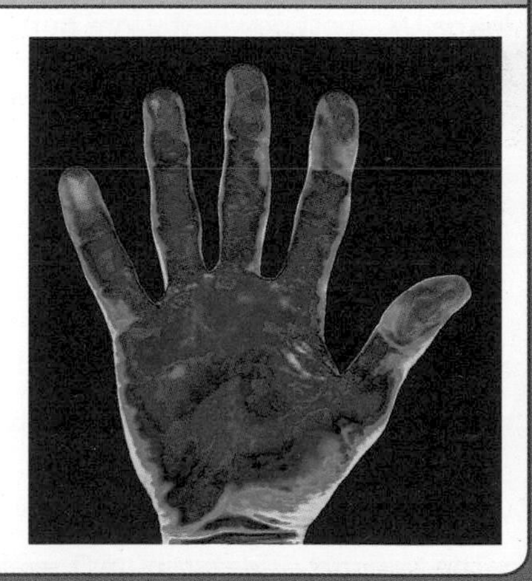

The use of waves to provide evidence that the universe is expanding

> Scientists observe the universe by collecting radiation data from across the electromagnetic spectrum.

> Current evidence suggests the universe is expanding and matter and space began from a small initial point, expanding violently and rapidly from a 'big bang'.

Power stations

> **You will find out:**
> - how power stations generate electricity
> - energy transfer taking place inside a power station
> - how the efficiency of power stations can be improved

Going with a bang!

Lake Kivu, in northwest Rwanda, is in an area with a lot of volcanic activity. Highly flammable gases from the volcanic activity, and from bacteria, are building up at the bottom of Lake Kivu. Scientists fear there may be a massive explosion, destroying local villages. In 2003, engineers began pumping out the gas and burning it to generate electricity. Electricity for most of Rwanda could be generated from this gas.

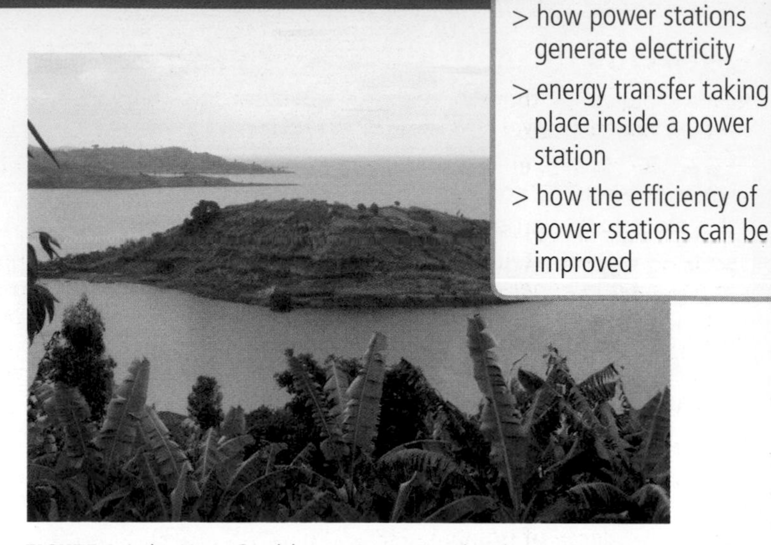

FIGURE 1: Lake Kivu. Could a power station be the answer to this area's problems?

How does a power station work?

In the UK, most power stations burn **fossil fuels** such as coal, natural gas or oil. Figure 2 shows the main stages in one of these power stations. Look at the diagram and work out where each of these stages happens.

- Fuel is burned to heat water.
- The hot water changes to steam.
- The steam drives round a **turbine**.
- Steam condenses back into water and is returned to the boiler.
- The turbine turns and, in turn, makes the **generator** turn.
- The generator generates electricity.
- The **step-up transformer** increases the voltage of the electricity to the very high voltages needed for the **National Grid**.
- The National Grid distributes the electricity to homes and businesses.

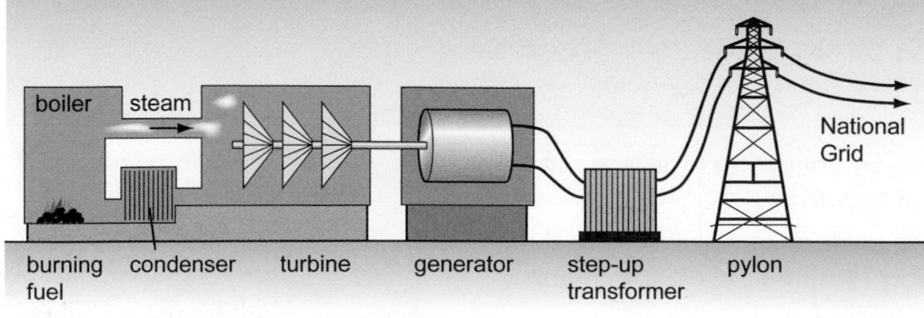

FIGURE 2: From fossil fuel to electricity for your home.

In some modern gas-fired power stations, the burning gas heats air instead of water. The hot air is pressurised and used to drive round a turbine.

Coal, natural gas and oil are all **non-renewable** fuels. One day they will run out. How long they last depends on (a) how fast we use them, and (b) whether or not any new supplies are found.

Did you know?

A large power station may burn a trainload of coal every half-hour.

QUESTIONS

1 Which part of the power station generates electricity?

2 Explain what happens in the boiler of a power station.

3 Describe what 'non-renewable' fuels are. Give three examples.

Energy changes

Fossil fuels

You already know that when energy is transferred, some heats the surroundings and is 'wasted'. Power stations are no different.

In a power station, energy is wasted at every stage (Figure 3). Its efficiency is only about 35%.

Most new fossil fuel power stations are 'combined cycle gas turbine power stations'. The first stage uses burning natural gas to drive round one turbine, connected to a generator. Waste energy from this gas turbine is used to heat water. The steam drives a steam turbine. This type of power station is just over 50% efficient. The most efficient fossil fuel power stations are combined heat and power stations: the waste energy heats local houses and businesses. These are usually 70% to 80% efficient.

Remember

$$\text{efficiency} = \frac{\text{useful energy output}}{\text{total energy input}}$$

Nuclear power stations

Nuclear power stations work in the same way as fossil fuel power stations. However, instead of burning fossil fuels, they use a nuclear reaction (the radioactive decay of uranium-235 and plutonium-239) to produce the energy to change water to steam. Uranium and plutonium atoms split up into smaller atoms (this is nuclear fission). Some of the energy stored in the uranium and plutonium atoms is released and used to heat the water.

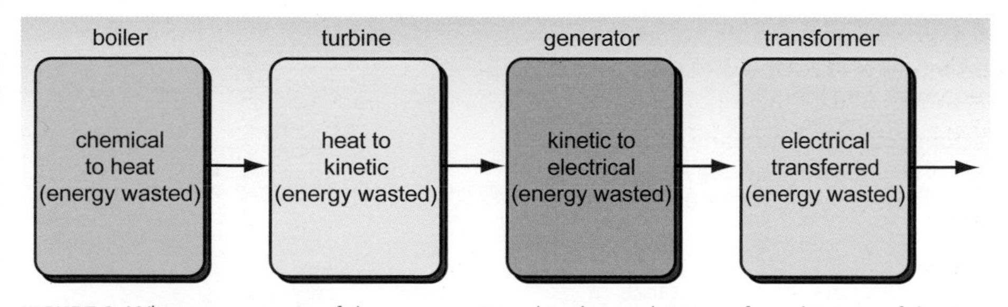

FIGURE 3: What percentage of the energy, stored in the coal, is transferred into useful energy, in a power station?

QUESTIONS

4 Explain why it might be more accurate to describe a coal-burning power station as a 'heat station'.

5 Explain why a combined cycle gas turbine power station is more efficient.

What should we burn?

Large amounts of coal are present in deep underground seams that are inaccessible to mining. In 2002, some engineers suggested that it would be possible to burn this coal. The resulting gases, trapped at the surface, could fuel gas turbine power stations. Critics say the resulting fires, in the coal seams, would probably be impossible to put out. Also, the burning coal would release massive amounts of carbon, possibly causing unstoppable global warming, affecting the whole planet.

FIGURE 4: What gases are released from a fossil fuel power station?

6 Discuss the plan to burn coal in deep underground seams. What extra information do you need?

Renewable energy

You will find out:

- > what are renewable energy resources
- > how different renewable resources can be used to generate electricity

Energy everywhere?

One day, fossil fuels will run out. Fortunately, many renewable resources can be used to generate electricity. Renewable resources are always available (such as solar energy) or can be replaced as they get used (such as fuels from plants). Scientists are finding new ways to use the energy from these resources.

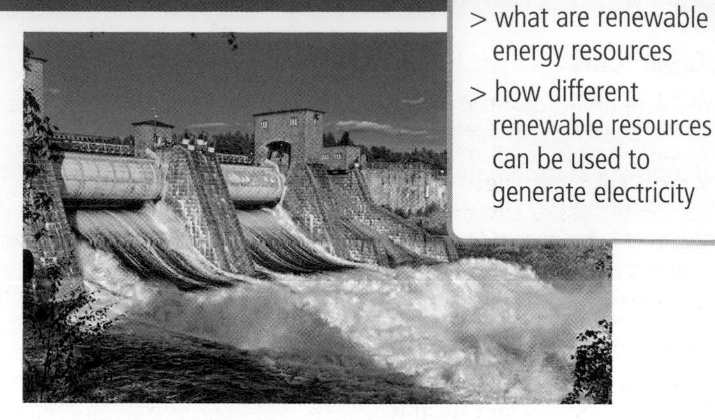

FIGURE 1: Generating electricity from moving water.

Three renewable resources

Energy from waste

'Energy from waste' power stations burn waste that cannot be recycled. This can be domestic, industrial or agricultural waste. Much of this waste is **biomass**: waste woodchips, used animal bedding, nutshells or meat production waste. Ash (the residue from burning) is used as aggregate or fertilisers.

Other power stations burn **methane gas**. The methane comes from landfill sites, sewage or **biofuels**. Biofuels are specially made from plant materials such as willow saplings or fermented sugar cane.

Wind turbines

The world's largest wind farms, in California, have thousands of **wind turbines**. Small wind farms may have only two or three.

Wind turns the turbine blades, making the turbine turn. The turbine is connected to and turns a **generator**. Electricity is generated.

Hydroelectric power

Hydroelectric power stations use falling water to turn turbine blades. A dam holds river water in a reservoir. Gates open to allow water to flow through turbines into the river below the dam. The turbines turn generators which generate electricity. The flow rate of the water controls the speed at which blades spin. The faster they spin, the more electricity is generated.

FIGURE 2: What is the difference between a wind turbine and a wind farm?

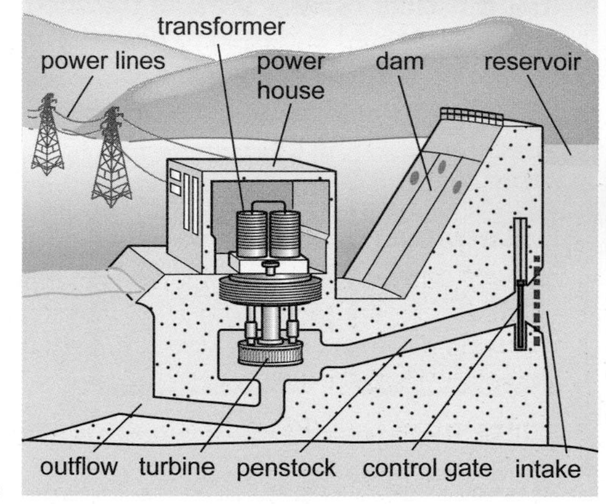

FIGURE 3: Hydroelectric power station.

Did you know?

30% of the electricity generated from renewable resources comes from 'energy from waste' power stations.

QUESTIONS

1 What is burned in an 'energy from waste' power station?

2 Explain how a wind turbine produces electricity.

3 Explain how electricity is generated in a hydroelectric power station.

... renewable energy KS4

Four more renewable resources

Tidal power

Tidal power stations are often built across estuaries. A wall is built across the estuary. As the tide comes in, water flows through turbines in the wall. The turbine blades spin and turn a generator, which generates electricity. A gate stops water flowing out again. As the tide falls, the gate is opened. Water again flows through the turbines, but in the opposite direction. This generates more electricity. Tidal power stations work well only where the difference between high tide and low tide is 15 metres or more.

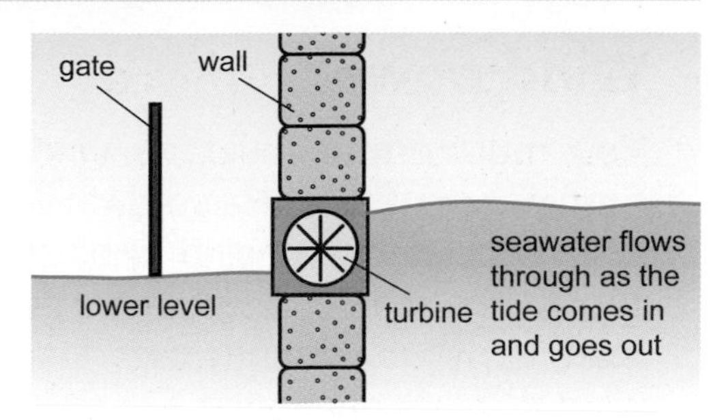

FIGURE 4: Using tides to generate electricity

Wave power

Wave power generators vary in design. In one type, tubes of air (like a chimney) dip into the sea. The tube is open to the sea at the bottom, with a turbine at the top. When a wave hits the wave generator, the water inside the tube rises. This forces the air through the turbine, so the blades spins. The turbine is connected to a generator. As the water level falls, air is sucked back down through the turbine again, so the turbine continues to spin.

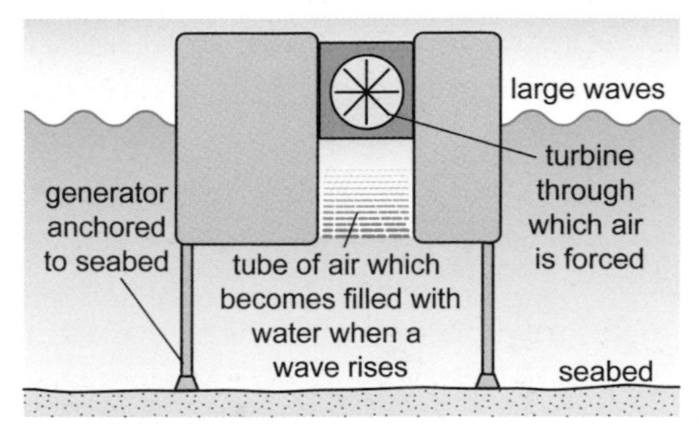

FIGURE 5: Using waves to generate electricity

Geothermal power stations

Rocks deep underground are hot. In volcanic areas, steam sometimes rises to the surface. **Geothermal** power stations use it to drive a turbine. In other places, water is pumped through boreholes into the hot rock. Steam comes back and is used to drive a turbine.

Solar power stations

In 2007, Europe's first commercial **solar power** station, near Seville in Spain, began generating electricity. Hundreds of computer-controlled mirrors reflect sunlight onto a tower, to heat water. The steam produced drives a turbine.

Photovoltaic cells (PV cells) change sunlight into electricity directly.

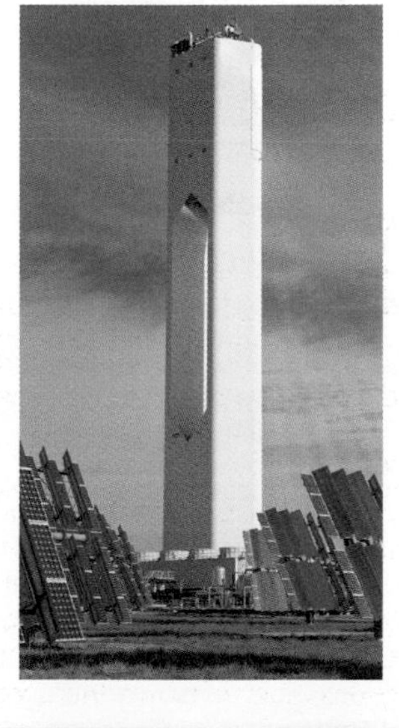

FIGURE 6: By 2013, the Seville solar power station will supply electricity for 180 000 homes.

QUESTIONS

4 Describe the similarities and differences between tidal power stations and wave power generators.

5 Energy transfers in a geothermal power station generate electricity. Describe the basic principle of this.

Storing energy

What happens when the wind is not blowing or there is no sunshine? The electricity that can be generated from wind or solar energy is much less.

In 2010, engineers from Cambridge announced a possible solution – giant 'gravel batteries'. When more electricity is generated than is needed, it is used to heat and pressurise gas. This gas is then pumped though gravel.

The hot gravel (about 500 °C) stores energy. When the energy is needed, cooler gas is passed through the gravel. The gas heats up and can drive a turbine.

QUESTIONS

6 Suggest some reasons why gravel is a good material to use in 'giant batteries'.

Electricity and environment

You will find out:

- > benefits and disadvantages of different ways of electricity generation
- > limitations on where different types of power station can be used

Wind power

How much power could the wind supply? Scientists at Harvard University have calculated that wind turbines could generate far more electricity than the world currently uses. In 2010, wind turbines generated 175 GW – 2% of the total electricity generated worldwide – so many wind farms would need to be built.

FIGURE 1: Turbines in the UK's first offshore wind farm, off the north coast of Wales.

Comparing power stations

Electricity is essential to modern daily life. Different ways of generating electricity have good and bad points.

Good points

> Burning fossil fuels is relatively cheap and can generate large amounts of electricity.

> Wind turbines can generate electricity in remote areas. They are cheaper than running National Grid cables to these areas.

> Hydroelectric power stations, across large rivers, supply very large amounts of electricity.

> Tidal and wave power generators have low running costs.

> Geothermal power stations are small. They do not need fuel or other deliveries. They have only a small impact on the environment.

Harmful effects

> Burning fossil fuels produces gases that cause **global warming.**

> Burning waste produces toxic gases that may cause cancer or birth defects.

> Wind farms (large numbers of turbines) are ugly (in the eyes of some people) and may harm birds.

> Building hydroelectric power stations floods valleys. This may destroy wildlife habitats.

> Tidal and wave power can change water flow. That disrupts harbours or shipping and may destroy wildlife habitats.

> Geothermal power stations might release dangerous gases from deep in Earth's crust.

FIGURE 2: Waste steam is coming out of the chimneys at this geothermal power station.

QUESTIONS

1 Explain how burning fossil fuels harms the environment.

2 State two types of power station that can damage wildlife habitats.

3 Give one advantage and one disadvantage of wind turbines.

Q ... renewable vs non-renewable energy

More effects of generating electricity

Air pollution and climate change

There are two types of **air pollution**.

> Visible – Smoke is an example of **particulate matter**. It can be harmful to health and worsen asthma and other breathing problems.

> Invisible – Burning fossil fuels produces **carbon dioxide**. The carbon dioxide escapes into the atmosphere. Increasing amounts trap more infrared radiation, causing Earth's surface to warm up very slightly. This is global warming, or **climate change**. Scientists are developing **carbon capture** technologies – ways of trapping the carbon dioxide before it escapes into the atmosphere. One method is to store the gas deep underground or under the sea, in old oil or gas fields.

Location matters

Wires and pylons cost money to construct and put in place. The longer the wires, the more energy is wasted by heating the surroundings. Power companies try to generate electricity near to where it is needed.

> Fossil fuel power stations have to be near a river for water, and near railways or roads.

> Wind turbines work best where it is windy.

> Hydroelectric power stations only work where it is hilly.

> Geothermal power stations need to be on rock that is easy to drill through.

Size matters

Some people generate electricity on a small scale, for their own use. They might install small wind turbines or put solar cells on the roof. The National Grid buys any surplus electricity. The householder can buy electricity from the Grid when they need to.

Remember

Solar cells are photovoltaic cells (PVs). They absorb sunlight and generate electricity directly.

FIGURE 3: Why is it expensive to bring the National Grid to places like this?

QUESTIONS

4 Explain what 'global warming' means.

5 Explain why burning renewable fuels does not cause global warming.

Not so green hydroelectric power

A study in Brazil estimated that one hydroelectric dam caused three times as much global warming as a power station burning fossil fuels. The reason is methane.

Methane is a greenhouse gas. Its effect is 21 times stronger than carbon dioxide.

The reservoir for a hydroelectric power station is made by damming and flooding an area.

> Dead plants decay at the bottom of the reservoir, where there is no oxygen, and produce methane. The methane is released into the atmosphere when water passes through the dam's turbines.

> Water levels in the reservoir change with the seasons. Plants grow on the banks of the reservoir, then are flooded and rot. More methane is produced.

Did you know?

The UK's green energy strategy is to ensure that at least 20% of our electricity is generated from renewable energy resources by 2020.

QUESTIONS

6 Explain how hydroelectric dams contribute to global warming.

Q ... air pollution AND climate change ...hydroelectric problems

Making comparisons

You will find out:

> factors to consider when deciding on the best type of power station

> how the flexibility, cost and reliability of different types of power station vary

Nuclear power

Supporters say nuclear power is cheap, reliable and does not produce the pollution that fossil fuels do. Critics say that, although only a tiny amount of waste is produced, that waste is extremely dangerous.

Did you know?

In 2009, nuclear power stations produced 14% of the world's electricity.

FIGURE 1: Loading fuel rods into a nuclear reactor. The reactor is immersed in the blue water.

Which power station?

Choosing power stations

Modern fossil fuel power stations are designed to be efficient. They recycle some wasted energy. This means that they burn less fuel to generate the same amount of electricity. A modern power station is about 50% efficient. Older power stations are about 35% efficient.

Fossil fuels are **non-renewable**. One day, fossil fuels will run out and alternative types of power stations will be needed.

Some **renewable** energy power stations are less efficient than fossil fuel power stations. Others are more efficient. However, efficiency is less important when the energy resource is renewable.

Other factors are important, such as:

> building costs and running costs

> reliability of the energy resource

> ability to generate different amounts of electricity as and when it is needed.

QUESTIONS

1 Give an example of a renewable energy power station.

2 Explain what 'a more efficient power station' means.

3 List three things that may matter more than efficiency, for a renewable energy power station.

Electricity without power stations

About a third of the world's population does not have mains electricity or batteries. **Solar cells** use electromagnetic radiation from the Sun to produce electricity. People in developing countries can use solar cells to generate electricity for:

> lamps

> telephones

> computers

> water-purifying machinery

> fridges to keep vaccines cold

> other appliances to improve the quality of life.

FIGURE 2: How can solar cells improve the quality of life for people in the developing world?

Costs and reliability

Changing demand

Power stations cannot store the electricity they generate. Instead, they change the amount of electricity that is generated to meet demand at a given time. For example, demand for electricity goes up in a cold spell in winter. Demand also increases in the morning and evening, when people are making meals. Fossil fuel and hydroelectric power stations can adjust the amount of electricity they generate more easily than wind, tidal or wave power generators can.

FIGURE 3: Why does the demand for electricity, at half-time in a World Cup final, suddenly increase?

The cost of generating electricity

There are two costs:

> Capital cost – how much it costs to build the power station.

> Operating cost – how much it costs to run the power station once it is built.

At present, most types of renewable energy power stations – especially tidal and geothermal – have high capital costs. Their operating costs are low, compared with fossil fuel power stations.

In future, fossil fuel prices will continue to rise. Meanwhile, technologies for power stations using renewable energy resources are improving. The total cost of energy from fossil fuels, compared with renewable sources, will change.

Reliability

All power stations need an energy resource, but some are more reliable than others.

> Fossil fuels and nuclear fuels can be stored and always be available.

> Wind turbines rely on wind. They do not operate at all at low wind speeds; they work best at wind speeds over 50 kph. Wind turbines are only a good choice in areas with steady, high winds.

> Hydroelectric power stations rely on a fast flow of water. They need reliable rainfall, to maintain the water level in the reservoir.

QUESTIONS

4 Suggest how power stations cope with varying demands for electricity.

5 Explain the difference between capital cost and operating cost.

6 Do you think geothermal energy is reliable or not? Give your reasons.

Electricity from sewage

Sewage is a fuel for electricity generation.

One method uses the methane that sewage treatment produces.

In the United States, researchers are developing quite a different method. Bacteria, in a generator, oxidise liquid organic waste. This releases electrons and protons. The electrons are attracted to a positive anode, and flow round a circuit connected to the generator. The protons move through the liquid to a negatively charged, central cathode. At the cathode, the protons combine with oxygen, from the air, and electrons from the cathode, to produce water.

QUESTIONS

7 Suggest potential limitations and advantages of a sewage electricity generator.

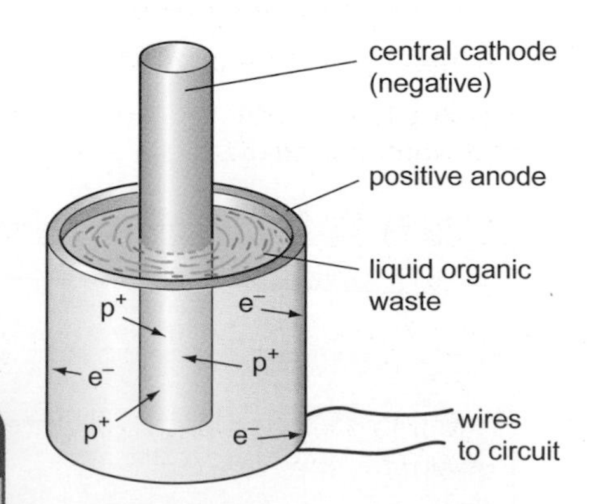

FIGURE 4: What factors might affect the amount of electricity produced by this generator?

The National Grid

You will find out:
> the National Grid distributes electricity
> how the voltage changes throughout the National Grid
> advantages of using very high voltages to distribute electricity

Are power lines safe?

In 2005, a study by the University of Oxford and the owners of the National Grid showed that children born within 600 metres of overhead power lines had a 20% increased risk of developing childhood leukaemia. Those born within 200 metres had a 70% increased risk. Another UK Childhood Cancer Study showed there was no increased risk. Which report is correct?

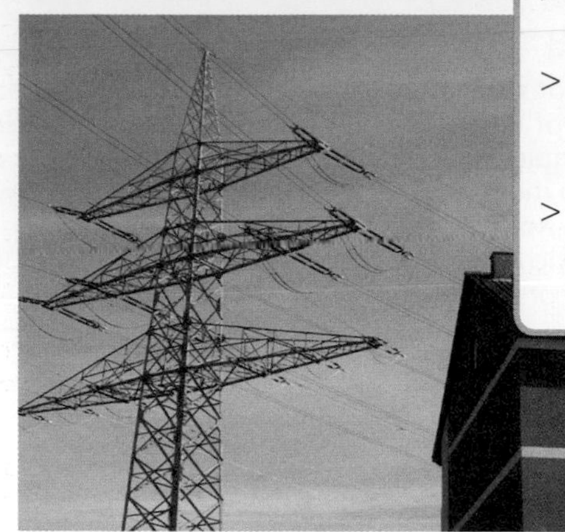

FIGURE 1: Are there invisible risks?

What is the National Grid?

Mains electricity is generated in power stations. The network of cables that carries electricity to local areas is called the **National Grid**. Electricity companies then run smaller cables to your school and home.

Electricity from power stations carries very large amounts of energy. This is because one power station may supply the energy for thousands of homes and businesses. It makes the cables extremely dangerous. You may have seen warning signs saying 'Danger Overhead Cables'. Near canals or lakes the sign is to warn against accidentally touching cables with fishing rods – an electric **current** would flow down the rod and kill the fisherman.

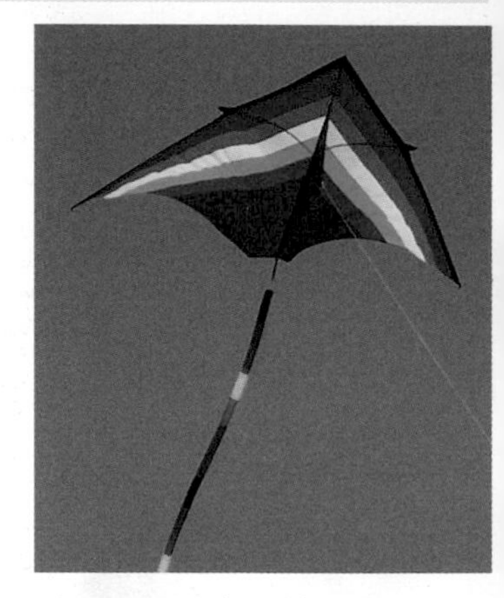

FIGURE 2: Explain why it is dangerous to fly a kite near overhead cables.

Changing voltage

> At the power station, a **step-up transformer** increases the **voltage** of the electricity. A high voltage reduces the cost of transferring energy. Most energy is **distributed** around the country at 400 000 V.

> **Step-down transformers**, in large sub-stations, reduce the voltage to 132 000 V.

> More step-down transformers, in smaller sub-stations, reduce the voltage to 230 V. There is usually a small sub-station for every few streets.

> Home appliances use 230 V, but **transformers** can reduce the voltage even more, such as in a mobile phone charger.

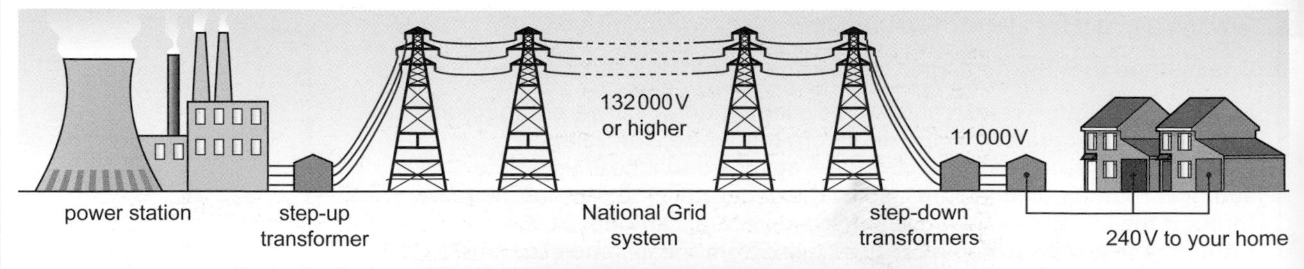

FIGURE 3: The National Grid distributes electricity throughout the UK.

QUESTIONS

1 What is the National Grid?

2 Explain why the electricity, as it leaves power stations, has a high energy.

3 Describe what happens at an electricity sub-station.

Q ... National Grid GCSE

High voltage

Why use such high voltages?

Electrical power is the energy that an electric current transfers each second.

This power depends on the size of the current and the voltage.

power = current × voltage
or $P = I \times V$

The National Grid distributes large amounts of power and so P, in the equation, is high. The current can be kept low as long as the voltage is high. You may remember that, if the current is small, the wires carrying it can be thinner. Thinner wires contain less metal, so are cheaper.

A huge current does not come into homes when the voltage is changed down to 230 V. The single current from the power station splits into lots of smaller, **parallel currents**, all taking electricity to different houses.

Heat losses

Some of the energy carried by an electric current is wasted. The wires become warm. The amount transferred in heating the wire increases as the current increases. By keeping the current from the power station low, it reduces wasted energy. About 10% of the energy from power stations is wasted in the National Grid.

When high voltage cables are laid underground, they need special casings to keep them cool. It is one reason why underground cables are expensive.

Remember
Very high voltages transmit very large amounts of energy with relatively small energy losses.

QUESTIONS

4 If the voltage from a power station doubled, how would the current change?

5 Give one advantage of thin wires, as well as cost.

6 If 10% of the energy is wasted, what is the efficiency of the National Grid?

FIGURE 4: What advantages are there in using high voltages in these cables?

Choosing cables

Most of the connecting wires you have used in circuits are copper, with plastic insulation around them. Copper is a better electrical conductor than most metals, so a smaller proportion of the energy flowing through it is wasted. In the National Grid, overhead power transmission cables are aluminium reinforced with steel, not copper.

QUESTIONS

7 Explain why the National Grid cables, on pylons, are not insulated and whether or not this affects safety.

8 Suggest some factors that affect the choice of putting cables overhead or underground.

FIGURE 5: A cable is a bundle of wires.

Q ... electricity transformers ... National Grid cables

Preparing for assessment: Planning an investigation

To achieve a good grade in science, you not only have to know and understand scientific ideas, but you need to be able to apply them to other situations and investigations. These tasks will support you in developing these skills.

Investigating what affects the energy output of a solar cell

A student is investigating the factors that affect the amount of energy output by a solar cell. She lists possible factors: position of the Sun, intensity of sunlight, atmospheric conditions, cloud cover, surface area of cell and type of cell.

As she cannot control the Sun or the atmosphere, the student decides to use a light source in the lab, and the same solar cell (with a digital voltmeter showing the energy output by the cell).

She decides to vary the distance from the light source to the cell, to vary the amount of light falling on the cell.

These were the student's preliminary results:

Distance (cm)	20	40	60	80	100	120	140	160
Voltage (V)	9.20	3.05	1.91	1.51	1.33	1.23	1.17	1.13

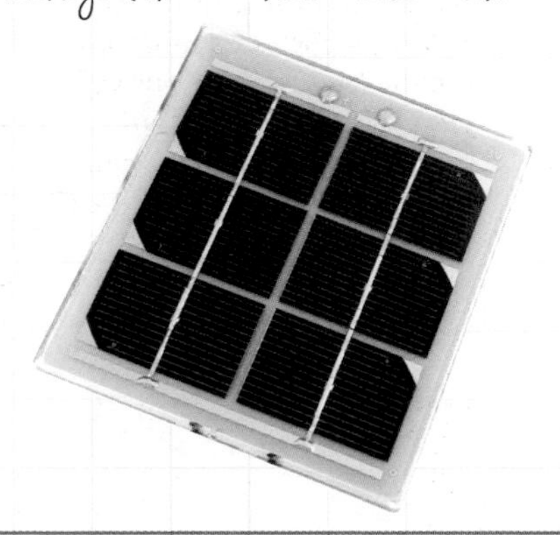

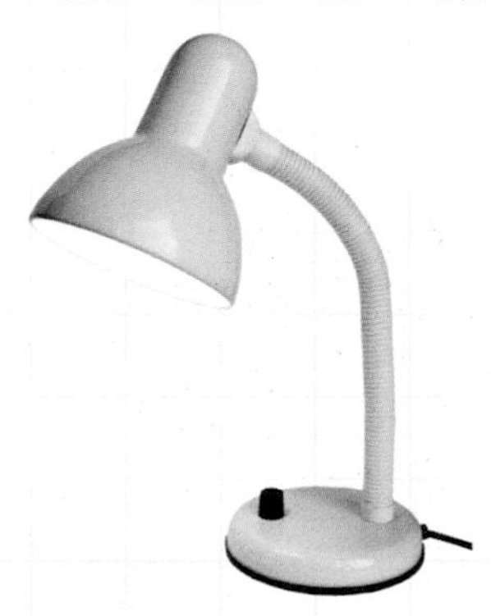

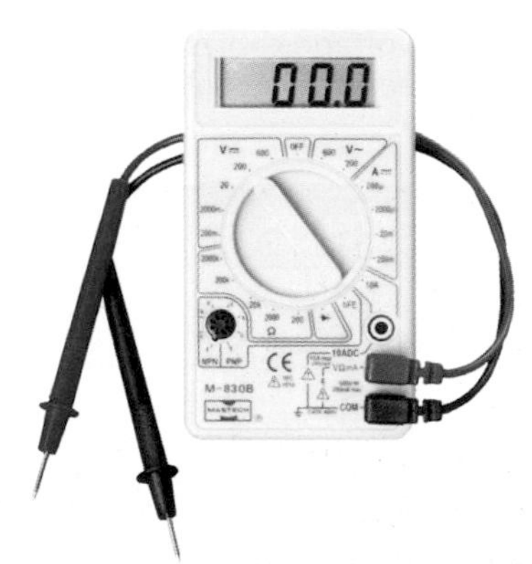

Planning

The student carried out some research in library books and on the internet. She found information about how light intensity varies with distance: as the distance increases, the intensity decreases.

1. Before the light is switched on, should the student expect the solar cell to show a reading of zero? Explain your answer.

> Think about whether there could be other sources of light also hitting the solar cell.

2. Identify the independent and dependent variables in the investigation.

> Remember that the dependent variable changes as a result of changes to the independent variable.

3. Suggest at least two variables that should be kept the same in this investigation. How could this be done?

> Some variables may be difficult to control. Think of the ones that the student should be able to control and how they should be controlled. Think about those that would be difficult to control – is there a way of compensating for them?

Processing and evaluating data

When the student carries out the full investigation, she will need a carefully drawn results table.

4. Prepare a suitable results table. Think about the following:

> Would you do each reading once?

> Would you repeat each reading? If yes, how many times?

> How many rows does the table need? How many columns does it need?

> What should go in the column headings?

Taking several readings increases the accuracy and reliability of the test.

In which column do the independent variable readings usually go?

If a measurement for a dependent variable is repeated, there should be a column for each reading, and, usually, extra columns for calculated values, such as a mean value.

Extending the investigation

The student wanted to extend the investigation, to use coloured filters to alter the wavelength of light hitting the solar cell. Suggest any difficulties there might be.

How could the student be sure that the same amount of light is hitting the solar cell for each of the different colours? Could this be done using different colour filters? Should the student think about a different way of doing this?

Connections

How Science Works

> Planning an investigation

> Assess and manage risks when carrying out practical work

> Select and process primary and secondary data

> Analyse and interpret primary and secondary data

Science ideas

P1.1 Energy (pages 194–195)

P1.5 Using waves (pages 252–253)

What are waves?

You will find out:
> waves transfer energy
> there are different types of waves

Ocean waves

A wave is not really the water itself. It's what happens to the water. Look at a big picture of the sea from a long way away. You can see that the seawater does not move forward, backwards or sideways as a wave goes past. It just goes up and down.

FIGURE 1: What's moving here and in which direction?

Types of waves

Mechanical waves transfer energy

Tie the end of a piece of rope to a table leg. Hold the other end and move the rope up and down. A wave moves along the rope. Like waves in the sea, it is an example of a **mechanical wave**. Sound is another example, with air doing what water and the rope did.

Electromagnetic waves transfer energy

There is another group of waves called **electromagnetic waves** or **electromagnetic radiation**. Infrared, visible light and ultraviolet are examples. TV, telephone and radio signals are all sent using electromagnetic waves.

> Mechanical waves need a **medium** to travel through – water or air, for example.

> Electromagnetic waves can pass through a **vacuum**.

Did you know?

As a surfer rides a wave, surfer and wave move forward, but not the water – the water is only moving up and down.

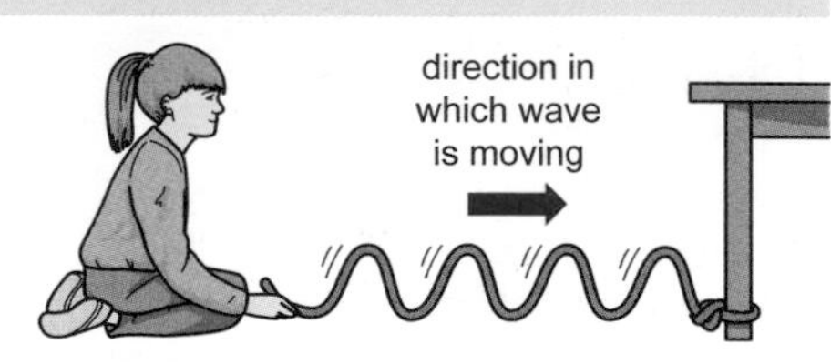

FIGURE 2: Making a mechanical wave.

QUESTIONS

1 Give some examples of mechanical waves.

2 Describe in your own words something that all waves have in common.

3 Explain why sound cannot travel through a vacuum.

Transferring energy

All waves travel outwards from a **source**. All waves transfer energy.

It is often easy to see the source for mechanical waves. When you move the end of a slinky to and fro, energy from your moving hand transfers to the slinky. This energy moves along the slinky as a wave.

Sound waves also carry energy away from a moving object. That is why, if you stop supplying energy to a loudspeaker or amplifier, the cone stops moving and the sound dies away.

Transverse and longitudinal

Moving the end of the slinky in and out makes a **longitudinal wave**. Tying thread to one loop of the slinky helps you to see that the loop moves back and forth in the same direction as the wave. Moving the end of from side to side makes a **transverse wave**. The thread moves at right angles to the direction the wave travels along the slinky.

FIGURE 3: Energy is transferred as the ripples spread outwards. Where does the energy come from?

> Longitudinal waves **oscillate** in the same direction as the wave travels.

> Transverse waves oscillate at right angles to the direction the wave travels.

Sound waves are longitudinal waves. There are sections where the particles are squashed together (**compression**) and sections where the particles are spread out (**rarefaction**).

Water waves are transverse waves: the surface of the water goes up and down as the wave travels along.

All electromagnetic waves are transverse waves. They pass through a vacuum. In other words, they do not need a medium. Electromagnetic waves oscillate at right angles to the direction the wave is travelling. At any point along the wave, the signal will get stronger, then weaker, then stronger and so on as the wave goes past.

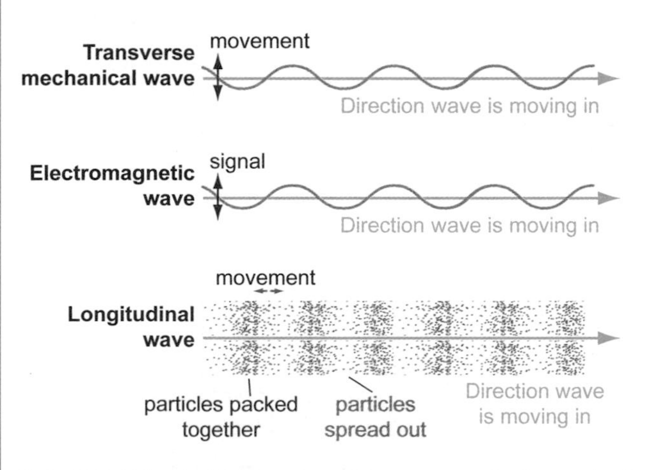

FIGURE 5: What is the direction of energy transfer for each of these waves?

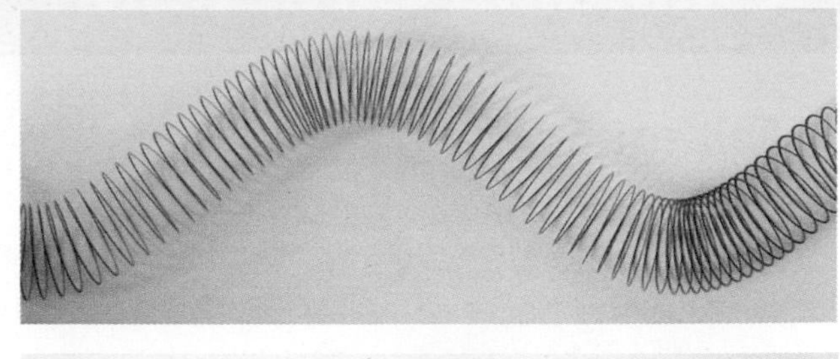

FIGURE 4: Transverse and longitudinal waves. Which is which?

QUESTIONS

4 Imagine you are listening to a live concert, broadcast over a TV channel. Describe how waves have transferred energy to bring the music to your ears.

5 Use ideas of energy to suggest why the water waves die away after a stone is dropped into water.

6 Suggest how the size of a water wave might be related to the energy it transfers. Give an example of evidence that supports your idea.

Earthquakes

Earthquakes happen when tectonic plates grind past one another or collide. Mechanical waves, called seismic waves, transfer energy away from this epicentre of the earthquake.

Two main types of waves spread outwards from the earthquake epicentre. The primary waves are longitudinal pressure waves that travel outwards at the speed of sound. They are detected as low frequency sound waves, often too low frequency to hear. Most of the damage from an earthquake is caused by the transverse secondary waves that travel more slowly. Primary waves can travel through oceans and through the Earth's core, but secondary waves only travel through the solid crust.

QUESTIONS

7 Primary waves, from an earthquake, can be detected thousands of kilometres away from the epicentre. Explain what this shows about the way that sound waves travel in different materials.

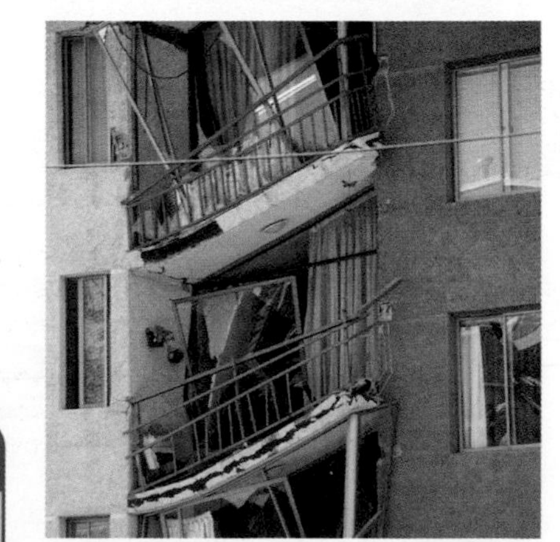

FIGURE 6: The earthquake that caused this damage measured 8.8 M_w. Find out what scales are used to measure earthquakes.

Changing direction

You will find out:

- waves can be reflected, refracted and diffracted
- waves change direction when they are refracted

Mirages

You can often see real mirages over roads, in very hot weather. They look like shimmering pools of water on the road, even though the road is dry. Very hot air just above the road surface bends light rays so that you see an image of the sky at ground level. Our brains 'know' that an image of the sky at ground level means water, so you 'see' water.

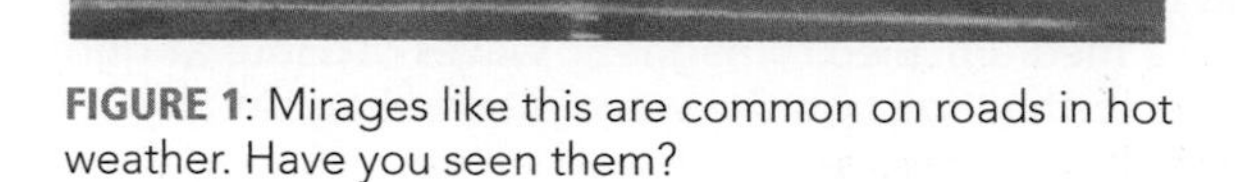

FIGURE 1: Mirages like this are common on roads in hot weather. Have you seen them?

Reflection and diffraction

Reflection

Without **reflection** things could not be seen. Objects reflect light waves towards our eyes. Smooth, shiny surfaces give a clear **image**. Mirrors are an example.

Other waves also reflect. Sound waves reflect from hard surfaces, but less well from soft surfaces.

Move the end of a slinky back and forth once: you will see a mechanical wave travel to the other end of the slinky and then reflect back. A ripple tank can show how water waves reflect from objects.

When a sound wave is reflected in certain places, you hear an **echo**.

The **law of reflection** says:

When any wave reflects from a surface, the angle at which it is reflected is the same as the angle that it strikes the surface.

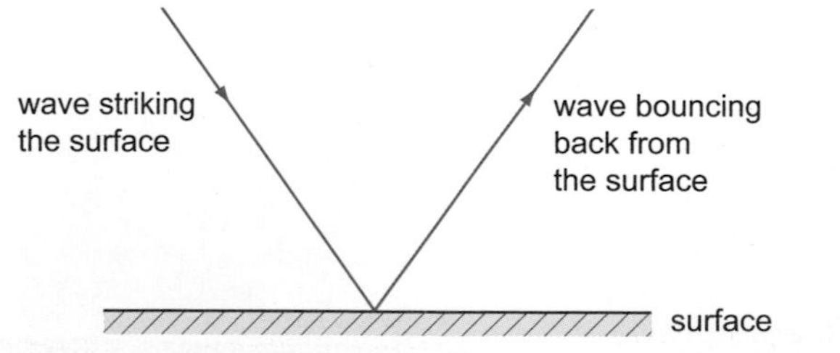

Diffraction

Diffraction happens to waves when they pass through gaps or round corners:

- As a wave passes close to an obstacle, the part of the wave nearest the obstacle is curved.
- A gap acts like two obstacles, so the wave curves at each side of the gap.

Diffraction is greatest when the gap is the same size as the **wavelength** of the wave. The wave at the far side of the gap becomes semi-circular.

Light has a very small wavelength. It goes through very small gaps without any diffraction.

Radio waves have a very long wavelength. They can be diffracted by nearby mountains.

FIGURE 2: How can you tell that these water waves are being diffracted, not reflected?

QUESTIONS

1 Draw a diagram to show how light reflects from an object to our eyes.

2 Describe how you could use a slinky to show reflection of both transverse and longitudinal waves.

3 Draw diagrams to show how (a) a light wave (b) a sound wave curve when they are close to an obstacle.

Refraction

Different waves travel at different speeds. Light travels much, much faster than sound. That is why, during a storm, you hear thunder after you see the lightning.

Waves also travel at different speeds through different mediums. Light waves travel slower in air than in a vacuum, and slower in glass than in air.

The **normal** is an imaginary line drawn at right angles to the boundary between two materials. When a wave passes from one material to a different material where it travels more slowly, it is **refracted** towards the normal.

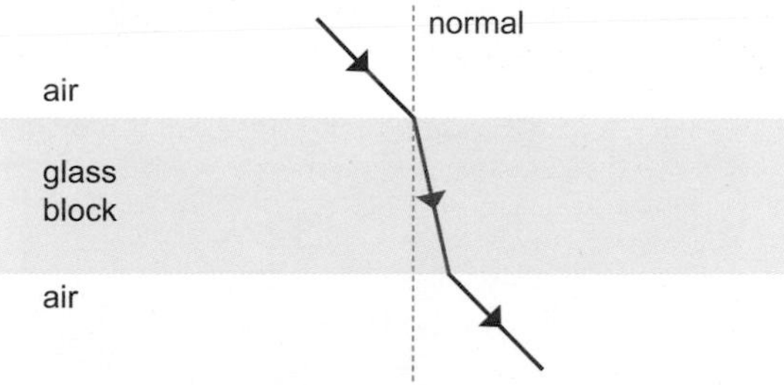

FIGURE 3: Light is refracted towards the normal as it passes from air into glass. You can also see that some light is reflected.

FIGURE 4: Do you think a pool of water will look shallower or deeper than it really is?

Sound is also refracted when the air temperature changes. At night time, sound is often refracted, by the atmosphere, down towards Earth. You can hear things that are further away.

QUESTIONS

4 Explain why a beam of light is not refracted, if it hits a block of glass at right angles.

5 Look at the light travelling through a glass block, in Figure 3. Which way is the light refracted when it passes from glass to air?

6 Draw a diagram to show how sound, refracted towards Earth, means that you can hear things from further away, at night.

Sunrise and sunset

Light from the Sun is refracted in all directions by the atmosphere. Light waves, carrying different colours, are refracted different amounts. Blue light is refracted most and red least. So, at sunrise and sunset, when the sunlight has to travel through the greatest thickness of atmosphere, only the red end of the spectrum gets through to Earth's surface – the sky looks red.

FIGURE 5: Refraction of light from the Sun by Earth's atmosphere means that, when you see the Sun like this, it is already below the horizon.

QUESTIONS

7 Turner, the artist, painted very red sunsets following volcanic eruptions. Dust in the atmosphere has an effect on the refraction of light. Explain how these are connected.

Sound

You will find out:

- > how sounds are caused, and how they travel
- > how to describe sounds
- > how to compare sounds using oscilloscope traces

Sound or noise?

Would you call church bells, on Sunday morning, 'noise' or a pleasant 'sound'? People usually call sounds they don't like noise, but they disagree about which sounds are noise. Noise pollution is sounds that are loud enough or go on long enough to cause damage to hearing. Can you think of some situations where it is important to be able to measure sound?

FIGURE 1: Why does this noise monitor need to wear ear defenders?

Looking and listening

Sounds are caused by things vibrating, such as a tuning fork, a loudspeaker in a music system or your vocal chords when you talk. Vibrations make pockets of air particles vibrate. Sound transfers energy as a wave of vibrating air pockets. You hear sounds when the vibrating air pockets make your ear drum vibrate.

sound source

compression

rarefaction

FIGURE 2: The sound source vibrates, making pockets of air particles vibrate.

Sound waves are longitudinal mechanical waves – air particles vibrate backwards and forwards parallel to the direction the sound travels.

Like all mechanical waves, sound cannot pass through a vacuum. The experiment in Figure 3 shows this. As air is pumped out of the bell jar, the sound from the ringing bell gets quieter and quieter until you can no longer hear it.

Describing sounds

To describe a sound, you need to know its **amplitude** and **frequency**.

> **amplitude** (volume) – how loud it is (very loud sounds can damage your hearing).

> **frequency** (pitch) – how high or low it is. Frequency is measured in cycles per second (Hertz, Hz). Humans can hear frequencies in the range 20–20 000 Hz.

wires

to vacuum pump

bell jar

electric bell

FIGURE 3: How can you tell that the bell still produces sound when the air is pumped out?

QUESTIONS

1 Explain what causes sound.

2 What is a longitudinal wave?

3 Give an example of a high amplitude, low frequency sound.

Sound is a wave

Sound can be reflected, diffracted and refracted, just like all other waves.

An **echo** is a reflection of a sound. It is heard later because it has reflected off a surface, such as a hill or building, and travels further than the sound itself. It is always fainter than the sound, because some of the energy is absorbed by the reflecting surface. Sometimes you may hear a double echo.

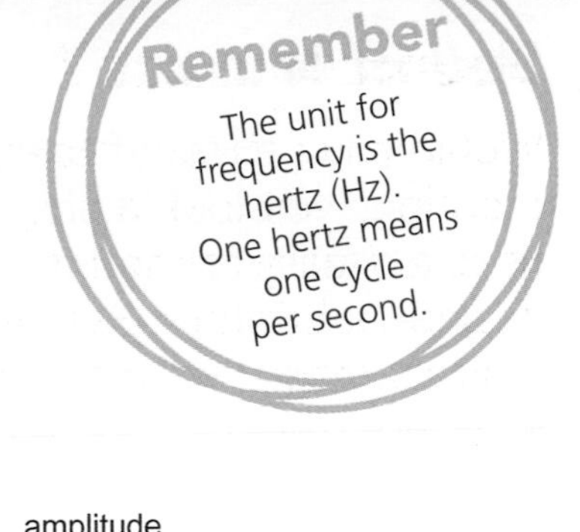

Wavelength

Like all waves, a sound wave is characterised by its **wavelength, frequency** and **amplitude**. In a fixed period of time:

> the shorter the wavelength

  > the more energy it can transfer

  > the more cycles (number of waves per second) and, therefore, higher frequency

> the longer the wavelength

  > the less energy it can transfer

  > the fewer cycles and, therefore, lower frequency.

Frequency is the number of waves a source produces per second. It is also the number of waves that pass a certain point each second.

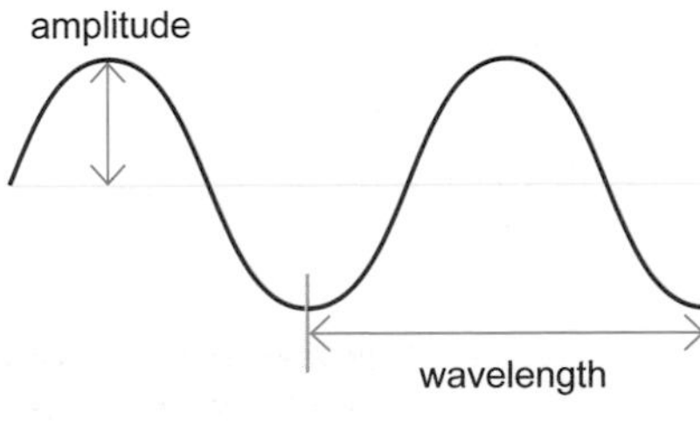

FIGURE 4: Representing a sound wave as a graph of amplitude against time.

QUESTIONS

4 Explain compression and rarefaction in sound waves.

5 Draw a diagram to represent a sound wave that has a lower frequency and is louder than the one shown in Figure 4.

Using an oscilloscope to compare sounds

An oscilloscope measures the amplitude and frequency of sound waves. It produces a trace. The vertical height of the trace shows amplitude. The horizontal width of the trace shows frequency.

Look at Figure 5. Trace A has higher amplitude than trace B. It is a louder sound. The horizontal axis of each trace represents time. You can see that trace C completes two cycles in the same time as trace D completes one cycle. Trace C will complete twice as many cycles in one second as trace D. So, trace C is for a sound with twice the frequency of the sound making trace D.

The 'time base' of an oscilloscope shows how many seconds each horizontal division of the oscilloscope represents. You can use this to calculate the exact frequency of a sound.

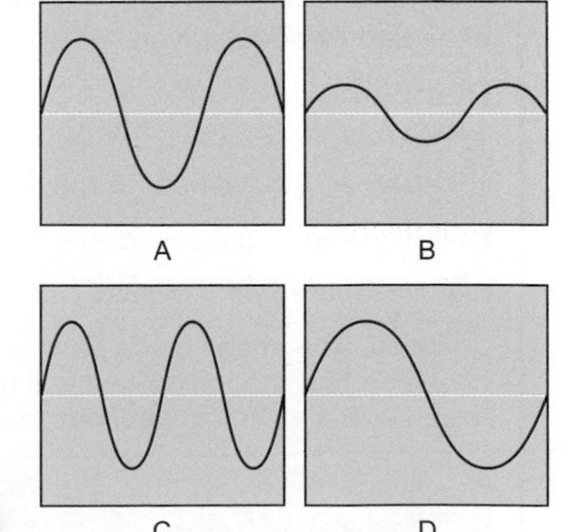

FIGURE 5: What would the trace look like, for a sound that is getting louder?

QUESTIONS

6 Describe the oscilloscope trace for a sound that gets quieter, then louder.

7 An oscilloscope trace shows five complete cycles in 0.1 second. What is the frequency of the sound wave?

Light and mirrors

You will find out:

- > how to construct ray diagrams for the images formed by mirrors
- > how to describe the images formed by plane and curved mirrors
- > about uses of concave and convex mirrors

Hall of Mirrors

Ever been in a Hall of Mirrors? The mirrors can make you look much fatter or thinner, or much taller or shorter. They can even change just part of your reflection so you look as though you have very long legs, or a very long neck. How do they do that?

FIGURE 1: You are not what you see.

Describing light rays

The **ray diagram** in Figure 2 has these important features:

> The **normal** is a line drawn perpendicular the point where the **incident ray** strikes the mirror.

> The incident ray is the ray of light going towards the mirror. The angle between this ray and the normal is the **angle of incidence**.

> The **reflected ray** is the ray of light going away from the mirror. The angle between this ray and the normal is the **angle of reflection**.

Whenever light reflects from a mirror, the law of reflection says:

The angle of incidence equals angle of reflection.

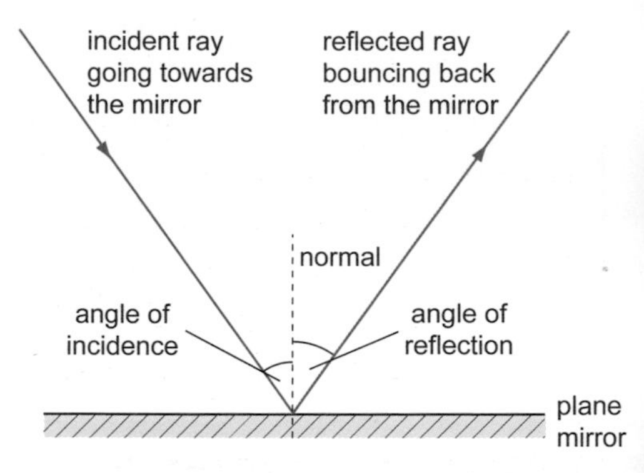

FIGURE 2: Ray diagram showing a ray of light hitting a plane (that is, flat) mirror.

Describing images

Look at yourself in a **plane** (flat) **mirror**. Your image appears:

> the same way up as you – it is **upright** and not inverted (upside down)

> the same size as you – it is not magnified

> the same distance behind the mirror as you are in front of it

> with left and right reversed – it is **laterally inverted**.

It is the same for any object.

Of course, the image is not really where it appears to be. If you put a screen behind the mirror you will not see an image of you on the screen. It is a **virtual image,** not real.

FIGURE 3: Try to describe this reflection, using correct scientific language.

QUESTIONS

1 Name six things that should be included in a ray diagram.

2 Describe or define each item in your answer to question 1.

3 Write down all four characteristics that describe the image formed by a plane mirror. Use the correct scientific language.

What happens if the mirror is curved?

You do not need to know about curved mirrors for the exam, but looking at how they work will help you better understand reflection. The science is the same as for plane mirrors. Figure 4 shows how parallel rays of light change direction, when they hit a curved mirror.

> A **concave** (converging) mirror has a reflecting surface that curves inwards, like the bowl of a spoon.

Parallel rays hitting it come together, or converge.

> A **convex** (diverging) mirror has a reflecting surface that curves outwards, like the back of a spoon.

Parallel rays hitting it spread out, or diverge.

> The law of reflection also applies to curved mirrors.

The normal is drawn at right angles to where the incident ray strikes the mirror.

> When parallel incident rays are reflected from a smoothly curved mirror, the focus is the point:

- > the reflected rays pass through, or
- > where the reflected rays appear to have come from.

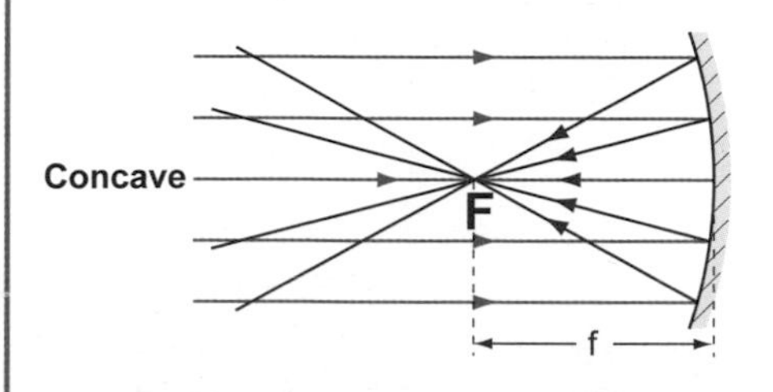

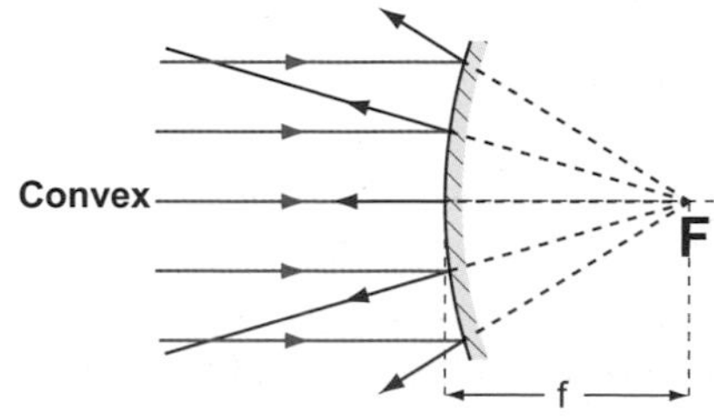

FIGURE 4: Reflection in curved mirrors.

Remember

For any mirror, the angle of incidence is always equal to the angle of reflection.

QUESTIONS

4 Describe what the 'focus' of a curved mirror means.

5 Jan's eye is 3 cm wide, but in her make-up mirror it looks 3.6 cm wide. What is the magnification of her mirror?

6 Discuss with a partner how the images in a Hall of Mirrors are formed.

Using curved mirrors

The image in a convex mirror is always upright and smaller than the object. This makes convex mirrors useful for driving mirrors or shop security mirrors – you can see a much larger area than using a plane mirror.

If the object is far away from a concave mirror, the image is inverted and smaller than the object. This makes concave mirrors useful for collecting and focusing a signal, for example, a satellite dish or telescope. If the object is closer, its image is upright and larger than the object. Make-up mirrors are concave.

FIGURE 5: A convex mirror in a clothes shop.

QUESTIONS

7 Explain, with a ray diagram, why a car wing mirror is convex. Some drivers fit extra 'blind spot' mirrors. Find out how these help with safety.

Using waves

You will find out:

> waves transfer energy

> some of the many uses of electromagnetic waves

How energy travels through air

To get a good TV picture, you have to be in line with a transmission mast that sends the signal. In a deep valley with hills all round, you may not get a good TV picture. That tells us two things. The wave that carries the signal travels in a straight line. Also, the signal can move through the air, but obstacles, such as hills, can interfere with the transmission of the wave.

FIGURE 1: The signal from the transmitter comes in a straight line to these aerials.

Invisible waves

Electromagnetic radiation is invisible. It travels in transverse waves at the speed of light, transferring energy.

It has a range – a **spectrum** – of energies. Radiation from each part of the **electromagnetic spectrum** has uses:

Increasing Energy ↓

Radio waves are low energy.

They are used to send radio, TV and GPS signals. Bluetooth signals enable mobile phones to communicate with landline phones or with laptops. Bluetooth signals are very weak radio signals.

Microwaves have many everyday uses because of their low energy.

They are used in microwave ovens, communication systems and speed traps. Mobile phones, very long-distance phone communication and satellite television use microwave signals reflected off satellites in space.

Infrared radiation. You feel this as heat.

An electric bar heater emits infrared radiation. TV remote controllers also use infrared, but you will not feel the warmth coming from them, because they use very weak signals.

Visible light enables us to see.

This includes seeing with spectacles, microscopes and telescopes. It is also used for photography.

Ultraviolet radiation tans us.

Sun-beds use UV light to tan us artificially.

X-rays have high energy.

They are used to produce X-ray images of bones and teeth.

Gamma rays have very high energy.

Gamma rays can kill cancer cells.

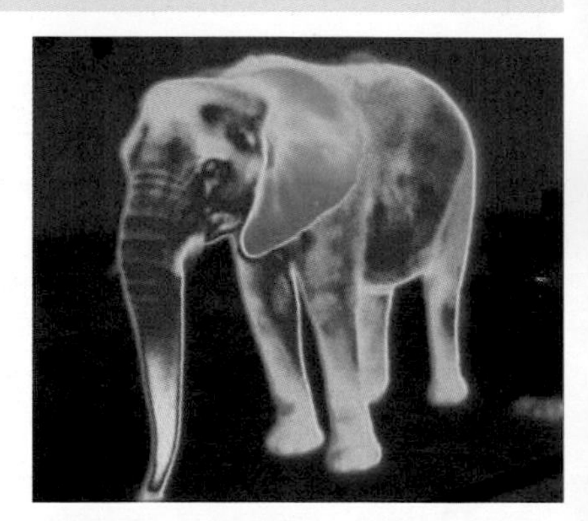

FIGURE 2: A thermogram reveals infrared radiation that you cannot see.

Did you know?

Light travels at 300 000 000 metres per second. Nothing can travel faster.

QUESTIONS

1 List the different types of radiation. Give an example of how each is used.

2 Which are your top three most useful applications?

... electromagnetic radiation

More about radiation and its uses

Like waves in all parts of the electromagnetic spectrum, radio waves have a range of energies. The lowest energies transmit local radio and police and ambulance messages. The highest energies are for TV.

Anything warm emits infrared radiation. Infrared cameras detect it and produce thermogram images. A person can be 'seen' when trapped in a collapsed building or breaking into a house at night. Thermograms reveal arthritis and cancer because these parts of the body are warmer.

Visible light comes from the Sun, fluorescent lamps and some lasers. It is 'visible' because you can see where it comes from and the objects that reflect it.

Ultraviolet (UV) radiation comes from the Sun and very hot objects, such as welding equipment. Sunscreens block some of the Sun's harmful UV radiation that can damage skin.

X-rays and gamma rays are dangerous to body cells. They must be used carefully. X-ray images show when bones are broken, teeth decaying or lungs diseased.

Radioactive substances, including uranium, produce gamma rays. They are used to sterilise food by killing bacteria, to kill cancer cells, and in industry to check for faults and cracks in pipelines.

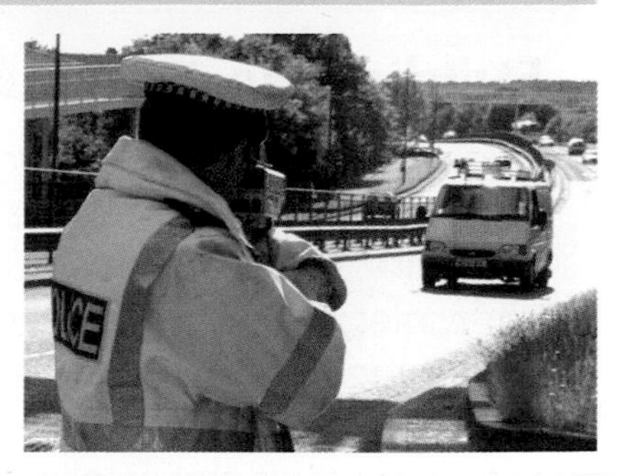

FIGURE 3: What types of waves are used to detect speeding motorists?

QUESTIONS

3 Explain how infrared rays are used to find people trapped in collapsed buildings.

4 As a group, discuss the safety measures we must take when using gamma rays.

5 Suggest a property of gamma rays that makes them suitable for sterilising food.

FIGURE 4: What is UV radiation revealing about this forged banknote?

Security

Airport security ensures millions of passengers fly in safety. You put your baggage on a conveyor belt, where it is X-rayed. Airport staff see an image of the object on a screen. Passport control staff use UV radiation to identify illegal passports.

An infrared burglar alarm senses intruders from their body heat. A laser beam can also be used as a burglar alarm, acting as an invisible light 'trip wire'. If the beam is broken it sets off the alarm.

FIGURE 5: A gun is detected in a suitcase.

QUESTIONS

6 Describe ways that airport security ensures passenger safety.

7 A valuable diamond necklace is on display at an exhibition. Suggest how you might protect the necklace.

The electromagnetic spectrum

You will find out:

> electromagnetic radiation travels as waves

> wavelength, frequency and energy are connected by the wave equation

Building up the radiation picture

In 1666, Isaac Newton shone white light through a prism and revealed the seven colours of the visible spectrum. In the 19th century, infrared rays and ultraviolet rays were discovered. On 11 July 1962, *Arthur* received the first live television broadcasts from America that had travelled as electromagnetic radio waves via the satellite, *Telstar*. Now there are thousands of satellites in orbit around the world, transmitting many kinds of communications.

FIGURE 1: Why does Arthur point at the horizon? Think about the UK and US are on a globe.

Electromagnetic radiation

Electromagnetic (EM) radiation travels as waves. It carries energy from place to place.

Electromagnetic waves move in straight lines. They travel at the same speed in a vacuum – often called the **speed of light**.

All EM radiation transfers energy through space at the same speed.

Different types of EM radiation transmit different amounts of energy. The **electromagnetic spectrum** shows the different types of EM radiation in order of the amount of energy they transfer.

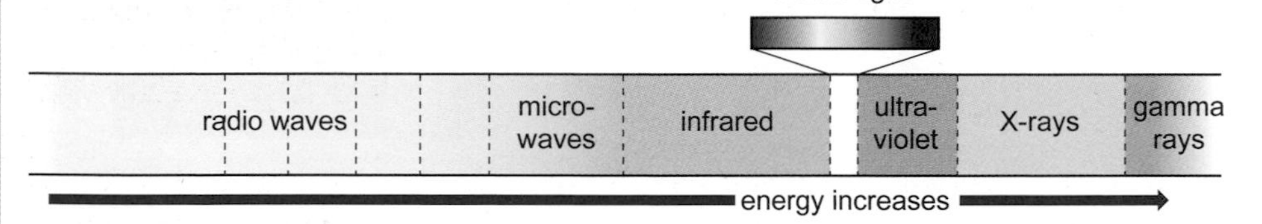

FIGURE 2: The electromagnetic spectrum. Remembering the order will help you to understand the way different radiations behave.

Did you know?

The electromagnetic spectrum stretches from radio waves to high-energy gamma rays, but it is not in separate parts – it is continuous.

QUESTIONS

1 How does electromagnetic radiation travel, how fast does it go and what does it carry?

2 Describe what the electromagnetic spectrum shows.

3 Which type of radiation has the highest energy.

Energy and wavelength

Electromagnetic radiation travels as transverse electromagnetic waves. Key terms are wavelength, frequency and amplitude.

Wavelength is the distance between a point on one wave and the same point on the next wave.

For EM radiation, the points on the wave indicate signal strength. Maximum amplitude is maximum signal strength.

The shorter the wavelength of EM radiation, the more energy the wave transfers. Frequency is the number of waves passing a particular point in a second. It is measured in cycles per second, hertz (Hz).

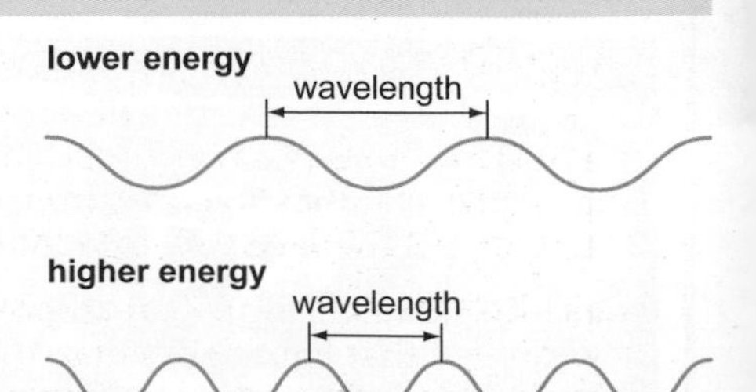

FIGURE 3: Wavelength, frequency and energy are linked.

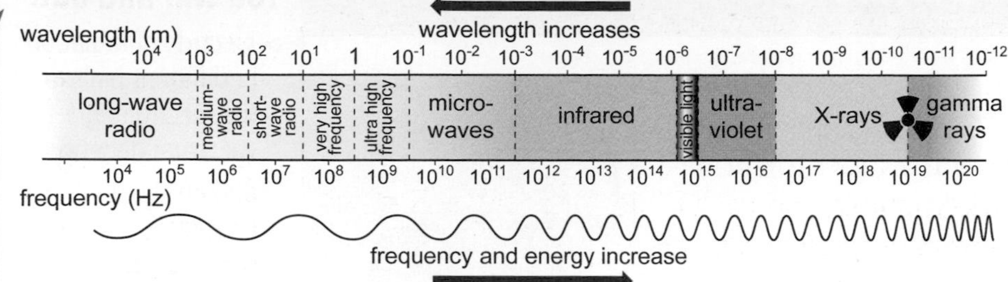

FIGURE 4: The electromagnetic spectrum.

Waves that carry a lot of energy have a high frequency and a short wavelength.

You can see in Figure 4 that, as the wavelength gets longer, the frequency gets less. Frequency and wavelength are connected by the **wave equation**:

$$v = f \times \lambda$$

where

v = the speed of light (300 000 000 m/s)

f = frequency (hertz)

λ = wavelength (metres)

Remember

Wavelengths in the EM spectrum range from about 10^{-15} metres to over 10^4 metres.

Detecting electromagnetic radiation

The different types of electromagnetic radiation merge into each other. Think about heating a small bar of metal in a Bunsen burner. At first, it gets hot and you feel the infrared radiation being emitted. Keep heating and it begins to glow red, then orange, then yellow: different colours of visible light are being emitted. Our Sun emits mainly visible light and UV radiation, but it is so hot that it also emits X-rays and gamma rays.

QUESTIONS

4 List the different types of electromagnetic radiation in descending order of frequency.

5 Describe the features and characteristics of electromagnetic radiation.

6 Research for and add some more radiation detectors to those in Table 1.

TABLE 1: Detectors of electromagnetic radiation.

Wave type	gamma rays	X-rays	ultraviolet	visible light	infrared	microwaves	radio waves
Wavelength	10^{-12} m	10^{-10} m	10^{-7} m	0.0005 mm	0.1 mm	1–10 cm	10–1000+ m
Sources	radioactive nuclei	X-ray tube	Sun very hot object	Sun hot object LEDs	warm or hot object	radar microwave oven	radio transmitter
Detectors	Geiger–Muller tube	Geiger–Muller tube	photographic film fluorescent chemical	eyes photographic film	skin thermometer	aerial mobile phone	aerial TV radio

Light: wave or particle?

The Ancient Greeks had two opposing theories about how we see things. Empedocles said that light rays from our eyes touch any object we look at. Plato thought that every object radiated light rays that entered our eyes.

Francesco Grimaldi, in 1665, first spoke of light as a form of wave. Isaac Newton said that light behaved as if it were a particle. Einstein showed that light could 'knock' electrons off a metal surface – a particle property. He called light 'particles' photons. Louis de Broglie brought the two ideas together: light has the properties of both particles and waves. This idea became known as **wave–particle duality**.

QUESTIONS

7 Research the wave–particle duality of light and summarise your findings.

Dangers of radiation

You will find out:

- hazards of radiation in different parts of the electromagnetic spectrum depend on wavelength and frequency
- there are ways to protect ourselves from these hazards

Microwave claims

Mobile phones give out microwave radiation, when they are being used. Some scientists claim that the radiation can damage cells in our brains, if using a mobile for long periods of time. Some say that the microwaves given out by phone masts are stronger and even more harmful. Other scientists dispute both claims. There still has not been enough research to be absolutely certain about either.

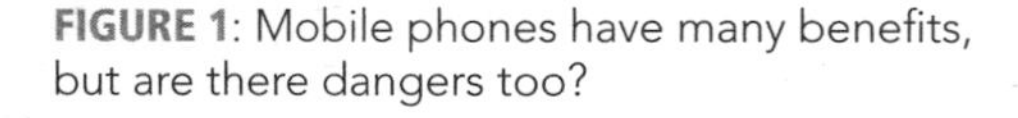

FIGURE 1: Mobile phones have many benefits, but are there dangers too?

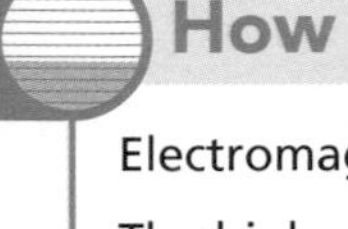

How dangerous is electromagnetic radiation?

Electromagnetic radiation brings many benefits, but there is a downside.

The higher the energy of electromagnetic radiation, the more damage it does to our bodies. Very high energy radiation can **ionise** molecules in cells. This damages the cells or even kills them. High energy radiation penetrates deeper into our bodies too, so it does more internal damage.

- Radio waves are harmless.
- Microwaves are strongly absorbed by water. They pass through the skin and can heat up the cells beneath.
- Infrared radiation burns skin and tissues beneath.
- Visible light, if it is very bright, can damage the eyes.
- Ultraviolet radiation causes sunburn and can cause skin cancer.
- X-rays can cause severe burns to the skin and can damage cells, causing cancer.
- Gamma rays damage cells and can cause cancer.

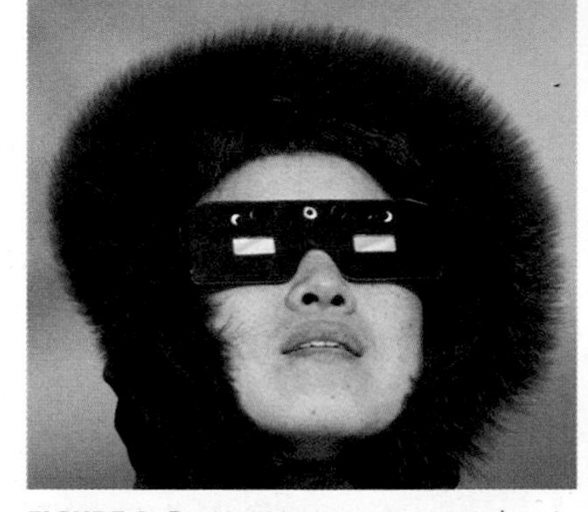

FIGURE 2: Protecting your eyes makes it safe to view a solar eclipse.

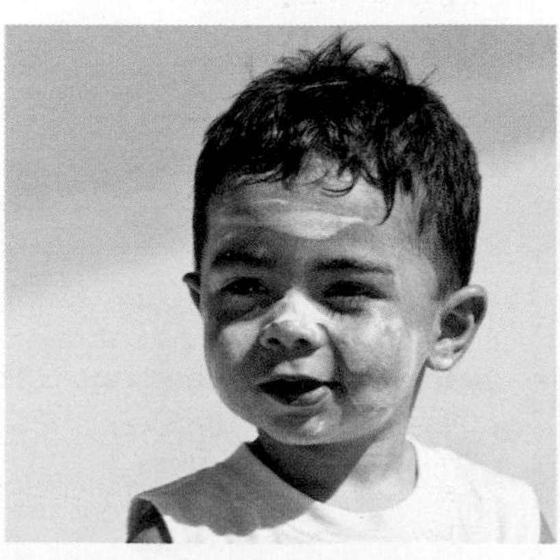

FIGURE 3: Do you know what the '30' means, in Factor 30 sun cream?

QUESTIONS

1 List the radiation in the electromagnetic spectrum, in order of increasing energy.

2 List the radiation in the electromagnetic spectrum, in order from the most damaging to the least damaging.

3 Suggest ways that you can protect yourself from the UV radiation in sunlight.

... radiation hazards GCSE

Protecting from the dangers

Microwave ovens cook by heating the water molecules in food. Human bodies are mostly water and so could be affected by the microwaves from an oven. To stop this radiation getting out, the glass oven door has a fine metal mesh inside it. The oven only starts when the door is closed.

Infrared radiation from a bonfire keeps you warm. Getting too close would cause serious burns to your skin and the soft tissue beneath.

Very bright visible light can damage the cells of the eye's retina, causing blindness.

Ultraviolet radiation produces essential vitamin D in our bodies, but prolonged exposure may cause sunburn or even skin cancer. Sun creams contain chemicals that absorb the UV radiation from the Sun, helping to protect the skin.

X-rays pass through soft body tissue easily, but are partially absorbed by denser substances, such as bone. A hospital radiographer, who operates the X-ray equipment, stands behind a lead–glass screen while the X-ray is taken. Depending on the strength of the X-rays used, a patient may have a lead apron to protect parts of the body not being X-rayed.

Strong X-rays and gamma rays ionise cell molecules, killing the cells. Gamma rays are used to kill cancer cells. They are finely focused so that they do not kill healthy cells. Radiation doses are carefully controlled, for the safety of both patients and hospital staff. Lead or concrete shielding is needed to stop gamma rays.

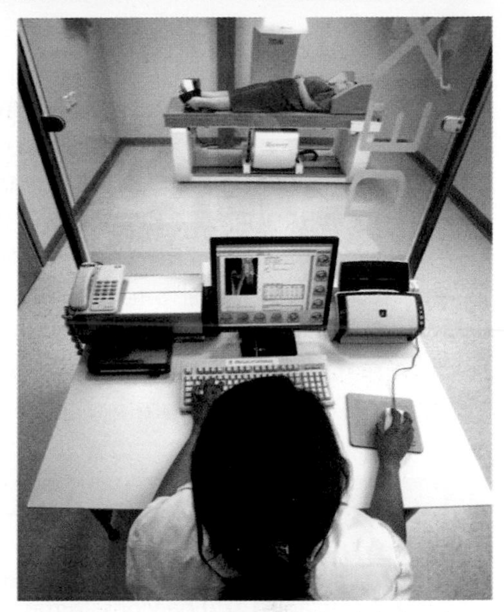

FIGURE 4: The radiographer stands behind a lead screen to protect herself.

QUESTIONS

4 Explain why it is dangerous to stay out in the hot summer sun for too long. Does it matter what part of the world you are in?

5 Describe how radiographers protect themselves from dangerous gamma and X-rays.

6 Read the introduction to this page again. Suggest a reason why some scientists fear that using mobile phones may be harmful.

Did you know?

Some air crew wear radiation badges. If exposed to too much cosmic radiation, they are grounded until their levels return to normal.

Mutations

Ultraviolet, X-rays and gamma rays are called **ionising radiations**. They have enough energy to penetrate cells and kill or damage them.

DNA molecules, in the nucleus of a cell, may be damaged if ionising radiation reaches them. The cell may die or it may continue to divide, but with a damaged genetic code. Sometimes, these mutated cells keep dividing out of control and form a cancerous **tumour**.

The chance of ionising radiation causing cancer depends on the type and intensity of the radiation, and how much is absorbed.

QUESTIONS

7 Some people wear radiation badges at work. Find out which types of jobs these are. What information do the badges give?

... radiation safety GCSE ... ionising radiation

Telecommunications

You will find out:

> how electromagnetic waves are used for communication

Communications

There used to be a delay of several seconds between speaking and getting a reply, when making long-distance telephone calls. It was due to the time taken for the signal to travel along the cable and back. Nowadays many signals are relayed via satellites orbiting around Earth. The signals travel so fast that there is no delay.

FIGURE 1: Satellites receive and transmit sound and visual messages.

Sending information long distances

The earliest 'long-distance' communications were semaphore (using flags to send messages) and beacons. In the late 1800s, telephone technology began: sending electrical signals along cables.

Soon after, people started experimenting with radio waves. This led to radio and television broadcasts. Now, all **telecommunications** use some type of electromagnetic radiation.

> Radio waves are used for local radio, emergency services communications and television.

> Microwaves are used for mobile phones and satellite communications. Satellite TV comes to homes this way.

> Infrared and visible radiation transmits information along fibre optic cables – this is what cable TV uses.

FIGURE 2: Radio waves help keep emergency and security services in touch.

QUESTIONS

1 Describe the types of radiations used in telecommunications. What do you know about their wavelengths?

2 Name one thing that the types of electromagnetic radiation used for communications have in common.

3 Give the type of signals transmitted through fibre optic cables.

The technological age

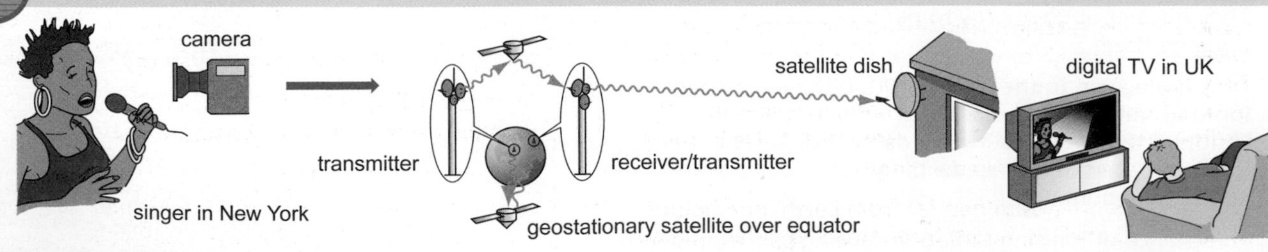

FIGURE 3: How a satellite TV programme reaches you.

Modern technology

There have been two main changes in communications over recent years. They have got faster and it is possible to send much more information at one time.

Communications satellites enable signal to be sent from one place on Earth to another. Electromagnetic waves carry signals to a satellite and back down to a different place, very quickly. Live television broadcasts from the other side of the world really do show things almost as they are happening. Phone messages sent via satellite mean that you can have a 'live conversation' with someone via email, *Skype*™ or text messages.

Bluetooth uses low-power radio waves to enable signals to be sent over very short distances. Your mobile phone can connect to your laptop, to download photos for example, or your mobile phone can connect to a hands-free device.

The internet uses the telephone system. Signals may be sent along cables as infrared or visible radiation, or as microwaves from mast to mast, or via satellites.

Interactive TV uses infrared signals to communicate between your remote control and the TV, which is connected to the internet.

Wi-Fi networks use low-power radio waves to communicate between computers in a house or office, or to connect computers into a central system, at places such as stations, hotels or city centres.

QUESTIONS

4 Look at Figure 3. Describe how the diagram would be different for a cable TV programme.

5 Suggest the main ways in which communications satellites and fibre optic cables have changed telecommunications.

6 Describe how bluetooth systems can be used.

7 Find out how the Global Positioning System works.

Space technology

Communications satellites and telephone masts are used for much more than sending TV programmes or telephone conversations.

Modern phones can connect to the internet by communicating with nearby telephone masts using microwaves. Each individual telephone mast sends out its own specially coded signal. The phone can tell where it is by measuring the strength of signals from different telephone masts.

Modern phones can also connect directly, via satellites, into the **Global Positioning System (GPS)**. This takes more power and makes the phone battery go flat more quickly.

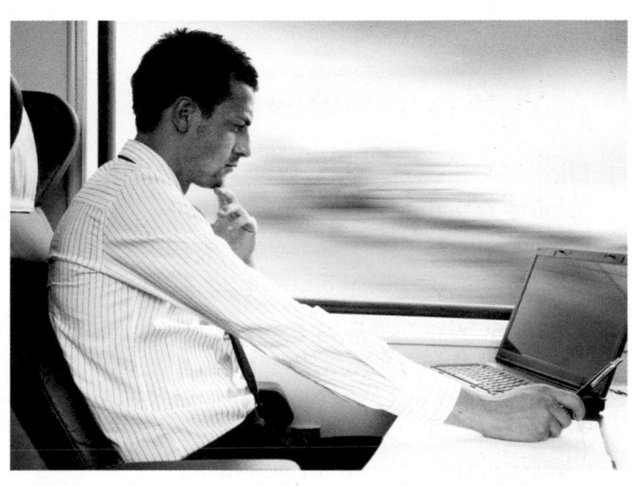

FIGURE 4: A laptop dongle lets you surf the net while you are on the move.

Did you know?

The first radio broadcast was in 1916 and the first television broadcast was in 1936.

Communications satellites

Geostationary satellites (also called geosynchronous satellites) are used for all kinds of telecommunications. They stay above the same point on the ground, usually over the equator, and appear to hang in space. In reality, they are orbiting at the same rate as Earth turns on its axis – one orbit every 24 hours.

The satellites are positioned far from Earth, at a height of 36 000 kilometres, about three times the diameter of Earth away.

QUESTIONS

8 Explain why telecommunications satellites need to be in geostationary orbit.

9 Research other types of artificial satellites and their purpose. Are they all geostationary? Explain your answer.

Cable and digital

You will find out:

> optical fibres are used for communications

> digital signals have advantages over analogue

Digital cameras

When you use a digital camera, perhaps in your phone, light rays from the scene reach millions of tiny areas on a memory card, in the camera. Each area develops a different electrical charge. The camera instantly scans the charges and produces a sequence of digital signals.

FIGURE 1: Press 'view' to convert digital signals into an image.

What is digital and cable?

Digital

'Digital' is to do with the way the information is sent and stored.

You might have heard news stories about 'analogue TV' being switched off. Some old TV sets will not work and others will need re-tuning. 'Analogue' is how TV signals and other information used to be sent. It is an electromagnetic wave that carries information in the form of tiny changes in the strength and frequency of the wave.

Digital signals are sent in 'pulses'. These short bursts of signal are switched on and off extremely rapidly.

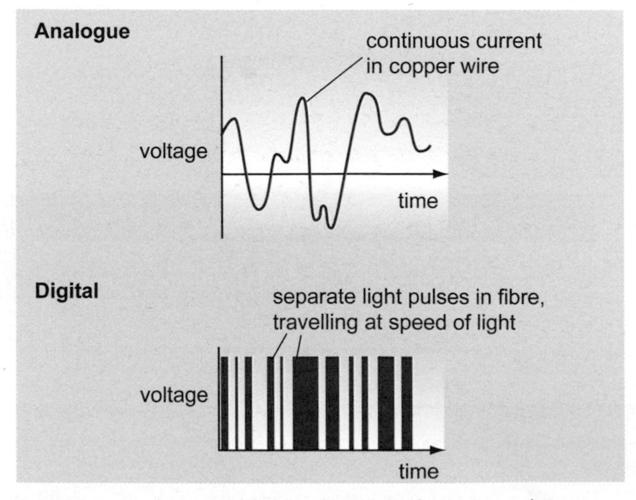

FIGURE 2: Analogue and digital signals shown as voltages.

Cable

Fibre optic cables are thin, flexible strands of glass. They are also called **optical fibres**. They carry light waves:

> Light is shone in one end of the fibre optic cable.

> It reflects backwards and forwards off the inside walls of the glass, in a zigzag.

> It reaches the other end of the cable.

Messages can be sent though the cable as digital 'pulses' of either visible light or infrared radiation.

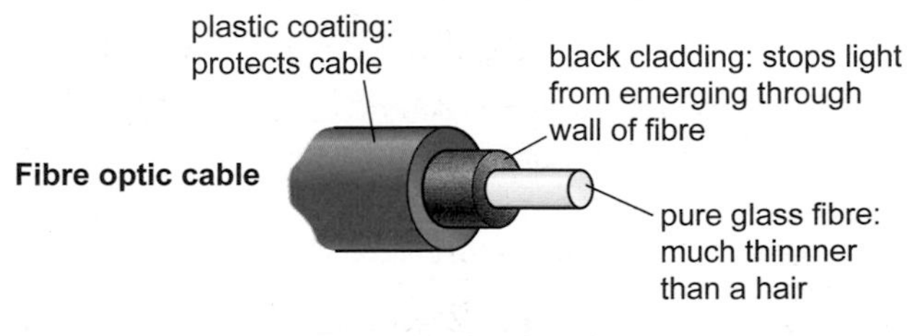

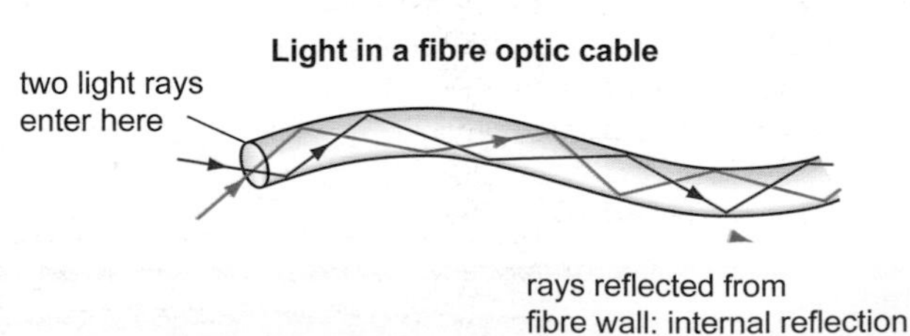

FIGURE 3: Fibre optic cables carry information fro phones, cable TV and computers.

QUESTIONS

1 Explain the difference between analogue signals and digital signals.

2 Describe what optical fibres are made from. How do they work?

Q ... digital vs cable GCSE

Why use digital?

Analogue signals carry different information on different waves. Part of the information is in the signal strength.

Transmitted information can be poor quality because:

> the different waves interfere with each other

> there is a maximum distance that signals can travel

> boosting the signal (to travel further) affects the information.

Digital signals are pulses of light or infrared. They are either 'on' or 'off'.

The quality is better because:

> signals do not interfere with each other

> signal quality is not affected by distance

> signals be made stronger without losing information.

Without the interference, more signals can be sent down one cable at the same time.

QUESTIONS

3 Describe why digital signals are better quality than analogue signals.

4 Suggest why broadband is used with fibre optic cables, but not with old metal cables.

5 Explain the optical properties that make optical fibres so good for transmitting signals.

Cable and broadband

Light or infrared signals travel faster through optical fibres than electrical signals along metal cables. Optical fibres:

> transmit information fast, with good quality

> can carry many signals.

However, a computer must control when signals are sent, to avoid overloading the cable with too many signals.

Broadband is a way of controlling when each is sent, allowing signals to be sent close together. This makes it possible to download films, for example, that contain a lot of information.

Fibre optic cables have uses in medicine, in endoscopes. Endoscopes are used for internal examinations and to help surgeons perform keyhole surgery.

> A fibre optic cable carries light inside a patient's body.

> A second cable carries the light back, so that a doctor can see any problems.

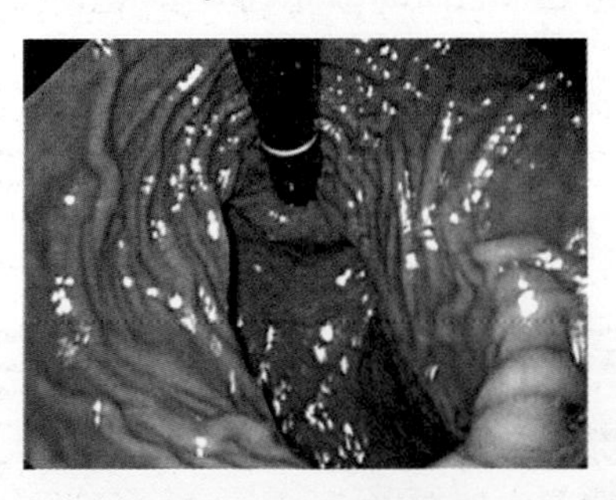

FIGURE 4: Inside the stomach leading to the duodenum, seen through an endoscope.

Total internal reflection

Figure 5 shows how light rays, inside an optical fibre, behave at the boundary between glass and air. When the angle of incidence reaches the critical angle of 42°, the ray grazes the boundary. At greater angles, there is total internal reflection – the light ray is reflected back into the glass. The light is reflected back and forth along the fibre, even if there is no shielding around the outside of the fibre.

QUESTIONS

6 Suggest a reason why the signals, sent along a fibre optic cable, are laser signals that contain a single wavelength of light or infrared radiation.

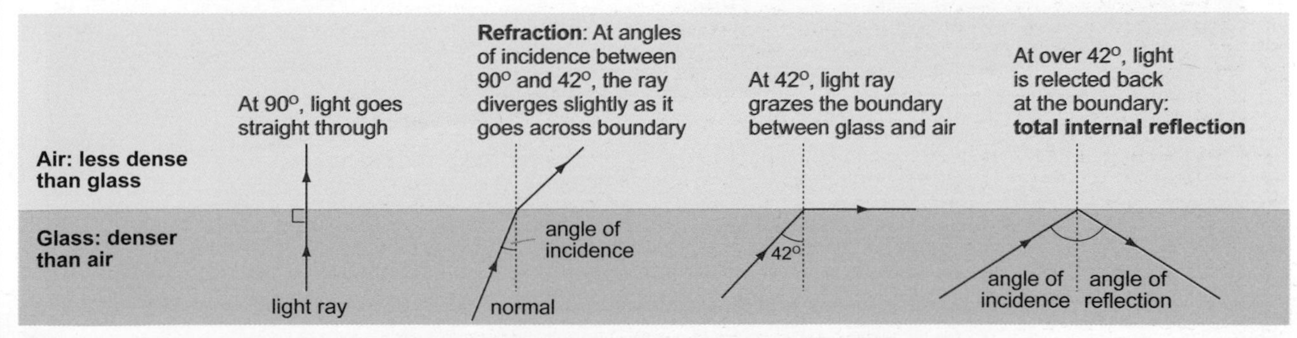

FIGURE 5: Behaviour of light at a glass–air boundary. Light travels slightly faster in air than in glass.

Preparing for assessment: Applying your knowledge

To achieve a good grade in science, you not only have to know and understand scientific ideas, but you need to be able to apply them to other situations and investigations. These tasks will support you in developing these skills.

Communicating in space

In 1968, the Apollo 8 spacecraft orbited the Moon with three astronauts on board. For 20 hours, the spacecraft kept orbiting the Moon. Whenever the spacecraft was behind the Moon, its radio signals were blocked from Earth for about 45 minutes. Mission Control, in the United States, could only wait for the spacecraft to re-emerge and make contact again. During this time, the astronauts could still speak to each other and take colour photographs like the one below (the first image of Earth, taken from Apollo 8). As part of the mission, astronauts sent television pictures showing Earth seen from space for the very first time.

In 1998, the first module of the International Space Station was launched. This giant science laboratory orbits Earth. For over ten years, astronauts have lived in the International Space Station, carrying out experiments and essential maintenance. Inside the space station, the astronauts can speak to each other normally. On space walks outside the space station, astronauts use audio systems fitted inside their space helmets, allowing radio communication with each other. Microwaves are used to transmit information between the Space Station and Earth. Unlike the astronauts orbiting the Moon in 1968, scientists on the International Space Station are in constant contact with Earth. The Space Station orbits the Earth several times a day, so communication links in both the United States and Russia are needed for continuous coverage. The astronauts orbit nearly 300 km above Earth's surface. They can be many hundreds of kilometres from Mission Control, but messages sent using radio links are received almost immediately.

Task 1

Describe some ways that astronauts communicate with different groups of people. Explain how they use waves to do this.

Task 2

Describe some differences and similarities between the communications during the Apollo 8 mission and a mission in the International Space Station.

Task 3

Explain how instructions from Mission Control reach astronauts on the International Space Station and are then passed on to astronauts outside the Space Station.

Task 4

Discuss why astronauts and Mission Control need to use different communication methods in different situations.

Maximise your grade

Grade	Answer includes showing that you…
	know that radio waves, microwaves and sound waves are used for communication.
E	know that sound waves cannot travel through a vacuum.
	know that electromagnetic waves can travel through a vacuum.
	know that radio waves and microwaves travel at the same speed.
C	know that electromagnetic waves travel at the speed of light.
	can compare the use of different types of wave for communication in different situations.
A 	can evaluate the most suitable type wave used for communication with and in spacecraft.

Searching space

You will find out:

- > how different types of telescope are used to make discoveries about the universe
- > telescopes collect radiation from across the electromagnetic spectrum

Galileo Galilei

Galileo may have invented the first optical telescope, in about 1609 (although other scientists may have got there earlier). Galileo's telescope could only magnify eight times, but he was able to observe moons orbiting around Jupiter. He also saw that the planet Venus had phases just like the phases of our own Moon. These observations led him to support the new theory that the Sun was at the centre of the solar system – a very unpopular theory at the time.

FIGURE 1: Galileo using his telescope to observe space.

Telescopes

The earliest telescopes were optical telescopes. Optical telescopes magnify visible light coming from distant stars and planets. The first telescopes used lenses to magnify the image. Later, scientists added concave mirrors, to make reflecting optical telescopes. Mirrors can be large. They collect more light and give brighter, clearer images than lenses on their own.

In the 1930s, scientists discovered radio signals coming from space. They built radio telescopes to study these signals. Radio telescopes have a dish to collect radio waves. The dish focuses waves onto an antenna, above the middle of the dish. A large dish collects more radio waves than a small dish. A radio telescope with a large dish can study fainter objects.

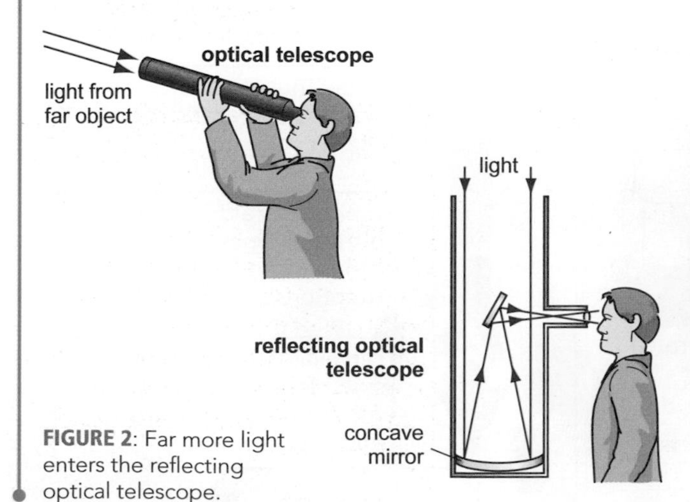

FIGURE 2: Far more light enters the reflecting optical telescope.

FIGURE 3:The Arecibo radio observatory in Puerto Rico can 'see' pulsars. Pulsars emit a pulse of radio waves about every second.

QUESTIONS

1 What is the name for a telescope that you look through with your eyes?

2 Give two reasons why you can see better in the dark with binoculars than just with your eyes.

3 Suggest what 'fainter objects' means.

... telescopes GCSE

Looking further

Earth-based telescopes

Over time, larger, more sophisticated telescopes, with complex systems of lenses, were built. Their greater magnification allows astronomers to make more and more detailed observations of the universe.

Observing the sky, from Earth, has problems:

> Moisture, in the atmosphere, absorbs and scatters some of the incoming radiation – on cloudy nights, optical telescopes are useless.

> Dust particles and pollution in Earth's atmosphere have the same effect.

> Weather patterns cause patches of air that are warmer or cooler than the surrounding air. Light is refracted by these patches. This distorts the image formed by a telescope.

All this makes extremely accurate observations, from ground-based telescopes, impossible. Telescopes are built at the tops of mountains, where the atmosphere is thinner, cleaner and drier.

Space telescopes

In 1990, the ***Hubble* telescope** was launched into orbit around Earth, outside the atmosphere. When it began transmitting back to Earth, the images were the clearest that astronomers had ever seen. It 'looks' deeper into space than ever before. Astronomers have seen new galaxies, many thousands of light-years away. *Hubble* sends back computer data that are converted to images on a computer screen. The colours that you see, on images from deep space, are not actual colours. Rather, they represent temperature, the intensity of radio waves or other electromagnetic radiation given off.

Hubble's successor, the *James Webb Space Telescope* (JWST), is due to be launched in 2014. It is expected to give even better images.

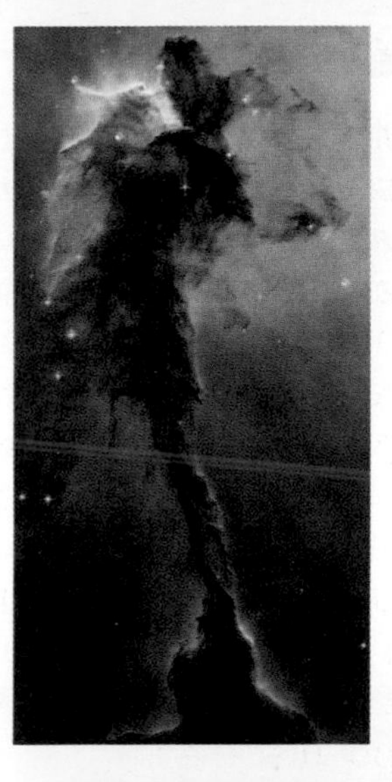

FIGURE 4: The fairy gas cloud, in the Eagle nebula, taken by the *Hubble* telescope.

QUESTIONS

4 Explain why many telescopes are built high up on top of mountains.

5 Explain why the *Hubble* telescope is so important.

6 Find out more about discoveries made with the *Hubble* space telescope.

Seeing the invisible

Objects in space emit radiation across the electromagnetic spectrum, but most wavelengths are scattered or absorbed and re-emitted by the atmosphere.

Astronomers can detect infrared radiation – their instruments compensate for infrared radiation from Earth and its atmosphere.

Astronomers can detect ultraviolet radiation, X-rays and gamma rays – telescopes are mounted on satellites orbiting above Earth's atmosphere.

TABLE 1: Using radiation to 'see' objects in space

Radiation	Objects 'seen' in space
gamma ray	neutron stars
X-ray	neutron stars
ultraviolet	hot stars, quasars
visible	stars
infrared	red giants
far infrared	protostars, planets
radio	pulsars

Did you know?

'Optical' comes from the Greek word *optikos* meaning *to see*.

QUESTIONS

7 Find out more, with an internet search, about telescopes that detect radiation from different parts of the electromagnetic spectrum. Putting 'telescope+electromagnetic+spectrum' into a search engine is a good place to start.

... radiation AND space GCSE

Waves and movement

You will find out:

> the observed wavelength and frequency of a wave change when the wave source is moving relative to an observer

Extreme weather warning

Knowing which way a severe storm is travelling is critical, if you are nearby. Meteorologists track a storm by sending out radio waves towards it. If the storm is approaching, the reflected waves have a shorter wavelength. If moving away, the wavelength increases. The data is converted into pictures of wind speed and direction – and you can download these, in real time, onto your mobile phone.

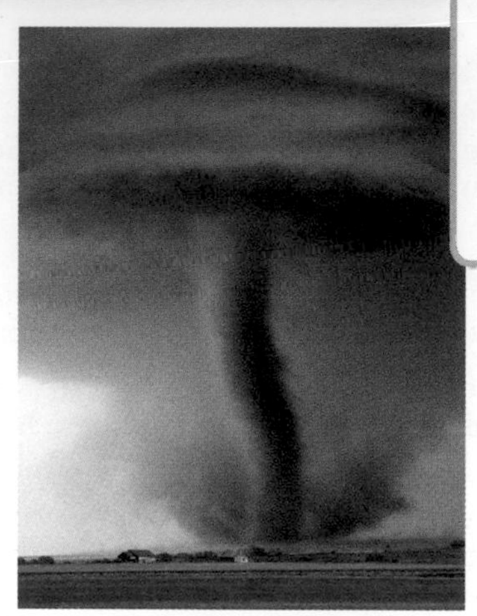

FIGURE 1: You may want to track this twister in case it is coming your way.

Changing sounds

The faster something vibrates, the higher the pitch of the sound it makes. High pitch means higher frequency of the sound waves.

A high frequency sound has more complete cycles passing a particular point, every second, than a lower frequency wave (Figure 2). For lots of complete cycles to go past a point in one second, each cycle must be very short. A high frequency sound has a short wavelength.

Moving sounds

Think about an emergency vehicle driving past you at high speed, with its siren on. The pitch of the siren seems to change as the vehicle goes past.

In Figure 3, sound waves travel out in all directions from the siren. When the siren is stationary, the waves are the same distance apart in every direction, so people at A and at B hear the same sound.

When the siren is moving, the waves in front of it are bunched up (Figure 4) – the source of the sound is always 'catching up' the sound it sent out earlier. The waves behind the siren are spread out – the source of the sound is always moving away from the sound it sent out earlier.

> As the siren gets closer, you hear sound with a *higher frequency*, and *shorter wavelength*, than the actual siren sound.

> As the siren moves away, you hear sound with a *lower frequency*, and *longer wavelength*, than the actual siren sound.

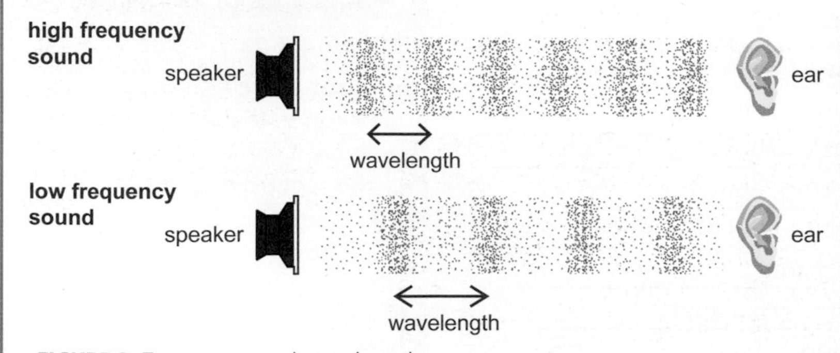

FIGURE 2: Frequency and wavelength.

FIGURE 3: When the siren is stationary, both people hear the same sound.

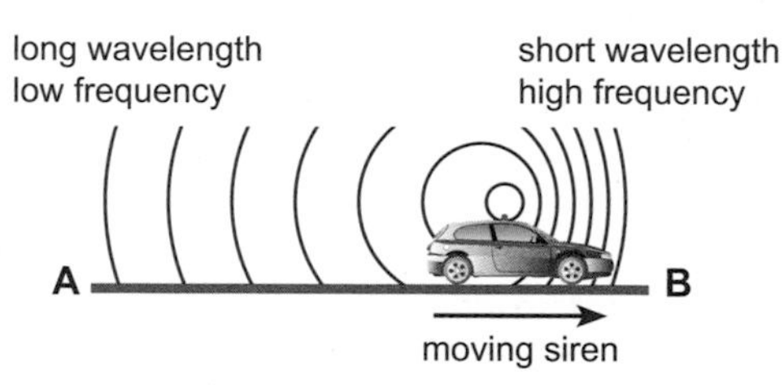

FIGURE 4: When the siren is moving, what will people at A and at B hear?

QUESTIONS

1 Explain why a low frequency sound has a longer wavelength.

2 When the speed of a wave does not change, increasing the frequency makes the wavelength shorter. Use the wave equation to explain why.

3 A stationary emergency vehicle, behind you, turns on its siren. As it moves off, the pitch of its siren sounds lower. How you can tell, without looking, whether it is moving away or towards you?

Q ... sound wavelength

Doppler and light

The observed change in frequency and wavelength, when the source of radiation is moving, is called the **Doppler effect**. It is true for all waves, not just sound. The effect is not usually noticeable for light sources – light waves travel so much faster than any light source on Earth that the difference is too tiny to notice.

Stars and galaxies may be moving through space at thousands of kilometres per second. That is fast enough to make a difference to the way we see the light coming from them.

If the light source were moving towards us, the waves would seem 'bunched up'. This makes the light look bluer (a shorter wavelength) than the light that is being emitted.

If the light source were moving away from us, the waves would seem 'spread out'. This makes the light look redder (a longer wavelength) than the light that is being emitted.

It should be simple to tell whether stars are moving towards or away from us, but it is not. Scientists cannot be sure what colour the light from each star 'really' is, so it is not easy to tell if it looks to have changed!

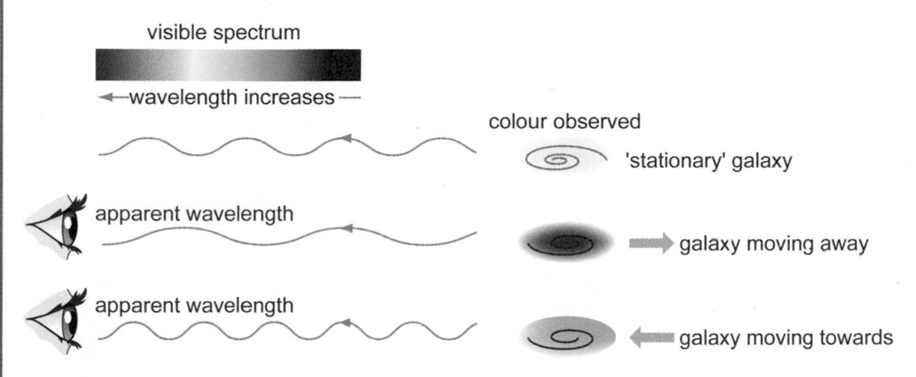

FIGURE 5: If things moved fast enough, their colour would appear to change.

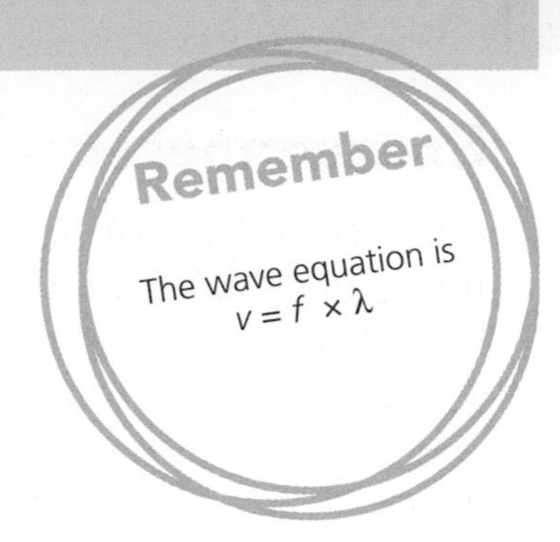

QUESTIONS

4 Explain what happens to light waves when their source is moving away from you.

5 The sound of an emergency vehicle siren changes as it drives past you, but the colour of the flashing light does not seem to change. Use the Doppler effect, and what you know about light and sound, to explain why.

Mystery object

Using a radio telescope, astronomers detect a very strange object. It sends out radio waves, but the frequency changes regularly. The frequency increases and then decreases on a regular cycle. Once in each cycle, the object seems to stop radiating for a brief period.

QUESTIONS

6 Use your knowledge of the Doppler effect to give a possible explanation for what the astronomers are observing. Hint: not all bodies in the universe give off radiation. There might be something present that astronomers cannot detect.

Did you know?

The Andromeda Galaxy is moving towards us. Scientists can tell because of its blue shift. As it is 2 500 000 light-years away, we will not collide for billions of years!

Origins of the universe

You will find out:

> light from distant galaxies is red-shifted
> how the red-shift and the cosmic microwave background radiation give evidence for the origin of the universe

Spectrometer

Astronomers use spectrometers to study radiation from stars. Radiation enters a 'telescope' and is focused onto a diffraction grating. This splits the light into a spectrum of different wavelengths (similar to a prism, but better). A second 'telescope' gives a magnified view of the spectrum produced, so that the light can be studied.

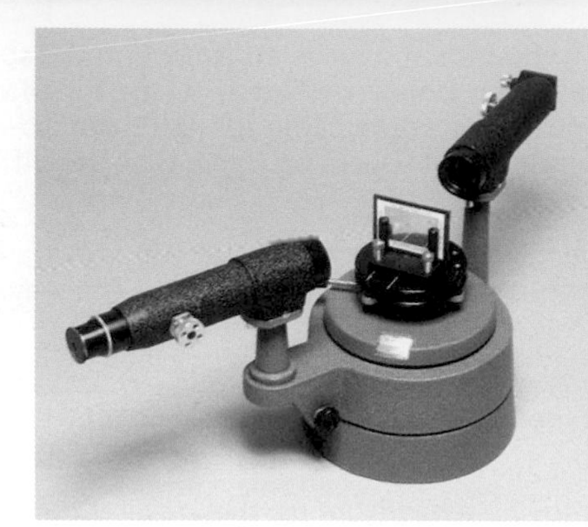

FIGURE 1: A simple spectrometer. Astronomers use more sophisticated ones to study radiation from stars.

Red-shift and Big Bang

The red-shift

Astronomers have used spectrometers to study the light from many thousands of galaxies. They found that galaxies radiate different amounts of different wavelengths of light. A **spectrometer** produces a spectrum. It has bright bands and dimmer bands. Each bright band is a wavelength that the galaxy radiates.

Edwin Hubble discovered that different galaxies produced the spectra. The bright and dimmer bands are in the same order and distance apart.

Hubble found that the pattern was sometimes shifted towards the red end of the visible spectrum. This is called the **red-shift**. For galaxies that were further away, the pattern was shifted further towards the red end of the visible spectrum – the red-shift was bigger.

These observations show that:

> All galaxies, in whatever direction we look, are moving away from us.

> The galaxies that are further away from us are travelling faster.

FIGURE 2: The redder the galaxy, the faster it is travelling away from Earth.

The Big Bang

The **Big Bang theory** is the most popular theory about the origin of the universe. It says that all galaxies are moving away from each other; everything in the universe started from a very small point and exploded rapidly outwards. What happened before the Big Bang? Scientists say simply, "There was no before".

Did you know?

'Dark matter' cannot be detected by telescopes because it does not emit radiation. Scientists think that 'dark matter' makes up 80% of the universe.

QUESTIONS

1 Name the instrument that astronomers use, to look at the spectra from stars.

2 Explain why a red shift tells astronomers about a galaxy.

3 Suggest what astronomers would think, if they found a galaxy with a blue shift.

CMBR and the Big Bang theory

Cosmic microwave background radiation

Seen through an optical telescope, the region between stars and galaxies looks completely black. Look with radio telescopes and a very faint glow of microwave radiation, coming from everywhere in space, can be detected. This is the **cosmic microwave background radiation (CMBR)**. Scientists think it is the 'glow' of radiation left over from the Big Bang. No other theory has managed to explain why there is a cosmic microwave background radiation.

Theory or fact?

While scientists know the universe is expanding, they do not know how it started. There are many theories. At the moment, the Big Bang theory best explains the observations that scientists have made. So far, it is the only one that explains CMBR.

The theory says that, shortly after the Big Bang, there was a super-hot 'fog' of plasma. The plasma filled all of, what was then, a much smaller universe. As it cooled, most of this plasma formed into atoms, then stars and galaxies. Some was left over. Once stable atoms had formed, they could no longer absorb this left-over radiation. This radiation is what astronomers 'see' today as the CMBR.

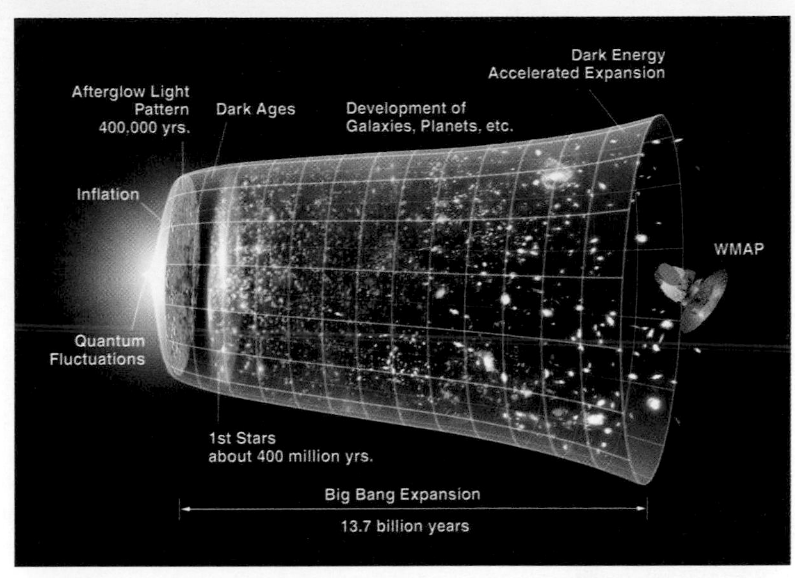

FIGURE 3: Scientists think that the universe may have evolved like this.

QUESTIONS

4 Describe what is meant by the cosmic microwave background radiation.

5 Explain why the Big Bang is a theory and not fact.

When will the universe stop expanding?

The simple answer is, no one knows. Galaxies hurtle outwards because of the enormous energy they received at the Big Bang, just as matter hurtles outwards from any explosion. Yet, the galaxies are also attracted towards each other by the force of gravity. If this force is big enough, it will eventually slow down all the galaxies and cause them to move towards each other again.

Scientists do not know exactly how strong the pull due to gravity is. The force depends on how much matter there is in the universe. Scientists do not know this – they cannot 'see' all matter because it does not all give off radiation.

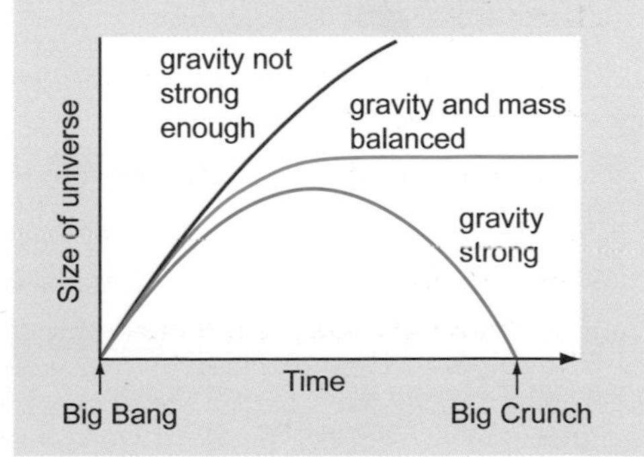

FIGURE 4: Different fates for the universe.

QUESTIONS

6 Imagine there was a new theory that claimed to explain the origins of the universe, better than the Big Bang theory. Discuss the process that scientists would go through, to decide whether or not to adopt the new theory.

Checklist P1.4–1.5

To achieve your forecast grade in the exam you will need to revise

Use this checklist to see what you can do now. Refer back to the relevant topic in this book if you are not sure. Look across the three columns to see how you can progress.

Remember that you will need to be able to use these ideas in various ways, such as:

> interpreting pictures, diagrams and graphs

> applying ideas to new situations

> explaining ethical implications

> suggesting some benefits and risks to society

> drawing conclusions from evidence that you are given.

Look at pages 276–297 for more information about exams and how you will be assessed.

To aim for a grade E	To aim for a grade C	To aim for a grade A
Describe energy sources that heat water in a power station. Explain how power stations use steam to generate electricity. Explain why fossil fuels are non-renewable.	Understand how electricity is generated from fossil and nuclear fuels, biofuel and biomass. Describe how different types of power station have differing efficiency.	
Understand what renewable energy resources are. Explain that water and wind can drive turbines directly.	Describe how wind, wave, geothermal and solar power can be used to generate electricity. Explain the energy transfers that take place at a power station.	Describe the advantages and disadvantages of solar cells and of small-scale electricity production.
Describe useful and harmful effects of different ways to generate electricity.	Explain some of the effects that generating power has on society and the environment. Describe ways that carbon capture may be put into practice.	
Describe some of the factors that affect the choice of energy source for power stations.	Understand the factors that affect the choice of power station. Explain how power generation matches supply with demand.	Evaluate different ways of generating electricity. Evaluate different ways of matching supply with demand.
Describe how electricity is distributed from power stations to consumers via the National Grid. Name the essential parts of the National Grid.		
Recall that step-up and step-down transformers change the voltages along the National Grid.	Explain how and why the National Grid uses high voltages to distribute electricity.	Evaluate the use of overhead power lines and underground cables to distribute electricity.
Recall that waves transfer energy. Describe examples of transverse and longitudinal waves.	Understand that electromagnetic waves are transverse, sound waves are longitudinal and mechanical waves may be either. Understand that energy transfer is perpendicular to the oscillations in a transverse wave and parallel in a longitudinal wave. Explain the meaning of compression and rarefaction.	

To aim for a grade E	To aim for a grade C	To aim for a grade A
Recall that mechanical waves only travel through a medium. Explain the terms frequency, wavelength and amplitude.	Recognise how the wave equation connects wavelength, frequency and energy.	
Recall that all electromagnetic waves travel at the same speed through a vacuum. Describe electromagnetic waves as a continuous spectrum.	Describe the order of electromagnetic waves within the spectrum, in terms of energy, frequency and wavelength.	
Describe uses of electromagnetic waves.	Describe hazards of the radiations in different parts of the electromagnetic spectrum.	Evaluate the hazards of radiations in the electromagnetic spectrum.
Recall that waves can be reflected and diffracted. Construct ray diagrams using the normal and the angle of incidence equal to the angle of reflection. Describe the image produced in a plane mirror.	Explain how waves change direction when they are reflected, refracted or diffracted. Construct ray diagrams to show that the image produced in a plane mirror is virtual, upright and laterally inverted.	Understand that waves are not refracted if they are travelling along the normal.
Recall that sound waves are longitudinal and cannot travel through a vacuum.	Understand that sound waves cause vibrations in a medium, which are detected as sound.	
Describe sounds using amplitude and frequency. Explain that echo is reflected sound.	Understand how a wave's wavelength and frequency affect the energy transferred.	Understand how sounds can be compared using oscilloscope traces.
Recall that EM waves are used for communication. Describe a fibre optic cable.	Describe ways that EM waves are used for communications. Compare the use of different types of waves for communication. Describe how a fibre optic cable transmits a signal.	
Describe the change in the wavelength and frequency when a wave source is moving relative to an observer.	Explain the Doppler effect and how it can be used to detect whether an object is moving away from or towards an observer.	Interpret changes in observed wavelength and frequency due to the Doppler effect.
Describe red-shift as the increase in wavelength of light from distant galaxies.	Understand that the farther away a galaxy, the faster it is moving, and the bigger the observed increase in wavelength – red-shift.	
Explain that red-shift provides evidence that the universe is expanding and supports the Big Bang theory.	Explain that the cosmic microwave background radiation (CMBR) gives further evidence for the origin of the universe.	Explain the significance of CMBR. Understand that the Big Bang theory is currently the only theory that explains CMBR.

Exam-style questions: Foundation P1.4–1.5

In the examination, equations will be given on a separate equation sheet.
Write down the equation that you will use. Show clearly how you work out your answer.

AO1 **1.** What is the wavelength and amplitude of the wave shown in the diagram? [2]

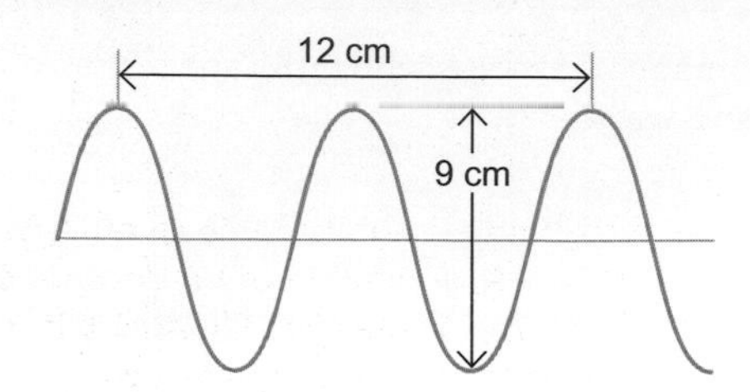

AO1 **2.** This is a list of different types of waves (in alphabetic order):

gamma, infrared, microwave, radio, sound, ultraviolet, visible, water, X–ray

(a) Write down any that are not in the electromagnetic spectrum. [1]

(b) Write the parts of the electromagnetic spectrum, in order of increasing wavelength. [2]

(c) Write the parts of the electromagnetic spectrum, in order of increasing frequency. [2]

(d) Explain why the electromagnetic spectrum cannot be written in order of increasing velocity. [1]

AO2 **3.** The diagram shows a light ray reflecting from a mirror.

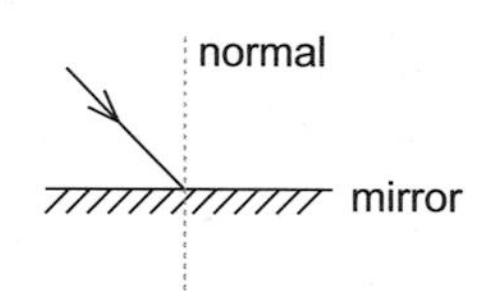

Complete the diagram and label the following:

incident ray, reflected ray, angle of incidence, angle of reflection [2]

AO3 **4.** A singer sings a note with a small amplitude and a long wavelength.

How would the note be heard?
Select one of these options:

(i) low and loud
(ii) low and quiet
(iii) high and loud
(iv) high and quiet [1]

5. (a) Fossil fuels are called non-renewable fuels.

AO1 **(i)** Name three fossil fuels. [1]

AO2 **(ii)** Explain why fossil fuels are called non-renewable fuels. [2]

(b) Some fuels come from renewable sources.

AO1 **(i)** Name three renewable sources of energy. [1]

AO2 **(ii)** Explain why these sources are called renewable. [2]

AO2 **6.** A transverse wave and a longitudinal wave each oscillates in a North–South direction.

(i) In which direction does the transverse wave move? [1]

(ii) In which direction does the longitudinal wave move? [1]

AO1 **7.** Which flowchart in Table 1 below shows the correct sequence for an electricity transmission system? [2]

Table 1

(i)	power station	→	step–down transformer	→	power lines	→	step–up transformer	→	electricity consumers
(ii)	step–up transformer	→	power station	→	step–down transformer	→	power lines	→	electricity consumers
(iii)	power station	→	step–up transformer	→	power lines	→	step–down transformer	→	electricity consumers

WORKED EXAMPLE – Foundation tier

This list shows different sources of energy (in alphabetical order):

biomass (trees and crops), coal, geothermal, hydroelectric, natural gas, nuclear, oil, solar, tidal, wave, wind

(a) Explain why electricity is not included in the list. [2]

This is a type of energy.

(b) Name the three sources of energy that depend on water. Explain why these are renewable energy sources. [4]

Hydroelectric Tidal Wave

They are renewable because the water is not used up.

(c) Complete this table, giving **one** advantage and **one** disadvantage for each of the energy resources: fossil fuels, nuclear, wind, solar, wave, hydroelectric. [6]

Source of Energy	Advantage	Disadvantage
fossil fuels	*Small amounts give a lot of energy.*	*Produce greenhouse gas.*
nuclear	*Does not produce greenhouse gas.*	*Produces radioactive waste.*
wind	*No fuel costs.*	*Does not work if there is no wind.*
solar	*No fuel costs. No pollution.*	*Does not work at night.*
wave	*No fuel costs. No pollution.*	*Needs waves of enough power.*
hydroelectric	*No fuel costs.*	*Needs to be where a dam can be built. Rotting vegetation produces greenhouse gas.*

How to raise your grade!

Take note of these comments – they will help you to raise your grade.

Electricity is not a source of energy. However, it should not be described as a 'type of energy'. Electricity is a means of transferring energy.

This is correct. The answer might have added that it is the movement of the water that is used to generate electricity. It might also have said that the supply of water and its movement will not run out.

If a question asks you to 'name' or 'state' something, you only need to write the answer. If it asks you to explain or justify something, include why you chose your answer.

The question only asks for one advantage and disadvantage, so that is all that the candidate should have written down.

If you can think of more than one advantage or disadvantage, choose the best one. If you write more than asked for and get them wrong, you will gain no marks. Try not to use the same answer in all the boxes.

Exam-style questions: Higher P1.4–1.5

In the examination, equations will be given on a separate equation sheet.
Write down the equation that you will use. Show clearly how you work out your answer.

AO2 **1.** The equation linking the velocity, wavelength and frequency of a wave is:

$v = f \times \lambda$

Complete the table, calculating the velocity, frequency or wavelength for different waves. [3]

Distance (m)	Time (s)	Velocity (m/s)
12	4	
	4.5	3
4		10
42	14	
39.6		13.2
	1.5	6

AO2 **2.** The diagram shows an object (an upright arrow) in front of a plane mirror.

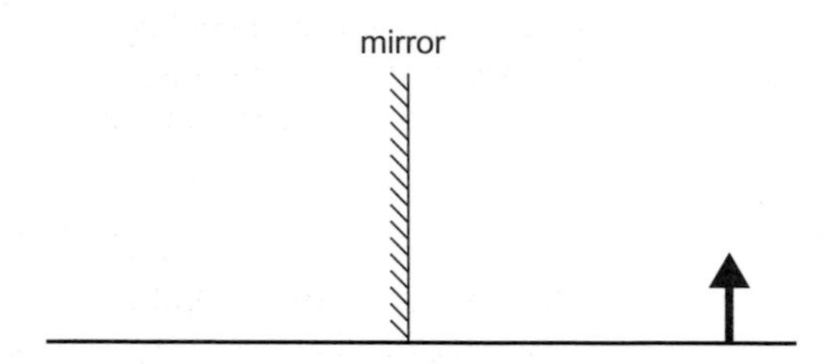

Draw a ray diagram to show:

(i) whether the image is upright or inverted

(ii) whether the image is real or virtual

(iii) the size of the image compared with the object

(iv) the distance of the image, behind the mirror, compared with the distance of the object in front of the mirror. [6]

AO3 **3. (a)** The spectrum of the light from distant galaxies shows that the light has been shifted to longer wavelengths (it has been red-shifted).

Does this suggest that the universe is expanding, contracting or not moving? Give your reasons. [2]

AO3 **(b)** The light from the Andromeda Galaxy (which is the nearest large galaxy to our Milky Way galaxy) is blue-shifted.

What does this suggest?

What might eventually happen to the two galaxies? [2]

AO3 **(c)** Student A says, "Red-shift means the universe is expanding, so red-shift proves the Big Bang happened".

Student B says, "Red-shift doesn't prove that the Big Bang happened".

Explain whose statement is correct. [2]

AO2 **4.** *In this question you will be assessed on using good English, organising information clearly and using specialist terms where appropriate.*

It is possible for a householder to install equipment to produce their electricity, for example solar cells or a wind-powered generator.

Outline the advantages and disadvantages for a householder in the UK of using wind or solar power as energy sources to produce their electricity. [6]

AO1 recall the science AO2 apply your knowledge AO3 evaluate and analyse the evidence

WORKED EXAMPLE – Higher tier

(a) Here are various components used in the National Grid (which is used to transfer electricity to our homes): **overhead power lines, power station, step-down transformer, step-up transformer.**

Complete the table by putting the components in the correct order and give the main function of each. [2]

Component	Main Function
Power station	*Generate electricity*
Step-down transformer	*Reduce current*
Overhead power lines	*Transmit electricity*
Step-up transformer	*Increase the current*

(b) When a step-up transformer is working, which of the following does it do?

(i) increase the current and increase the power

(ii) increase the current and increase the voltage

(iii) increase the voltage and decrease the power

(iv) increase the voltage and decrease the current [1]

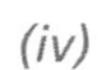

(iv)

(c) Apart from using overhead power lines, name one other method that could be used to transmit electricity from one location to another, and give one advantage and one disadvantage for this method. [2]

Using underground cables – Advantage (does not disfigure areas of natural beauty), Disadvantage (Expensive).

(d) To reduce energy loss, a step-up transformer is used to raise the voltage to a high value before electricity to carried from the power station to where it is needed. A step-down transformer is then used before the electricity is distributed to domestic houses. Explain why this method reduces energy waste by heating. [3]

A step-up transformer increases the voltage but it also decreases the current.

One of the main energy losses in electricity transmission wires is in heating the cables.

Heat is linked to the current – the higher the current, the more it heats up the wire.

How to raise your grade!

Take note of these comments – they will help you to raise your grade.

Step-up and step-down refers to voltage, not current. The transformers have been included in the wrong order.

This is correct. In an efficient transformer, the power is unchanged, so as the voltage increases, the current will decrease by a corresponding amount.

This answer is correct, but the advantage and disadvantage should not be written in brackets.

This answer is not complete. The candidate should explain that the low current heats up the cables less. Energy waste through heating is therefore reduced by using a high voltage in the National Grid.

Carrying out practical investigations in GCSE Science

Introduction

As part your GCSE Science course, you will develop practical skills and have to carry out investigative work in science.

Your investigative work will be divided into several parts:

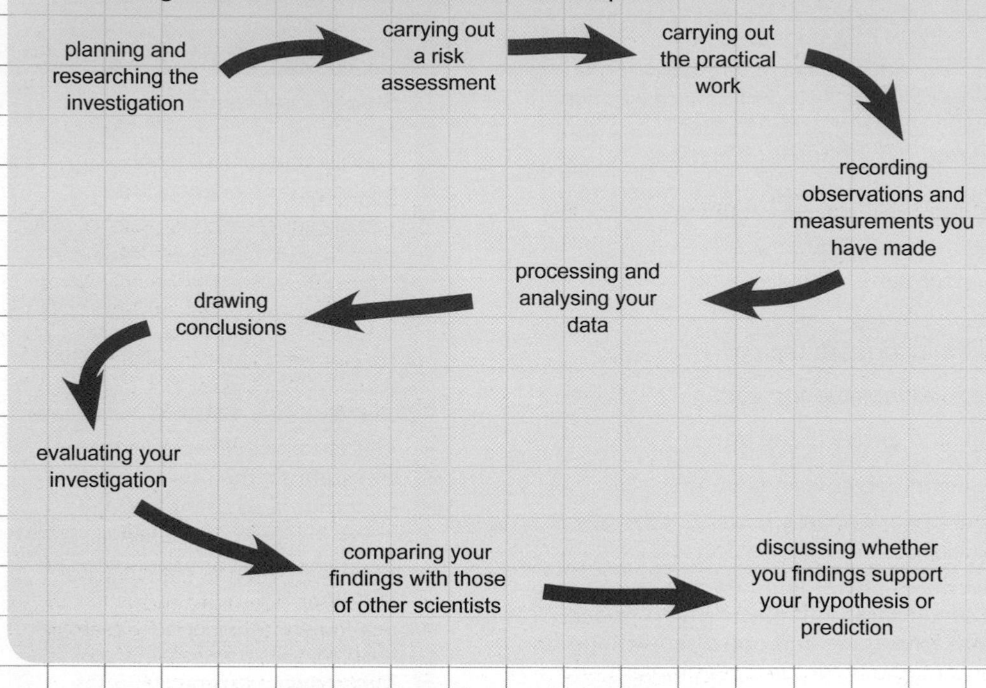

Planning and researching your investigation

A scientific investigation usually begins with a scientist testing an idea, answering a question, or trying to solve a problem.

You first have to plan how you will carry out the investigation.

Your planning will involve testing a **hypothesis**. For example, you might observe during a fermentation with yeast investigation that beer or wine is produced faster at a higher temperature.

So your hypothesis might be 'as the temperature increases, the rate of fermentation increases'.

In Science, you will be given a hypothesis to test. In Additional Science, or if you are doing Separate Sciences, you will have to produce or formulate a hypothesis.

To formulate a hypothesis you may have to research some of the background science.

First of all, use your lesson notes and your textbook. The topic you have been given to investigate will relate to the science that you have learned in class.

Also make use of the internet, but make sure that your internet search is closely focused on the topic that you are investigating.

Assessment tip

You will be given a hypothesis to investigate as part of your controlled assessment.

Definition

A **hypothesis** is a possible explanation that someone suggests to explain some scientific observations.

Assessment tip

If you are formulating a hypothesis, it is important that it's testable. In other words, you must be able to test the hypothesis in the school lab.

Assessment tip

In the planning stage, scientific research is important if you are going to obtain higher marks.

- The search terms you use on the internet are very important. 'Investigating fermentation' is a better search term than just 'fermentation', as it's more likely to provide links to websites that are more relevant to your investigation.
- The information on websites also varies in its reliability. Free encyclopaedias often contain information that hasn't been written by experts. Some question and answer websites might appear to give you the exact answer to your question, but be aware that they may sometimes be incorrect.
- Most GCSE Science websites are more reliable but, if in doubt, use other information sources to verify the information.

As a result of your research, you may be able to extend your hypothesis and make a **prediction** that's based on science.

Example 1

Investigation: Plan and research an investigation into the activity of enzymes

Your hypothesis might be 'when I increase the temperature, the rate of reaction increases'.

You may be able to add more detail, 'this is because as I increase the temperature, the frequency of collisions between the enzyme and the reactant increases'.

Assessment tip

Make sure that you make a detailed note of which sources you have used: for a book, the author's name and the title; for a website, the name of it.

Assessment tip

Higher tier

You are expected to be able to balance chemical equations. So, for example, if the enzyme is being used to decompose hydrogen peroxide to water and oxygen you should be able to balance the equation: $H_2O_2 \rightarrow H_2O + O_2$ to give $2H_2O_2 \rightarrow 2H_2O + O_2$

Choosing a method and suitable apparatus

As part of your planning, you must choose a suitable way of carrying out the investigation.

You will have to choose suitable techniques, equipment and technology, if this is appropriate. How do you make this choice?

You will have already carried out the techniques you need to use during the course of practical work in class (although you may need to modify these to fit in with the context of your investigation). For most of the experimental work you do, there will be a choice of techniques available. You must select the technique:

- that is most appropriate to the context of your investigation, and
- that will enable you to collect valid data, for example if you are measuring the effects of light intensity on photosynthesis, you may decide to use an LED (light-emitting diode) at different distances from the plant, rather than a light bulb. The light bulb produces more heat, and temperature is another independent variable in photosynthesis.

Your choice of equipment, too, will be influenced by the measurements that you need to make. For example:

- you might use a one-mark or graduated pipette to measure out the volume of liquid for a titration, but
- you may use a measuring cylinder or beaker when adding a volume of acid to a reaction mixture, so that the volume of acid is in excess to that required to dissolve, for example, the calcium carbonate.

Assessment tip

Technology, such as data-logging and other measuring and monitoring techniques, for example heart sensors, may help you to carry out your experiment.

Definition

The **resolution** of the equipment refers to the smallest change in a value that can be detected using a particular technique.

Assessment tip

Carrying out a preliminary investigation, along with the necessary research, may help you to select the appropriate technique to use.

Variables

In your investigation, you will work with independent and dependent variables.

The factors you choose, or are given, to investigate the effect of are called **independent variables**.

What you choose to measure, as affected by the independent variable, is called the **dependent variable**.

Independent variables

In your practical work, you will be provided with an independent variable to test, or will have to choose one – or more – of these to test. Some examples are given in the table.

Investigation	Possible independent variables to test
activity of yeast	> temperature > sugar concentration
rate of a chemical reaction	> temperature > concentration of reactants
stopping distance of a moving object	> speed of the object > the surface on which it's moving

Independent variables can be **categoric** or **continuous**.

> When you are testing the effect of different disinfectants on bacteria you are looking at categoric variables.

> When you are testing the effect of a range of concentrations of the same disinfectant on the growth of bacteria you are looking at continuous variables.

Range

When working with an independent variable, you need to choose an appropriate **range** over which to investigate the variable.

You need to decide:

✔ which treatments you will test, and/or
✔ the upper and lower limits of the independent variables to investigate, if the variable is continuous.

Once you have defined the range to be tested, you also need to decide the appropriate intervals at which you will make measurements.

The range you would test depends on:

✔ the nature of the test
✔ the context in which it is given
✔ practical considerations, and
✔ common sense.

Definition

Variables that fall into a range of separate types are called **categoric variables**.

Definition

Variables that have a continuous range are called **continuous variables**.

Definition

The **range** defines the extent of the independent variables being tested.

Example 2

1 Investigation: Investigating the factors that affect how quickly household limescale removers work in removing limescale from an appliance

You may have to decide on which acids to use from a range that you are provided with. You would choose a weak acid, or weak acids, to test, rather than a strong acid, such as concentrated sulfuric acid. This is because of safety reasons, but also because the acid might damage the appliance you were trying to clean. You would then have to select a range of concentrations of your chosen weak acid to test.

2 Investigation: How speed affects the stopping distance of a trolley in the lab

The range of speeds you would choose would clearly depend on the speeds that you could produce in the lab.

Concentration

You might be trying to find out the best, or optimum, concentration of a disinfectant to prevent the growth of bacteria.

The 'best' concentration would be the lowest in a range that prevented the growth of the bacteria. Concentrations higher than this would be just wasting disinfectant.

If, in a preliminary test, no bacteria were killed by the concentration you used, you would have to increase it (or test another disinfectant). However, if there was no growth of bacteria in your preliminary test, you would have to lower the concentration range. A starting point might be to look at concentrations around those recommended by the manufacturer.

Assessment tip

Again, it is often best to carry out a trial run or preliminary investigation, or carry out research, to determine the range to be investigated.

Dependent variables

The dependent variable may be clear from the problem that you are investigating, for example the stopping distance of moving objects. You may have to make a choice.

Example 3

1 Investigation: Measuring the rate of photosynthesis in a plant

There are several ways in which you could measure the rate of photosynthesis in a plant. These include:

- counting the number of bubbles of oxygen produced in a minute by a water plant such as *Elodea* or *Cabomba*
- measuring the volume of oxygen produced over several days by a water plant such as *Elodea* or *Cabomba*
- monitoring the concentration of oxygen in a polythene bag enclosing a potted plant, using a carbon dioxide sensor
- measuring the colour change of hydrogencarbonate indicator that contains algae embedded in gel.

Assessment tip

The value of the ***depend***ent variable is likely to ***depend*** on the value of the independent variable. This is a good way of remembering the definition of a dependent variable.

2 Investigation: Measuring the rate of a chemical reaction

You could measure the rate of a chemical reaction in the following ways:

- the rate of formation of a product

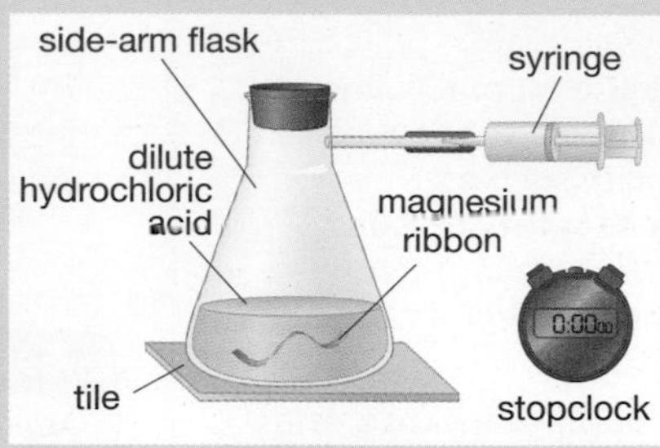

- the rate at which the reactant disappears

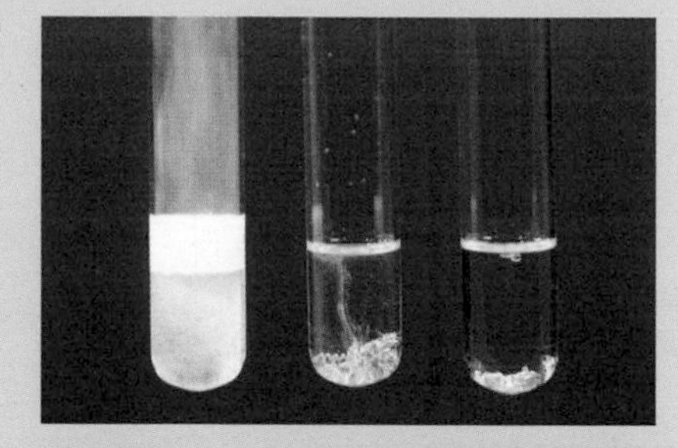

- a colour change

- a pH change.

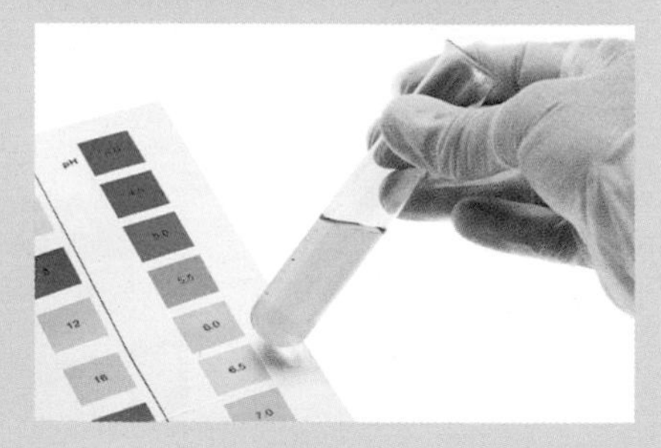

Control variables

The validity of your measurements depend on you measuring what you are supposed to be measuring.

Some of these variables may be difficult to control. For example, in an ecology investigation in the field, factors such as varying weather conditions are impossible to control.

Experimental controls

Experimental controls are often very important, particularly in biological investigations where you are testing the effect of a treatment.

Definition

Other variables that you are not investigating may also have an influence on your measurements. In most investigations, it is important that you investigate just one variable at a time. So other variables, apart from the one you are testing at the time, must be controlled, and kept constant, and not allowed to vary. These are called **control variables**.

Definition

An **experimental control** is used to find out whether the effect you obtain is from the treatment, or whether you get the same result in the absence of the treatment.

Example 4

Investigation: The effect of disinfectants on the growth of bacteria

If the bacteria do not grow, it could be because they have been killed by the disinfectant. The bacteria in your investigation may have died for some other reason. Another factor may be involved. To test whether any effects were down to the disinfectant, you need to set up the same practical, but this time using distilled water in place of the disinfectant. The distilled water is your control. If the bacteria are inhibited by the disinfectant, but grow normally in the dish containing distilled water, it is reasonable to assume that the disinfectant inhibited their growth.

Assessing and managing risk

Before you begin any practical work, you must assess and minimise the possible risks involved.

Before you carry out an investigation, you must identify the possible hazards. These can be grouped into biological hazards, chemical hazards and physical hazards.

Biological hazards include:

> microorganisms
> body fluids
> animals and plants.

Chemical hazards can be grouped into:

> irritant and harmful substances
> toxic
> oxidising agents
> corrosive
> harmful to the environment.

Physical hazards include:

> equipment
> objects
> radiation.

Scientists use an international series of symbols so that investigators can identify hazards.

Hazards pose risks to the person carrying out the investigation.

A risk posed by concentrated sulfuric acid, for example, will be lower if you are adding one drop of it to a reaction mixture to make an ester, than if you are mixing a large volume of it with water.

When you use hazardous materials, chemicals or equipment in the laboratory, you must use them in such a way as to keep the risks to absolute minimum. For example, one way is to wear eye protection when using hydrochloric acid.

Any action that you carry out to reduce the risk of a hazard happening is known as a 'control measure'.

Definition

A **hazard** is something that has the potential to cause harm. Even substances, organisms and equipment that we think of being harmless, used in the wrong way, may be hazardous.

Hazard symbols are used on chemical bottles so that hazards can be identified.

Definition

The **risk** is the likelihood of a hazard causing harm in the circumstances it's being used in.

Risk assessment

Before you begin an investigation, you must carry out a risk assessment. Your risk assessment must include:

- ✔ all relevant hazards (use the correct terms to describe each hazard, and make sure you include them all, even if you think they will pose minimal risk)
- ✔ risks associated with these hazards
- ✔ ways in which the risks can be minimised
- ✔ results of research into emergency procedures that you may have to take if something goes wrong.

You should also consider what to do at the end of the practical. For example, used agar plates should be left for a technician to sterilise; solutions of heavy metals should be collected in a bottle and disposed of safely.

Assessment tip

To make sure that your risk assessment is full and appropriate:

> Remember that, for a risk assessment for a chemical reaction, the risk assessment should be carried out for the products and the reactants.

> When using chemicals, make sure the hazard and ways of minimising risk match the concentration of the chemical you are using; many acids, for instance, while being corrosive in higher concentrations, are harmful or irritant at low concentrations.

Collecting primary data

- ✔ You should make sure that observations, if appropriate, are recorded in detail. For example, it is worth recording the appearance of your potato chips in your osmosis practical, in addition to the measurements you make.
- ✔ Measurements should be recorded in tables. Have one ready so that you can record your readings as you carry out the practical work.
- ✔ Think about the dependent variable and define this carefully in your column headings.
- ✔ You should make sure that the table headings describe properly the type of measurements that you have made, for example 'time taken for magnesium ribbon to dissolve'.
- ✔ It is also essential that you include units – your results are meaningless without these.
- ✔ The units should appear in the column head, and not be repeated in each row of the table.

Definition

When you carry out an investigation, the data you collect are called **primary data.** The term 'data' is normally used to include your observations as well as measurements you might make.

Definition

One set of results from your investigation may not reflect what truly happens. Carrying out repeats enables you to identify any results that don't fit. These are called **outliers** or **anomalous results**.

Definition

If, when you carry out the same experiment several times, and get the same, or very similar results, the results are **repeatable**.

Repeatability and reproducibility of results

When making measurements, in most instances, it is essential that you carry out repeats.

These repeats are one way of checking your results.

Results will not be repeatable of course, if you allow the conditions the investigation is carried out in to change.

You need to make sure that you carry out sufficient repeats, but not too many. In a titration, for example, if you obtain two values that are within 0.1 cm^3 of each other, carrying out any more will not improve the reliability of your results.

This is particularly important when scientists are carrying out scientific research and make new discoveries.

Definition

Taking more than one set of results will improve the **reliability** of your data.

Definition

The **reproducibility** of data is the ability of the results of an investigation to be reproduced by:

> using a different method and reaching the same conclusion

> someone else, who may be in a different lab, carrying out the same work.

Processing data

Calculating the mean

Using your repeat measurements you can calculate the arithmetical mean (or just 'mean') of these data. Often, the mean is called the 'average.'

Temperature (°C)	Number of yeast cells (mm^3)			Mean number of yeast cells (mm^3)
	Test 1	Test 2	Test 3	
10	1000	1040	1200	1080
20	2400	2200	2300	2300
30	4600	5000	4800	4800
40	4800	5000	5200	5000
50	200	1200	700	700

Definition

The **mean** is calculated by adding together all the measurements, and dividing by the number of measurements.

You may also be required to use formulae when processing data. Sometimes, these will need rearranging to be able to make the calculation you need. Practise using and rearranging formulae as part of your preparation for assessment.

Significant figures

When calculating the mean, you should be aware of significant figures.

For example, for the set of data below:

18	13	17	15	14	16	15	14	13	18

The total for the data set is 153, and ten measurements have been made. The mean is 15, and not 15.3.

This is because each of the recorded values has two significant figures. The answer must therefore have two significant figures. An answer cannot have more significant figures than the number being multiplied or divided.

Definition

Significant figures are the number of digits in a number based on the precision of your measurements.

Using your data

When calculating means (and displaying data), you should be careful to look out for any data that do not fit in with the general pattern.

It might be the consequence of an error made in measurement. But sometimes outliers are genuine results. If you think an outlier has been introduced by careless practical work, you should ignore it when calculating the mean. But you should examine possible reasons carefully before just leaving it out.

Definition

An **outlier** (or **anomalous result**) is a reading that is very different from the rest.

Displaying your data

Displaying your data – usually the means – makes it easy to pick out and show any patterns. It also helps you to pick out any anomalous data.

It is likely that you will have recorded your results in tables, and you could also use additional tables to summarise your results. The most usual way of displaying data is to use graphs. The table will help you decide which type to use.

Type of graph	When you would use the graph	Example
bar chart or bar graph	where one of the variables is categoric	'the diameters of the clear zones where the growth of bacteria was inhibited by different types of disinfectant'
line graph	where independent and dependent variables are both continuous	'the volume of carbon dioxide produced by a range of different concentrations of hydrochloric acid'
scatter graph	to show an association between two (or more) variables	'the association between length and breadth of a number of privet leaves' In scatter graphs, the points are plotted, but not usually joined.

If it is possible from the data, join the points of a line graph using a straight line, or in some instances, a curve. In this way, graphs can also help you to process data.

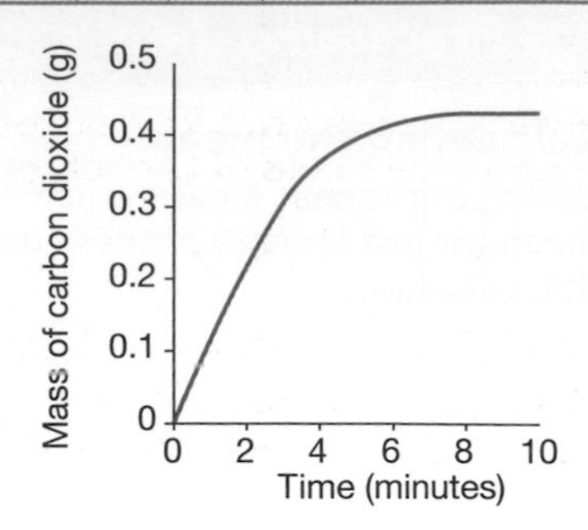

You can calculate the rate of production of carbon dioxide from the gradient of the graph.

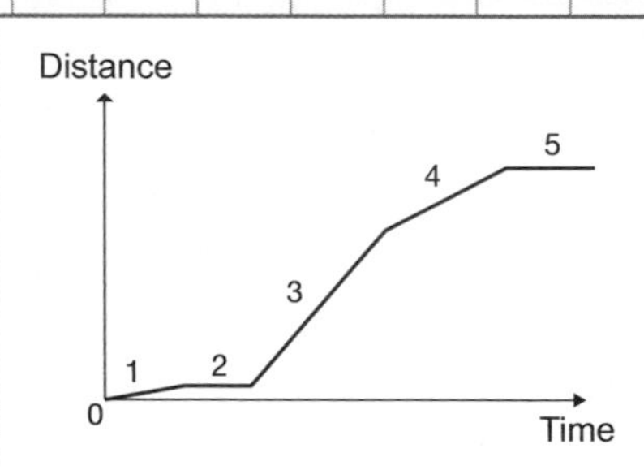

You can calculate the speed of the object from the gradient of the graph.

Assessment tip

Remember, when drawing graphs, plot the independent variable on the *x*-axis, and the dependent variable on the *y*-axis.

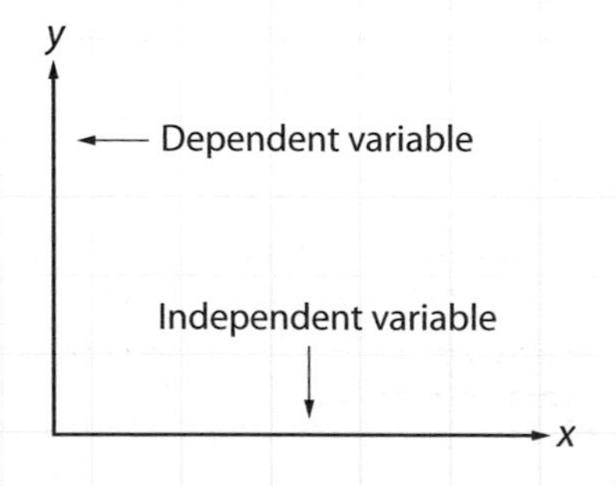

Conclusions from differences in data sets

When comparing two (or more) sets of data, you can often compare the values of two sets of means.

Example 5

Investigation: Comparing the effectiveness of two disinfectants

Two groups of students compared the effectiveness of two disinfectants, labelled A and B. Their results are shown in the table.

Disinfectant	Diameter of zone of inhibition (clear zone) (mm)										Mean dia (mm)
	1	2	3	4	5	6	7	8	9	10	
A	15	13	17	15	14	16	15	14	13	18	15
B	25	23	24	23	26	27	25	24	23	22	24

When the means are compared, it appears that disinfectant B is more effective in inhibiting the growth of bacteria. Can you be sure? The differences might have resulted from the treatment of the bacteria using the two disinfectants. But the differences could have occurred purely by chance.

Scientists use statistics to find the probability of any differences having occurred by chance. The lower this probability is, which is found out by statistical calculations, the more likely it is that it was (in this case) the disinfectant that caused the differences observed.

Statistical analysis can help to increase the confidence you have in your conclusions.

Assessment tip

You have learned about probability in maths lessons.

Definition

If there is a relationship between dependent and independent variables that can be defined, there is a **correlation** between the variables.

Drawing conclusions

Observing trends in data or graphs will help you to draw conclusions. You may obtain a linear relationship between two sets of variables, or the relationship might be more complex.

Example 6

Conclusion: The higher the concentration of acid, the shorter the time taken for the magnesium ribbon to dissolve.

Conclusion: The higher the concentration of acid, the faster the rate of reaction.

When drawing conclusions, you should try to relate your findings to the science involved.

> In the first investigation in Example 6, your discussion should focus on the greater possibility/increased frequency of collisions between reacting particles as the concentration of the acid is increased.

> In the second investigation in Example 6, there is a clear scientific mechanism to link the rate of reaction to the concentration of acid.

Sometimes, you can see correlations between data which are coincidental, where the independent variable is not the cause of the trend in the data.

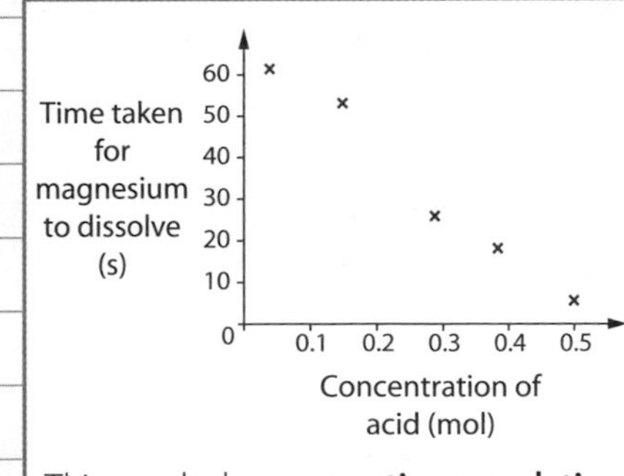

This graph shows **negative correlation**

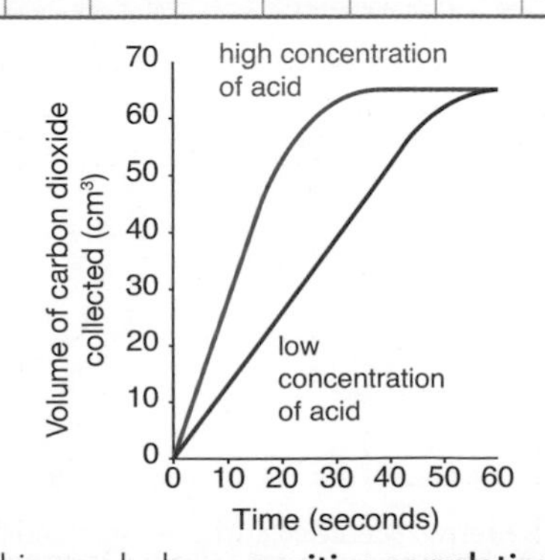

This graph shows **positive correlation**

Example 7

Studies have shown that levels of vitamin D are very low in people with long-term inflammatory diseases. But there's no scientific evidence to suggest that these low levels are the cause of the diseases.

Evaluating your investigation

Your conclusion will be based on your findings, but must take into consideration any uncertainty in these introduced by any possible sources of error. You should discuss where these have come from in your evaluation.

The two types of errors are:

- ✔ random error
- ✔ systematic error.

This can occur when the instrument that you are using to measure lacks sufficient sensitivity to indicate differences in readings. It can also occur when it is difficult to make a measurement. If two investigators measure the height of a plant, for example, they might choose different points on the compost, and the tip of the growing point to make their measurements.

They are either consistently too high or too low. One reason could be down to the way you are making a reading, for example taking a burette reading at the wrong point on the meniscus. Another could be the result of an instrument being incorrectly calibrated, or not being calibrated.

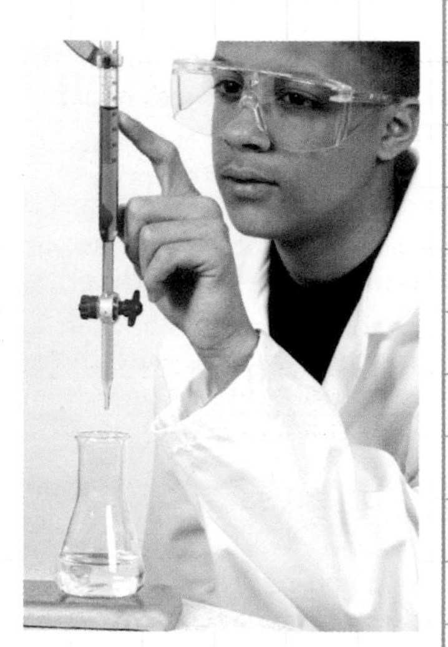

The volume of liquid in a burette must be read to the bottom of the meniscus

Accuracy and precision

When evaluating your investigation, you should mention accuracy and precision. If you use these terms, it is important that you understand what they mean, and that you use them correctly.

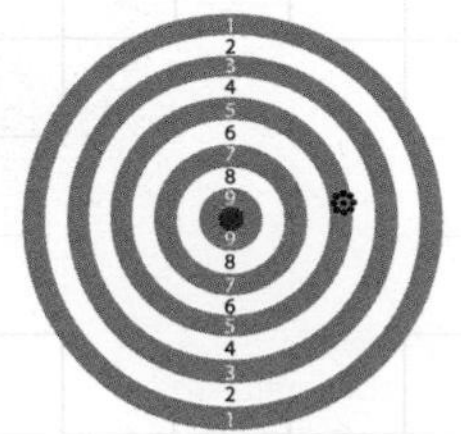

Precise but not accurate.

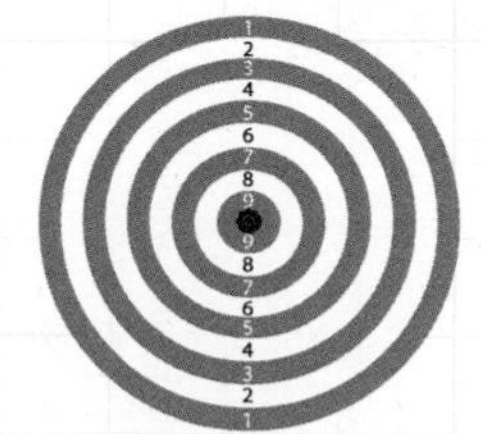

Precise and accurate.

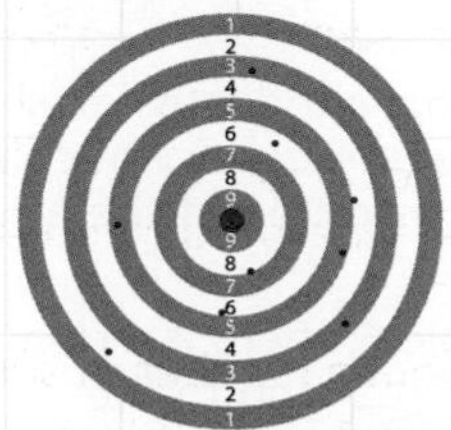

Not precise and not accurate.

The terms accuracy and precision can be illustrated using shots at a dartboard

Definition

Error is a difference between a measurement you make, and its true value.

Definition

With **random error**, measurements vary in an unpredictable way.

Definition

With **systematic error**, readings vary in a controlled way.

Assessment tip

A pH meter must be calibrated before use using buffers of known pH.

Assessment tip

What you should not discuss in your evaluation are problems introduced by using faulty equipment, or by you using the equipment inappropriately. These errors can, or could have been, eliminated, by:

> checking equipment

> practising techniques before your investigation

> taking care and patience when carrying out the practical.

Definition

When making measurements:

> the **accuracy** of the measurement is how close it is to the true value

> **precision** is how closely a series of measurements agree with each other.

Improving your investigation

When evaluating your investigation, you should discuss how your investigation could be improved. This could be by improving:

- ✔ the reliability of your data. For example, you could make more repeats, or more frequent readings, or 'fine-tune' the range you chose to investigate, or refine your technique in some other way.
- ✔ the accuracy and precision of your data, by using measuring equipment with a higher resolution.

In science, the measurements that you make as part of your investigation should be as precise as you can, or need to, make them. To achieve this, you should use:

- ✔ the most appropriate measuring instrument
- ✔ the measuring instrument with the most appropriate size of divisions.

The smaller the divisions you work with, the more precise your measurements. For example:

- ✔ In an investigation on how your heart rate is affected by exercise, you might decide to investigate this after a 100 m run. You might measure out the 100 m distance using a trundle wheel, which is sufficiently precise for your investigation.
- ✔ In an investigation on how light intensity is affected by distance, you would make your measurements of distance using a metre rule with millimetre divisions; clearly a trundle wheel would be too imprecise.
- ✔ In an investigation on plant growth, in which you measure the thickness of a plant stem, you would use a micrometer or Vernier callipers. In this instance, a metre rule would be too imprecise.

Using secondary data

As part of controlled assessment, you will be expected to compare your data – primary data – with **secondary data** you have collected.

One of the simplest ways of doing this is to compare your data with other groups in your class who have carried out an identical practical investigation.

In your controlled assessment, you will be provided with a data sheet of secondary data.

You should also, if possible, search through the scientific literature – in textbooks, the internet, and databases – to find data from similar or identical practical investigations so that you can compare the data with yours.

In addition, secondary data will be provided by AQA in section 2 of the ISA. To make full use of this, you may have to carry out further research yourself.

Ideally, you should use secondary data from a number of sources, carry out a full analysis of the secondary data that you have collected, and compare the findings with yours. You should critically analyse any evidence that conflicts with yours, and suggest and discuss what further data might help to make your conclusions more secure.

Definition

Secondary data are measurements/observations made by anyone other than you.

You should review secondary data and evaluate it. Scientific studies are sometimes influenced by the **bias** of the experimenter.

- ✔ One kind of bias is having a strong opinion related to the investigation, and perhaps selecting only the results that fit with a hypothesis or prediction.
- ✔ The bias could be unintentional. In fields of science that are not yet fully understood, experimenters may try to fit their findings to current knowledge and thinking.

There have been other instances where the 'findings' of experimenters have been influenced by organisations that supplied the funding for the research.

You must fully reference any secondary data that you have used, using one of the accepted referencing methods.

Referencing methods

The two main conventions for writing a reference are the:

- ✔ Harvard system
- ✔ Vancouver system.

In your text, the Harvard system refers to the authors of the reference, for example 'Smith and Jones (1978)'.

The Vancouver system refers to the number of the numbered reference in your text, for example '... the reason for this hypothesis is unknown.[5]'.

Though the Harvard system is usually preferred by scientists, it is more straightforward for you to use the Vancouver system.

Harvard system

In your references list a book reference should be written:

> Author(s) (year of publication). *Title of Book*, publisher, publisher location.

The references are listed in alphabetical order according to the authors.

Vancouver system

In your references list a book reference should be written:

> 1 Author(s). *Title of Book.* Publisher, publisher location: year of publication.

The references are number in the order in which they are cited in the text.

Assessment tip

Remember to write out the URL of a website in full. You should also quote the date when you looked at the website.

Do the data support your hypothesis?

You need to discuss, in detail, whether all, or which of your primary, and the secondary data you have collected, support your original hypothesis. They may, or may not.

You should communicate your points clearly, using the appropriate scientific terms, and checking carefully your use of spelling, punctuation and grammar. You will be assessed on this written communication as well as your science.

If your data do not completely match your hypothesis, it may be possible to modify the hypothesis or suggest an alternative one. You should suggest any further investigations that can be carried out to support your original hypothesis or the modified version.

It is important to remember, however, that if your investigation does support your hypothesis, it can improve the confidence you have in your conclusions and scientific explanations, but it can't prove your explanations are correct.

Your controlled assessment

The assessment of your investigation will form part of what is called controlled assessment. AQA will provide the task for you to investigate.

You may be able to work in small groups to carry out the practical work, but you will have to work on your own to write up your investigation.

The **Controlled Assessment** is worth 25% of the marks for your GCSE Science. It is worth doing it well!

Controlled Assessment is a two-part (sections 1 and 2) *Investigative Skills Assignment* (ISA) test. Before you start, you will be asked to:

> think about a hypothesis that you are given in Science or suggest a hypothesis as the result of your research in Additional or Separate Sciences

> make a risk assessment

> research the context of your investigation (for Section 2).

You are allowed to make brief notes about your research on one side of A4. You can use them to answer Sections 1 and 2 of the ISA paper.

Section 1 (45 minutes, 20 marks) consists of questions relating to your research.

Then you will carry out your experiment and record and analyse your results.

Section 2 of the ISA test (50 minutes, 30 marks) consists of questions related to your experiment. You will be given a sheet of secondary data by AQA. You should use it to select data to analyse and compare with the hypothesis. Finally you will be asked to suggest how ideas from your investigation and research could be used in a variety of ways.

How to be successful in your GCSE Science assessment

Introduction

AQA uses assessments to test how good your understanding of scientific ideas is, how well you can apply your understanding to new situations and how well you can analyse and interpret information you have been given. The assessments are opportunities to show how well you can do these.

To be successful in exams you need to:

✔ have a good knowledge and understanding of science
✔ be able to apply this knowledge and understanding to familiar and new
✔ situations
✔ be able to interpret and evaluate evidence that you have just been given.

You need to be able to do these things under exam conditions.

The language of the assessment paper

When working through an assessment paper, make sure that you:

✔ re-read a question enough times until you understand exactly what the examiner is looking for
✔ make sure that you highlight key words in a question. In some instances, you will be given key words to include in your answer.
✔ look at how many marks are allocated for each part of a question. In general, you need to write at least as many separate points in your answer as there are marks.

What verbs are used in the question?

A good technique is to see which verbs are used in the wording of the question and to use these to gauge the type of response you need to give. The table lists some of the common verbs found in questions, the types of responses expected and then gives an example.

Verb used in question	Response expected in answer	Example question
write down state give identify	These are usually more straightforward types of question in which you are asked to give a definition, make a list of examples, or the best answer from a series of options.	'Write down three types of microorganism that cause disease.' 'State one difference and one similarity between radio waves and gamma rays.'
calculate	Use maths to solve a numerical problem.	'Calculate the cost of supplying the flu vaccine to the whole population of the UK.'

estimate	use maths to solve a numerical problem, but you do not have to work out the exact answer	'Estimate the number of bacteria in the culture after five hours.'
describe	use words (or diagrams) to show the characteristics, properties or features of, or build an image of something	'Describe how antibiotic resistance can be reduced.'
suggest	come up with an idea to explain information that you are given	'Suggest why eating fast foods, rather than wholegrain foods, could increase the risk of obesity.'
demonstrate show how	use words to make something evident using reasoning	'Show how enzyme activity changes with temperature.'
compare	look for similarities and differences	'Compare the structure of arteries and veins.'
explain	to offer a reason for, or make understandable, information that you are given	'Explain why measles cannot be treated with antibiotics.'
evaluate	to examine and make a judgement about an investigation or information that you are given	'Evaluate the evidence for vaccines causing harm to human health.'

What is the style of the question?

Try to get used to answering questions that have been written in lots of different styles before you sit the exam. Work through past papers, or specimen papers, to get a feel for these. The types of questions in your assessment fit the three assessment objectives shown in the table.

Assessment objective	Your answer should show that you can...
AO1 recall the science	Recall, select and communicate your knowledge and understanding of science.
AO2 apply your knowledge	Apply skills, knowledge and understanding of science in practical and other contexts.
AO3 evaluate and analyse the evidence	Analyse and evaluate evidence, make reasoned judgements and draw conclusions based on evidence.

Assessment tip

Of course you must revise the subject material adequately. It is as important that you are familiar with the different question styles used in the exam paper, as well as the question content.

How to answer questions on: AO1 Recall the science

These questions, or parts of questions, test your ability to recall your knowledge of a topic. There are several types of this style of question:

- ✔ Fill in the spaces (you may be given words to choose from)
- ✔ Tick the correct statements
- ✔ Use lines to link a term with its definition or correct statement
- ✔ Add labels to a diagram
- ✔ Complete a table
- ✔ Describe a process

Example 8

a What is meant by the term *metabolic rate*?

Tick (✓) **one** box.

☐ the amount of energy a person uses each hour

☐ the amount of exercise a person does each day

☐ the amount of food a person eats each day

How to answer questions on: AO1 Recall the science in practical techniques

You may be asked to recall how to carry out certain practical techniques, either ones that you have carried out before, or techniques that scientists use.

To revise for these types of questions, make sure that you have learned definitions and scientific terms. Produce a glossary of these, or key facts cards, to make them easier to remember. Make sure that your key facts cards also cover important practical techniques, including equipment, where appropriate.

Example 9

Describe how to test the pH of a solution.

Assessment tip

Don't forget that mind maps – either drawn by you or by using a computer program – are very helpful when revising key points.

How to answer questions on: AO2 Apply skills, knowledge and understanding

Some questions require you to apply basic knowledge and understanding in your answers.

You may be presented with a topic that is familiar to you, but you should also expect questions in your Science exam to be set in an unfamiliar context.

Questions may be presented as:

- ✔ practical investigations
- ✔ data for you to interpret
- ✔ a short paragraph or article.

The information required for you to answer the question might be in the question itself, but for later stages of the question, you may be asked to draw on your knowledge and understanding of the subject material in the question.

Practice will help you to become familiar with contexts that examiners use and question styles. However, you will not be able to predict many of the contexts used. This is deliberate; being able to apply your knowledge and understanding to different and unfamiliar situations is a skill the examiner tests.

Practise doing questions where you are tested on being able to apply your scientific knowledge and your ability to understand new situations that may not be familiar. In this way, when this type of question comes up in your exam, you will be able to tackle it successfully.

Assessment tip

Work through the Preparing for assessment: Applying your knowledge tasks in this book as practice.

Example 10

Measles is an infectious disease caused by a virus. Today, most children are vaccinated against measles when they are very young.

The graph shows the number of measles cases per year in the USA between 1950 and 2000. The graph also shows when vaccination against measles was first introduced in the USA.

The use of the measles vaccine reduces the number of measles cases.

Explain why this graph alone does not prove that the use of the measles vaccine reduces the number of measles cases.

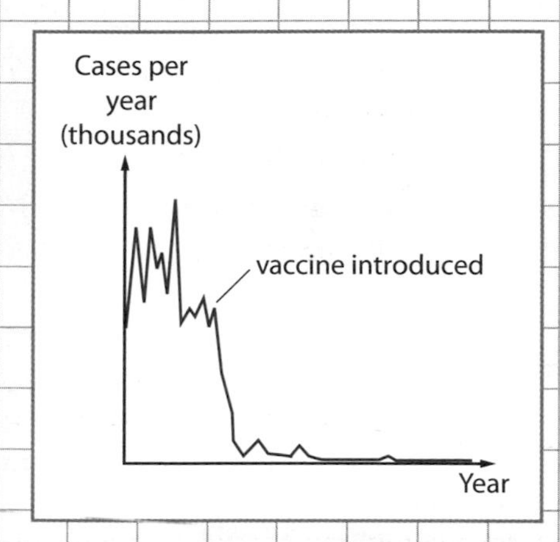

How to answer questions on: AO2 Apply skills, knowledge and understanding in practical investigations

Some opportunities to demonstrate your application of skills, knowledge and understanding will be based on practical investigations. You may have carried out some of these investigations, but others will be new to you, and based on data obtained by scientists. You will be expected to describe patterns in data from graphs you are given or that you will have to draw from given data.

Again, you will have to apply your scientific knowledge and understanding to answer the question.

Example 11

Look at the graph on the right showing the resistance of the bacterium, *Streptococcus pneumoniae,* to three different types of antibiotic.

a Which antibiotic does there seem to be least resistance to, even when it has been used before?

b Can you explain why this might be the case?

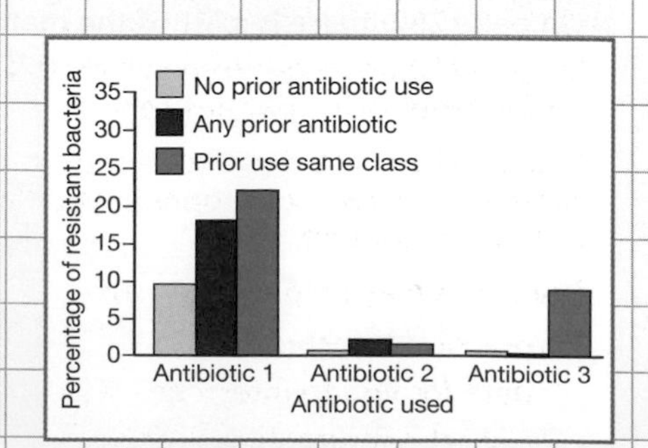

You will also need to analyse scientific evidence or data given to you in the question. It is likely that you will not be familiar with the material.

Analysing data may involve drawing graphs and interpreting them, and carrying out calculations. Practise drawing and interpreting graphs from data.

When drawing a graph, make sure that you:

- ✔ choose and label the axes fully and correctly
- ✔ include units, if this has not been done already
- ✔ plot points on the graph carefully – the examiner will check individual points to make sure that they are accurate
- ✔ join the points correctly; usually this will be by a line of best fit.

When reading values off a graph that you have drawn or one given in the question, make sure that you:

- ✔ do it carefully, reading the values as accurately as you can
- ✔ double-check the values.

When describing patterns and trends in the data, make sure that you:

- ✔ write about a pattern or trend in as much detail as you can
- ✔ mention anomalies where appropriate
- ✔ recognise there may be one general trend in the graph, where the variables show positive or negative correlation
- ✔ recognise the data may show a more complex relationship. The graph may demonstrate different trends in several sections. You should describe what's happening in each.
- ✔ describe fully what the data show.

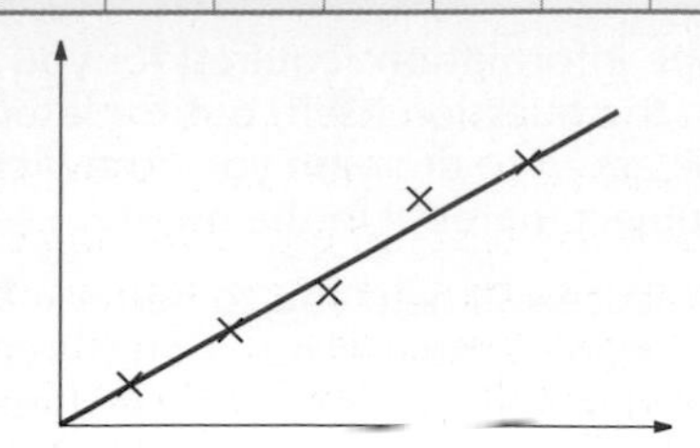

What type of line is drawn on this graph?

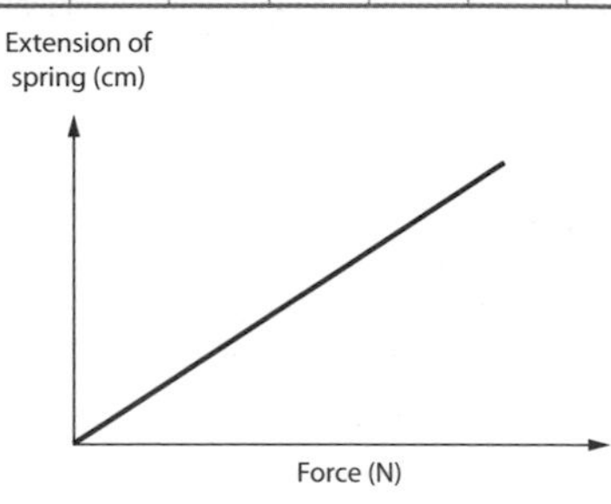

Make sure you know what type of relationship is shown in this graph

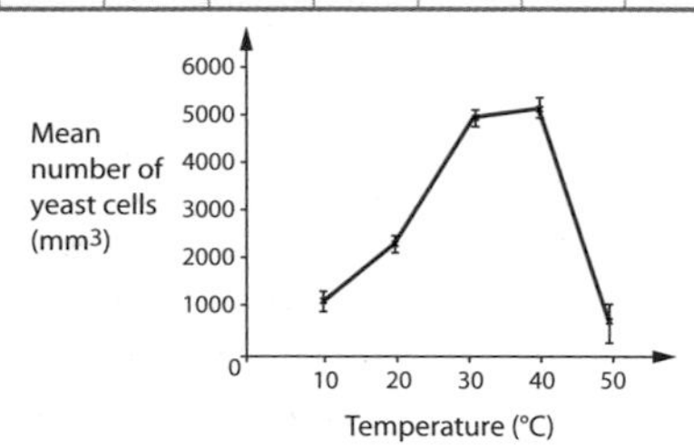

What type of relationship does this graph show?

How to answer questions needing calculations

- ✔ The calculations that you are asked to do may be straightforward, for example the calculation of the mean from a set of data.
- ✔ They may be more complex, for example calculating the yield of a chemical reaction.
- ✔ Other questions will require the use of formulae.

You will be given an equation sheet with the question paper.

On page 297, there is a list of the maths skills that you will need. Remember, these are the same skills that you have learned in maths lessons.

Example 12

Calculate the area on the agar plate, around the antibiotic disc, that is free from bacteria.

Use the formula:

area = πr^2

where $\pi = 3.14$

Assessment tip

When completing your calculation, make sure that you include the correct units.

Assessment tip

Check the specification, or with your teacher, to make sure that you know the formulae that you have to learn and remember.

Assessment tip

Remember, when carrying out any calculations, you should include your working at each stage. You may get credit for getting the process correct, even if your final answer is wrong.

How to answer questions on: AO3 Analysing and evaluating evidence

For these types of questions, in addition to analysing data, you must also be able to evaluate information that you are given. This is one of the hardest skills. Think about the validity of the scientific data: did the technique(s) used in any practical investigation allow the collection of accurate and precise data?

Your critical evaluation of scientific data in class, along with the practical work and controlled assessment work, will help you to develop the evaluation skills required for these types of questions.

Example 13

In the standard experiment testing of the inhibition of bacterial growth by a new antibiotic, explain why further investigation is required to confirm the effectiveness of the antibiotic.

You may be expected to compare data with other data, or come to a conclusion about its reliability, its usefulness or its implications. Again, it is possible that you will not be familiar with the context. You may be asked to make a judgement about the evidence or to give an opinion with reasons.

Example 14

A catalytic converter reduces nitrogen oxide emissions from a car engine to nitrogen. Evidence shows that the reaction needs the catalyst to be hot. Explain the effect of this on air quality in housing estates where many people use their car to commute to work, and suggest the possible implications for health.

Assessment tip

Work through the Preparing for assessment: Analysing and evaluating data tasks, in this book, as practice.

Assessment tip

Wherever possible, use as much data as you can in your answer, particularly when explaining trends or conclusions, so you can gain full marks. Try to use numbers and values rather than just trends in data or graphs. 'At 45 °C…' is always better than 'as the temperature rises it gets greater'.

The quality of your written communication

Scientists need good communication skills to present and discuss their findings. You will be expected to demonstrate these skills in the exam. You will be assessed in the longer-response exam questions that you answer. These questions are clearly indicated in each question paper. The quality of your written communication will also be assessed in your controlled assessment.

You will not be able to obtain full marks unless you:

- ✔ make sure that the text you write is legible
- ✔ make sure that spelling, punctuation and grammar are accurate so that the meaning of what you write is clear
- ✔ use a form and style of writing appropriate for its purpose and for the complexity of the subject matter
- ✔ organise information clearly and coherently
- ✔ use the scientific language correctly.

You will also need to remember the writing and communication skills that you have developed in English lessons. For example, make sure that you understand how to construct a good sentence using connectives.

Assessment tip

You will be assessed on the way in which you communicate science ideas.

Assessment tip

When answering questions, you must make sure that your writing is legible. An examiner cannot award marks for answers that he or she cannot read.

Revising for your Science exam

You should revise in the way that suits you best. It is important that you plan your revision carefully, and it is best to start well before the date of the exams. Take the time to prepare a revision timetable and try to stick to it. Use this during the lead up to the exams and between each exam.

When revising:

- ✔ Find a quiet and comfortable space in the house where you will not be disturbed. It is best if it is well ventilated and has plenty of light.
- ✔ Take regular breaks. Some evidence suggests that revision is most effective when you revise in 30 to 40 minute slots. If you get bogged down at any point, take a break and go back to it later when you're feeling fresh. Try not to revise when you are feeling tired. If you do feel tired, take a break.
- ✔ Use your school notes, textbook and possibly a revision guide. But also make sure that you spend some time using past papers to familiarise yourself with the exam format.
- ✔ Produce summaries of each topic.
- ✔ Draw mind maps covering the key information on a topic.
- ✔ Set up revision cards containing condensed versions of your notes.
- ✔ Ask yourself questions, and try to predict questions, as you are revising a topic.
- ✔ Test yourself as you go along. Try to draw key labelled diagrams, and try some questions under timed conditions.
- ✔ Prioritise your revision of topics. You might want to allocate more time to revising the topics you find most difficult.

Assessment tip

Try to make your revision timetable as specific as possible – don't just say 'science on Monday, and Thursday', but list the topics that you will cover on those days.

Assessment tip

Start your revision well before the date of the exams, produce a revision timetable, and use the revision strategies that suit your style of learning. Above all, revision should be an active process.

How do I use my time effectively in the exam?

Timing is important when you sit an exam. Do not spend so long on some questions that you leave insufficient time to answer others. For example, in a 60-mark question paper, lasting one hour, you will have, on average, one minute per question.

If you are unsure about certain questions, complete the ones you are able to do first, then go back to the ones you're less sure of.

If you have time, go back and check your answers at the end of the exam.

On exam day...

A little bit of nervousness before your exam can be a good thing, but try not to let it affect your performance in the exam. When you turn over the exam paper keep calm. Look at the paper and get it clear in your head exactly what is required from each question. Read each question carefully. Do not rush.

If you read a question and think that you have not covered the topic, keep calm – it could be that the information needed to answer the question is in the question itself or the examiner may be asking you to apply your knowledge to a new situation.

Finally, good luck!

Mathematical skills

You will be allowed to use a calculator in all assessments.

These are the maths skills that you need, to complete all the assessments successfully.

You should understand:

- ✔ the relationship between units, for example, between a gram, kilogram and tonne
- ✔ compound measures such as speed
- ✔ when and how to use estimation
- ✔ the symbols = < > ~
- ✔ direct proportion and simple ratios
- ✔ the idea of probability.

You should be able to:

- ✔ give answers to an appropriate number of significant figures
- ✔ substitute values into formulae and equations using appropriate units
- ✔ select suitable scales for the axes of graphs
- ✔ plot and draw line graphs, bar charts, pie charts, scatter graphs and histograms
- ✔ extract and interpret information from charts, graphs and tables.

You should be able to calculate:

- ✔ using decimals, fractions, percentages and number powers, such as 10^3
- ✔ arithmetic means
- ✔ areas, perimeters and volumes of simple shapes

In addition, if you are a higher tier candidate, you should be able to:

- ✔ **change the subject of an equation**

and should be able to use:

- ✔ **numbers written in standard form**
- ✔ **calculations involving negative powers, such as 10^{-1}**
- ✔ **inverse proportion**
- ✔ **percentiles and deciles.**

Some key physics equations

With the written papers, there will be an equation sheet. In order to make best use of the sheet, it will help if you practise using the following equations.

Equation	Meaning of symbol and its unit
specific heat capacity: $E = m \times c \times \theta$	E is energy transferred in joules m is mass in kilograms θ is temperature change in degrees Celsius c is specific heat capacity in J/kg °C
efficiency of a device: efficiency = $\frac{\text{useful energy out}}{\text{total energy in}}$ (× 100%) efficiency = $\frac{\text{useful power out}}{\text{total power in}}$ (× 100%)	← energy out and energy in are measured in the same units ← power out and power in are measured in the same units
amount of energy transfer: $E = P \times t$	E is energy transferred in kilowatt-hours P is power in kilowatts t is time in hours This equation may also be used when: E is energy transferred in joules P is power in watts t is time in seconds
cost of mains electricity used: *total cost* = E × *cost per kilowatt-hour*	*total cost* is in pence E is energy transferred in kilowatt-hours *cost per kilowatt-hour* is in pence per kilowatt-hour
at a reflecting surface: the angle of incidence = the angle of reflection	← angles are measured from the normal ← both angles are measured in the same units
wave equation: $v = f \times \lambda$	v is speed in metres per second f is frequency in hertz λ is wavelength in metres

When the evidence doesn't add up.

Sometimes people use what sound like scientific words and ideas to sell you things or persuade you to think in a certain way. Some of these claims are valid, and some are not. The activities on these pages are based on the work of Dr Ben Goldacre and will help you to question some of the scientific claims you meet. Read more about the work of Ben in his *Bad Science* book or at badscience.net.

How much to look younger?

There are many ways to make yourself look younger if you're an adult. These include the style and colour of your hair, the texture of your skin, your body shape and the clothes you wear. Manufacturers and retailers know this and recognise where there's money to be made.

Which of these do you think is more effective?

Are there other ways for adults to make themselves look younger?

How are these age-defying products promoted?

YOUNG SKIN FROM OLD?

Skin changes in appearance as people get older. These photographs show how older skin looks different to young skin.

> Examine the photographs. What are the differences?

> How might an anti-ageing skin cream work on the old skin? What would it need to do?

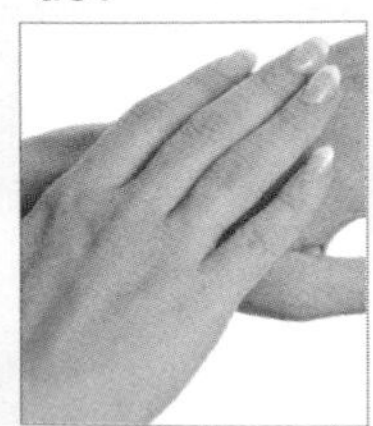

Young skin

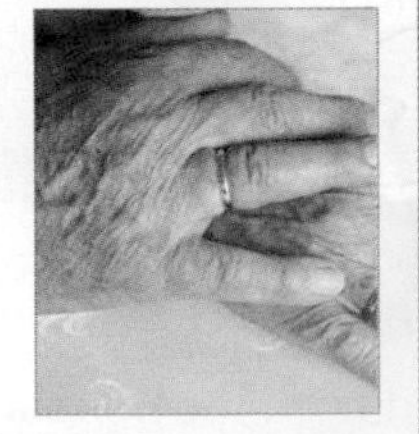

Old skin

THE SCIENCE BEHIND THE CLAIM

As you get older you may not like the appearance of wrinkles and crows feet. You can spend quite a lot of money on anti-ageing skin creams. Creams are advertised with appealing images and lavish claims, but do they really work?

One immediate gain from a cream is rehydration. Dried out skin doesn't look good so we can make it look better by moisturising it. This is easy and the active ingredients are really cheap. However something more is needed to make someone genuinely look younger.

These are three types of active ingredient commonly used:

> Alphahydroxy acids, such as vitamins A and C, are used to exfoliate the skin. Some of these work at high doses, but they are also irritants, so they can only be sold at low doses.

> Vegetable proteins, which are long chain molecules. As the cream dries on the skin the chain molecules tighten, applying tension and temporarily tightening it.

> Hydrogen peroxide, which is corrosive and will lightly burn the skin.

Why might someone who uses a cream with these ingredients think that it is working? Will the effects last?

We are learning to:

- find out how anti-ageing skin creams work
- examine claims that are made for them

NEW AND IMPROVED! ADVERTISER'S CLAIMS

Many anti-ageing skin creams are sold on the basis that there is a scientific reason that they work. Some claims are justified but others are pretty dubious, even if they look persuasive at first glance. You should think critically about what you are told by the advertisers.

- Claims sound more convincing if they are based on tests and if scientists have been involved. Powerful scientific words include *conclusive tests*, *laboratory*, *cleanse*, *purify* and *health*.
- They may claim to make you feel better, look younger, have more energy and be healthier. Some of the claims may be difficult to prove; they should have been tested, the full results published and independently checked in a scientific way. There are very few cases of anti-ageing skin creams being proved to get rid of wrinkles. Why do you think this is?
- Watch out for claims such as 'eight out of ten users said that…' if it's not clear what kind of people and how many were asked. What do you think ten company employees might say about their product and would this be representative of all their consumers?

THE PSYCHOLOGY OF COSMETICS

You can buy very cheap creams in the shops. You can even make your own skin cream using simple ingredients. If you did, it would be pretty good at moisturising, so your skin would feel soft and maybe a little smoother. It wouldn't, however, make you look younger for long.

Why do you think anti-ageing skin creams are sometimes quite expensive?

Do you think people who buy anti-ageing skin creams

a) genuinely believe that they make them look younger?

b) hope that they might but don't really believe it?

c) do it because it makes them feel good?

When the evidence doesn't add up.

Sometimes people use what sound like scientific words and ideas to sell you things or persuade you to think in a certain way. Some of these claims are valid, and some are not. The activities on these pages are based on the work of Dr Ben Goldacre and will help you to question some of the scientific claims you meet. Read more about the work of Ben in his *Bad Science* book or at badscience.net.

What we'll look like in the future

Here are some questions to get you thinking about the future of the human race.

> How do you think humans have evolved over the last few millions of years?

> What changes have taken place in the way we look and move?

> How might evolution affect us over the next few million years?

News stories that sound like they are about science can come from a number of different places. Sometimes they are fair reports of real scientific research, but sometimes they are just good stories.

The story opposite featured in a number of news reports, including The Times, where this version was printed, BBC, Daily Telegraph and The Sun. Your task is to work out if this story is good science or bad science.

You are going to investigate the predictions reported in this article. First you need to identify the predictions. This is the first one – that humans will evolve into two separate species. To decide whether you think this is good or bad science think about what you have learnt about evolution. Do you think this likely to happen?

From **THE TIMES**
October 17, 2006

The future ascent (and descent) of man

Within 100,000 years the divide between rich and poor could lead to two human sub-species

By Mark Henderson, Science Editor

The mating preferences of the rich, highly educated and well-nourished could ultimately drive their separation into a genetically distinct group that no longer interbreeds with less fortunate human beings, according to Oliver Curry.

Dr Curry, a research associate in the Centre for Philosophy of Natural and Social Science of the London School of Economics, speculated that privileged humans might over tens of thousands of years evolve into a "gracile" subspecies, tall, thin, symmetrical, intelligent and creative. The rest would be shorter and stockier, with asymmetric features and lower intelligence, he said.

THE BRAVO EVOLUTION REPORT

We are learning to:

- > apply ideas about cells and body systems to evaluate advice on how to focus on mental activities
- > examine scientific claims for accuracy
- > function effectively in a group, developing ideas collaboratively

THE SCIENCE BEHIND THE CLAIM

Let's look in more detail at some activities that some schools have used to try to improve students' concentration and learning. Your task is to work out which bits of science are good science and which are bad. To help you decide whether you think your activity is good or bad try discussing these questions:

> What *advice* is being given?

> What *claim* is being made?

> What *scientific ideas* are being used to justify that claim?

> What *scientific ideas* do you have that may tell you something about this topic?

> Is the advice *sound*?

This is the advice.

This is the claim.

Interlock the fingers of both your hands, holding your elbows out at the sides. This completes a circuit and allows positive energy to flow. Positive energy creates positive thoughts, stimulating the brain, stilling anxieties and clearing the way for a free flow of logical thought.

These are the scientific ideas used to back up this claim. Forming the arms in a loop creates no circuit that any kind of energy 'flows around' and 'positive energy' is a meaningless term.

So can you think of any reason why this might work? You know that regular exercise is good and could help to refocus on ideas and mental activities.

Is the advice sound? Well, it won't do you any harm and may even improve concentration, but not for the reasons claimed.

Now you have a go. Are these good or bad science?

Water is a vital ingredient of blood and blood is essential to transport oxygen to the brain. For the brain to work well you have to ensure your blood is hydrated. This needs water, little and often. The best way of rehydrating the blood taking oxygen to the brain is to hold water in the mouth for up to half a minute, thus allowing direct absorption.

Your carotid arteries are vital to supplying your brain with richly oxygenated blood. Ensure their peak performance by pressing your brain buttons. These are just below the collar bone, one on either side. Make 'C' shapes with forefinger and thumb to place over the brain buttons and gently massage.

WOULD YOU PAY MONEY FOR THIS?

Many products or services are sold on the basis that there is a scientific reason that they work. Some claims are justified but some are pretty dubious, even if they look persuasive at first glance. You should think critically about what you are told.

> Claims might sound more convincing if they use technical scientific terms, but sometimes they are used incorrectly, to make something sound scientific when it's not.

> Powerful scientific words include 'conclusive tests', 'energy', 'cleanse', 'purify' and 'health'.

> They may claim to make you feel better, look better, have more energy and be healthier.

> The claims may be true but they should have been tested, the full results published and independently checked.

Glossary

absorb an object absorbs energy when the energy from infrared radiation is transferred to the particles of the object, increasing the temperature of the object

adaptation the way in which an organism evolved to become better able to survive in its environment

addiction when a person becomes dependent on a drug

addition reaction reaction in which a C=C bond opens up and other atoms add on to each carbon atom

adrenaline hormone that helps to prepare your body for action

agar substance used to make jelly on which bacteria can be grown

aggregate stones, gravel or rock chippings used in the construction industry

alcohols a family of organic compounds containing an OH group, for example ethanol (C_2H_5OH)

alkali metal very reactive metal in Group 1 of the periodic table, for example sodium

alkanes a family of hydrocarbons: C_nH_{2n+2} with single covalent bonds – found in crude oil

alkenes a family of hydrocarbons: C_nH_{2n} with double covalent bonds (C=C), for example ethene (C_2H_4)

alloy mixture of two or more metals, with useful properties different from the individual metals

amino acids small molecules from which proteins are built

ampere (amp) unit used to measure electrical current

amplitude size of wave oscillations – for a mechanical wave, it describes how far the particles vibrate around their central position

angle of incidence angle between the ray hitting a mirror or lens and the normal

angle of reflection angle between a ray reflecting from a mirror and the normal

anode positive electrode

antibiotic therapeutic drug acting to kill bacteria which is taken into the body

antibiotic resistance ability of bacteria to survive in the presence of an antibiotic

antibody protein normally present in the body or produced in response to an antigen which it neutralises, thus producing an immune response

antiseptic substance that kills pathogens

antitoxin substance produced by white blood cells that neutralises the effects of toxins

antiviral therapeutic drug acting to kill viruses

appliance device that transfers the energy supplied by electricity into something useful

aqueous solution substance(s) dissolved in water

argon the most common noble gas – makes up nearly 1% of the air

atmosphere thin layer of gas surrounding a planet

atom the basic 'building block' of an element that cannot be chemically broken down

atomic number number of protons found in the nucleus of an atom

attraction force that pulls, or holds, objects together

auxin a plant hormone that affects rate of growth

axon long thread of cytoplasm in a neurone, carrying an impulse away from the cell body

bacteria single-celled microorganisms that can either be free-living organisms or parasites (they sometimes invade the body and cause disease)

balanced diet eating foods (and drinking drinks) that will provide the body with the correct nutrients in the correct proportions

balanced equation chemical equation where the number of atoms of each element is the same in the reactants as in the products

bauxite main ore of aluminium – impure aluminium oxide (Al_2O_3)

Big Bang theory theory that states that the universe originated from a point at very high temperature, and everything in the universe formed as energy and matter exploded outwards from that point and cooled down

biodegradable a biodegradable material can be broken down by microorganisms

biodiesel fuel made from plant oils such as rapeseed

biodiversity range of different living organisms in a habitat

biofuel fuel such as wood, ethanol or biodiesel – obtained from living plants or animals

biomass the mass of living material, including waste wood and other natural materials

blast furnace furnace for extracting iron from iron ore

blood plasma the liquid part of blood

Bluetooth low energy, short-range radio waves used to connect communications equipment such as mobile phones or laptops to other equipment nearby

boiling change of state from liquid to gas that happens at the boiling point of a substance

brass alloy of copper and zinc

brittle easily cracked or broken by hitting or bending

bronze alloy of copper and tin

brownfield site former industrial site that may be redeveloped for other uses

calcium carbonate compound with chemical formula $CaCO_3$ – main component of limestone

carbon capture technology that filters the carbon and carbon dioxide gas out of waste smoke and gases from power stations and industrial chimneys

carbon cycle the way in which carbon atoms pass between living organisms and their environment

carbon dioxide one of the gasses emitted from burning fossil fuels that contributes to global warming

carbon sinks carbon-containing substances, such as limestone and fossil fuels, which formed millions of years ago, removing carbon dioxide from the atmosphere

carbon-neutral fuel fuel grown from plants so that carbon dioxide is taken in as the plants are growing – this balances out the carbon dioxide released as the fuel is burned

carnivore an animal that eats other animals

cast iron iron containing 3–4% carbon – used to make objects by casting

casting making an object by pouring molten metal into a mould and allowing it to cool and solidify

catalytic converter ('cat') the section of a vehicle's exhaust system that converts pollutant gases into harmless ones

catalytic cracking cracking hydrocarbons by heating in the presence of a catalyst

cathode negative electrode

cell body the part of a nerve cell that contains the nucleus

cement substance made by heating limestone with clay – when mixed with water it sets hard like stone

central nervous system (CNS) collectively the brain and spinal cord

chalcopyrite common ore of copper – formula $CuFeS_2$

charge particles or objects can be positively or negatively electrically charged, or neutral: similar charges repel each other, opposite charges attract

chemical bond attractive force between atoms that holds them together (may be covalent or ionic)

chemical equation line of chemical formulae showing what reacts and what is produced during a chemical reaction

chemical reaction process in which one or more substances are changed into other substances – chemical reactions involve rearranging atoms and energy changes

cholesterol chemical needed for the formation of cell membranes, but that increases the risk of heart disease if there is too much in the blood

chromosome thread-like structure in the cell nucleus that carries genetic information

climate change changes in seasonal weather patterns that occur because the average temperature of Earth's surface is increasing owing to global warming

clone group of genetically identical organisms

coal solid fossil fuel formed from plant material – composed mainly of carbon

combustion process where substances react with oxygen, releasing heat

communications satellite artificial satellite that stays above the same point on Earth's surface as it orbits, and used to send communications signals around the world

community all the organisms, of all species, that live together in the same habitat at the same time

competition result of more than one organism needing the same resource, which is in short supply

compost partly rotted organic material, used to improve soil for growing plants

compound substance composed of two or more elements that are chemically joined together, for example H_2O

compression region of a longitudinal wave where the vibrating particles are squashed together more than usual

concrete mixture of cement, sand, aggregate and water

condensation change of state when a substance changes from a gas or vapour to a liquid: the substance condenses

conduction (electrical) transfer of energy when an electrical current passes through a material

conduction (thermal) transfer of energy through a substance when it is heated

conductor material that transfers energy easily

consumer an organism that feeds on other organisms

continental drift movement of continents relative to each other

continental plate tectonic plate carrying large landmass, though not necessarily a whole continent

convection heat transfer in a liquid or gas – when particles in a warmer region gain energy and move into cooler regions carrying this energy with them

convection current when particles in a liquid or gas gain energy from a warmer region and move into a cooler region, being replaced by cooler liquid or gas

coordination communicating between different parts of the body so that they can act together

core (of Earth) layer in centre of Earth, consisting of a solid inner core and molten outer core

cosmic microwave background radiation (CMBR) microwave radiation coming very faintly from all directions in space

covalent bond bond between atoms in which some of the electrons are shared

cracking oil refinery process that breaks down large hydrocarbon molecules into smaller ones

crust surface layer of Earth made of tectonic plates

current (electric) flow of electrons in an electric circuit

culture a population of microorganisms, grown on a nutrient medium

cuttings small pieces of a plant that can grow into complete new plants

decay (biological) the breakdown of organic material by microorganisms

dendrite a short thread of cytoplasm on a neurone, carrying an impulse towards the cell body

dendron long thread of cytoplasm on a neurone, carrying an impulse towards the cell body

diabetes disease in which the body cannot control its blood sugar level

diffraction change in the direction of a wave caused by passing through a narrow gap or round an obstacle such as a sharp corner

digital signal communications signal sent as an electromagnetic wave that is switched on and off very rapidly

distillation process for separating liquids by boiling them, then condensing the vapours

distribution the transmission of electricity from a power station to homes and businesses

DNA polymer molecule found in the nucleus of all body cells – its sequence determines genetic characteristics, such as eye colour

Doppler effect change in wavelength and frequency of a wave that an observer notices if the wave source is moving towards them or away from them

double covalent bond two covalent bonds between the same pair of atoms – each atom shares two of its own electrons plus two from the other atom

drug a chemical that changes the chemical processes in the body

drug dependency feeling that you cannot manage without a drug

earthquake shaking and vibration at the surface of the Earth resulting from underground movement or from volcanic activity

echo reflection of a sound wave

effector part of the body that responds to a stimulus

efficiency a measure of how effectively an appliance transfers the energy that flows into the appliance into useful effects

egg cell female gamete

electrical power a measure of the amount of energy supplied each second

electricity generator device for generating electricity

electrode solid electrical conductors through which the current passes into and out of the liquid during electrolysis – and at which the electrolysis reactions take place

electrolysis decomposing an ionic compound by passing an electric current through it while molten or in solution

electromagnetic (EM) radiation energy transferred as electromagnetic waves

electromagnetic spectrum electromagnetic waves ordered according to wavelength and frequency – ranging from radio waves to gamma rays

electromagnetic waves a group of waves that transfer energy – they can travel through a vacuum and travel at the speed of light

electron small particle within an atom that orbits the nucleus (it has a negative charge)

electronic configuration the arrangement of electrons in shells, or energy levels, in an atom

element substance made out of only one type of atom

embryo a very young organism, which began as a zygote and will become a fetus

embryo transplant taking an embryo that has been produced from one female's egg and placing it into another female

emit an object emits energy when energy is transferred away from the object as infrared radiation, decreasing the temperature of the object

emulsifier a substance that prevents an emulsion from separating back into oil and water

emulsion a thick, creamy liquid made by thoroughly mixing an oil with water (or an aqueous solution)

energy the ability to 'do work'

energy input the energy transferred to a device or appliance from elsewhere

energy levels the arrangement of electrons in atoms (shells)

energy output the energy transferred away from a device or appliance – it can be either useful or wasted

environment an organism's surroundings

enzyme biological catalyst that increases the speed of a chemical reaction

epidemic many people having the same infectious disease

essential oils oils found in flowers, giving them their scent – they vaporise more easily than natural oils from seeds, nuts and fruit

ethanol an alcohol that can be made from sugar and used as a fuel

evaporation change of state where a substance changes from liquid to gas at a temperature below its boiling point

evolution a change in a species over time

extremophile an organism that can live in conditions where a particular factor, such as temperature or pH, is outside the range that most other organisms can tolerate

fermentation process in which yeast converts sugar into ethanol (alcohol)

fertilisation fusion of the nuclei of a male and a female gamete

fertility drug hormone given to women to cause the ovaries to produce eggs

fibre optic cable glass fibre that is used to transfer communications signals as light or infrared radiation

finite resource material of which there is only a limited amount – once used it cannot be replaced

flammable catches fire and burns easily

food chain flow diagram showing how energy is passed from one organism to another

formula (for a chemical compound) group of chemical symbols and numbers, showing which elements, and how many atoms of each, a compound is made up of

fossil fuel fuel such as coal, oil or natural gas, formed millions of years ago from dead plants and animals

fractional distillation process that separates the hydrocarbons in crude oil according to size of molecules

fractionating column tall tower in which fractional distillation is carried out at an oil refinery

fractions the different substances collected during fractional distillation of crude oil

freezing change of state in which a substance changes from a liquid to a solid

frequency the number of waves passing a set point per second

FSH hormone produced by the pituitary gland that causes eggs to mature in the ovaries

fuel a material that is burned for the purpose of generating heat

fuel cell device that generates electricity directly from a fuel, such as hydrogen, without burning it

fungus (pl. fungi) living organisms whose cells have cell walls, but that cannot photosynthesise

gamete sex cell – a cell containing the haploid number of chromosomes, such as an egg or sperm

gamma ray ionising electromagnetic radiation – radioactive and dangerous to human health

gasohol mixture of gasoline (petrol) and alcohol (ethanol) used as a vehicle fuel

gene section of DNA that codes for a particular characteristic

genetic engineering changing the genes in an organism, for example by inserting genes from another organism

genetically modified organism that has had genes from a different organism inserted into it

geothermal power station power station generating electricity using the heat in underground rocks to heat water

gland organ that secretes a useful substance

global dimming gradual decrease in the average amount of sunlight reaching Earth's surface

Global Positioning System (GPS) navigation system using signals from communications satellites to find an exact position on the surface of Earth

global warming gradual increase in the average temperature of Earth's surface

gravitropism a growth response to gravity

green fuel fuel that does less damage to the environment than fossil fuels

greenhouse gas a gas such as carbon dioxide that reduces the amount of heat escaping from Earth into space, thereby contributing to global warming

group within the periodic table the vertical columns are called groups

HDL a type of cholesterol that does not appear to cause heart disease and may help to protect against it

heart disease blockage of blood vessels that bring blood to the heart

herbivore an animal that eats plants

hormones chemicals that act on target organs in the body (hormones are made by the body in special glands)

hot spot area of Earth's crust heated by rising currents of magma – Hawaii is above a mid-Pacific hot spot

hydrocarbon compound containing only carbon and hydrogen

hydroelectric power station power station generating electricity using the energy from water flowing downhill

hydrophilic water-loving (attracted to water, but not to oil) – used to describe parts of a molecule

hydrophobic water-fearing (attracted to oil, but not to water) – opposite of hydrophilic

hydroxide ion consisting of an oxygen and a hydrogen atom (written as OH^-)

hypothesis an idea that explains a set of facts or observations – a basis for possible experiments

image an image is formed by light rays from an object that travel through a lens or are reflected by a mirror

immiscible liquids that do not mix, but form separate layers, are immiscible

immune system a body system that acts as a defence against pathogens, such as viruses and bacteria

immunity you have immunity if your immune system recognises a pathogen and fights it

incident ray the ray of light hitting a mirror or lens

infrared radiation energy transferred as heat – a type of electromagnetic radiation

inoculating loop metal loop that is used to transfer microorganisms

insoluble not soluble in water (forms a precipitate)

insulator material that transfers energy only very slowly – thermal insulators transfer heat slowly, electrical insulators do not allow an electric current to flow through them

insulin hormone made by the pancreas that reduces the level of glucose in the blood

ion atom (or groups of atoms) with a positive or negative charge, caused by losing or gaining electrons

ionic bond chemical bond between two ions of opposite charges

ionise to cause electrons to split away from their atoms (some forms of EM radiation are harmful to living cells because they cause ionisation)

IVF *in vitro* fertilisation – the fertilisation of an egg by a sperm in a glass container

joule unit used to measure energy

kilowatt-hour the energy transferred in 1 hour by an appliance with a power rating of 1 kW (sometimes called a 'unit' of electricity)

kinetic theory model used to explain how energy is transferred by particles in a substance

lava magma that has erupted onto the surface of Earth

laterally inverted image left and right are reversed, when seen in a mirror

LDL a type of cholesterol that increases the risk of heart disease

leaching using a chemical solution to dissolve a substance out of a rock

LH hormone produced by the pituitary gland, which causes an egg to be released from an ovary

lichen small organism that consists of both a fungus and an alga

limestone type of rock consisting mainly of calcium carbonate

limewater calcium hydroxide solution

lithosphere the rocky, outer section of Earth, consisting of the crust and upper part of the mantle

longitudinal wave a wave in which the direction that the particles are vibrating is the same as the direction in which the energy is being transferred by the wave

low-grade ore ore containing only a small percentage of metal

lymphocyte type of white blood cell

magma molten rock found below Earth's surface

malnourished not having a balanced diet

mantle semi-liquid layer of the Earth beneath the crust

mass number number of protons and neutrons in the nucleus of an atom

mechanical wave wave in which energy is transferred by particles or objects moving, such as a wave on a string or a water wave

melting change of state of a substance from liquid to solid

menstrual cycle monthly hormonal cycle that starts at puberty in human females

menstruation monthly breakdown of the lining of the uterus leading to bleeding from the vagina

metabolic rate rate at which chemical reactions take place in the body

methane the simplest hydrocarbon, CH_4 – main component of natural gas

microorganism very small organism (living thing) that can be viewed only through a microscope – also known as a microbe

microwaves non-ionising radiation – used in telecommunications and in microwave ovens

mid-ocean ridge underwater mountain range formed by magma escaping from the seabed where continental plates are drifting apart

MMR vaccine for measles, mumps and rubella

molecule two or more atoms held together by covalent chemical bonds

molecular structure arrangement of atoms from which a molecule is made

molten made liquid by keeping the temperature above the substance's melting point

monomers small molecules that become chemically bonded to each other to form a polymer chain

mortar mixture of cement, sand and water

motor neurone nerve cell carrying information from the central nervous system to muscles

MRSA a form of the bacterium *Staphylococcus aureus* that is resistant to many antibiotics

mutation a change in the DNA in a cell

myelin sheath insulating layer around a nerve fibre

National Grid network that distributes electricity from power stations across the country

native (relating to metals such as gold) occurs in rocks as the element – not combined in compounds

natural gas gaseous fossil fuel formed from animals and plants that lived 100 million years ago – composed mainly of methane

natural oil oil produced by plants or fish

natural selection process by which 'good' characteristics that can be passed on in genes become more common in a population over many generations ('good' characteristics mean that the organism has an advantage which makes it more likely to survive)

nerve group of nerve fibres

neurone nerve cell

neutron small particle that does not have a charge – found in the nucleus of an atom

noble gas unreactive gas in Group 0 of the periodic table

non-renewable something that cannot be replaced when it has been used, such as fossil fuels and metal ores

normal line at right angles to a boundary, such as the line drawn for mirrors or glass blocks to help draw ray diagrams

nuclear power station power station generating electricity from the energy stored inside atoms – energy is released by the controlled splitting apart of large atoms (nuclear fission)

nucleus central part of an atom that contains protons and neutrons

nutrient medium liquid or jelly in which microorganisms can be grown

nutrient substance in food that we need to eat to stay healthy, such as protein

oceanic plate tectonic plate under the ocean floor – it does not carry a continent

oestrogen female hormone secreted by the ovary and involved in the menstrual cycle

OH group an oxygen atom bonded to a hydrogen atom and found in all alcohols

oil (crude) liquid fossil fuel formed from animals and plants that lived 100 million years ago

oil (from a plant) liquid fat obtained from seeds, nuts or fruit

opencast mining mining by digging out ore at the surface, rather than underground

optical fibre or cable glass fibre that is used to transfer communications signals as light or infrared radiation

oral contraceptive pills that prevent a woman releasing eggs

ore rock from which a metal is extracted, for example iron ore

organic compound a compound containing carbon and hydrogen, and possibly oxygen, nitrogen or other elements – living organisms are made up of organic compounds

oscillate vibration to and fro of particles in a wave

ovary organ (in females) which makes eggs

ovulation release of an egg from the ovary

oxidation process that increases the amount of oxygen in a compound – opposite of reduction

pandemic when a disease spreads rapidly across many countries – perhaps the whole world

Pangaea huge landmass with all the continents joined together before they broke up and drifted apart

pathogen harmful organism that invades the body and causes disease

payback time time taken for a type of domestic insulation to 'pay for itself' – to save as much in energy bills as it cost to install

period horizontal row in the periodic table

periodic table a table of all the chemical elements based on their atomic number

petrochemical substance made from petroleum

petroleum liquid fossil fuel formed from animals and plants that lived 100 million years ago

phagocytes white blood cells that surround pathogens and digest them with enzymes

photosynthesis process carried out by green plants where sunlight, carbon dioxide and water are used to produce glucose and oxygen

phototropism a growth response to light

photovoltaic cell device that converts the Sun's energy into electricity

phytomining using growing plants to absorb metal compounds from soil, burning the plants, and recovering metal from the ash

phytoremediation cleaning up contaminated soil by using growing plants to absorb harmful metal compounds

placebo 'dummy' treatment given to some patients, in a drug trial, that does not contain the drug being tested

plane mirror mirror with a flat surface

planet large ball of gas or rock travelling around a star – for example Earth and other planets orbit our Sun

plaque build-up of cholesterol in a blood vessel (which may block it)

plastics compounds produced by polymerisation, capable of being moulded into various shapes or drawn into filaments and used as textile fibres

plate boundaries edges of tectonic plates, where they meet or are moving apart

pollution presence of substances that contaminate or damage the environment

poly(ethene) plastic polymer made from ethene gas (also called polythene)

polymer large molecule made up of a chains of monomers

polymerisation chemical process that combines monomers to form a polymer: this is how polythene is formed

power amount of energy that something transfers each second and measured in watts (or joules per second)

power rating a measure of how fast an electrical appliance transfers energy supplied as an electrical current

power station place where electricity is generated to feed into the National Grid

producer organism that makes its own food from inorganic substances

products chemicals produced at the end of a chemical reaction

progesterone hormone produced by the ovary that prepares the uterus for pregnancy

protein molecule made up of amino acids (found in food of animal origin and also in plants)

proton small positive particle found in the nucleus of an atom

pyramid of biomass a diagram in which boxes, drawn to scale, represent the biomass at each step in a food chain

quarry place where stone is dug out of the ground

radio wave non-ionising radiation used to transmit radio and TV

rarefaction areas of a longitudinal wave in which the vibrating particles are spread out more than usual

rate of energy transfer a measure of how quickly something moves energy from one place to another

ray diagram diagrams showing how light rays travel

reactants chemicals that are reacting together in a chemical reaction

receptor nerve cell that detects a stimulus

red shift when lines in a spectrum are redder than expected – if an object has a red-shift it is moving away from the observer

reduction process that reduces the amount of oxygen in a compound, or removes all the oxygen from it – opposite of oxidation

reflected ray ray of light 'bouncing off' from a mirror or reflecting surface

reflection change of direction of a wave when it 'bounces off' from a surface

reflex action a fast, automatic response to a stimulus

reflex arc pathway taken by nerve impulse from receptor, through nervous system, to effector

refraction change of direction when a wave hits the boundary between two media at an angle, for example when a light ray passes from air into a glass block

renewable resource energy resource that is constantly available or can be replaced as it is used

repeatability consistent results are obtained when a person uses the same procedure a number of times

reproducibility consistent results are obtained when a number of people use the same procedure

resistant strain (of bacteria) a population of bacteria that is not killed by an antibiotic

respiration process occurring in living things in which oxygen is used to release the energy in foods

rod cell receptor cell in the eye that detects light

rutile an ore of titanium – impure titanium oxide (TiO_2)

Sankey diagram diagram showing how the energy supplied to something is transferred into 'useful' or 'wasted' energy

saturated fat solid fat, most often of animal origin, containing no C=C double bonds

saturated hydrocarbon hydrocarbon containing only single covalent bonds

secretion production and release of a useful substance

sensory neurone nerve cell carrying information from receptors to the central nervous system

shape memory alloy alloy that 'remembers' its original shape and returns to it when heated

shells electrons are arranged in shells (or orbits) around the nucleus of an atom

slag waste material produced during smelting of a metal – it contains unwanted impurities from the ore

smelting extracting metal from an ore by reduction with carbon – heating the ore and carbon in a furnace

solar cell device that converts the Sun's energy into electricity

solar panel panel that uses the Sun's energy to heat water

solar power station power station generating electricity using energy transferred by the Sun's radiation

solvent liquid in which solutes dissolve to form a solution

specific heat capacity a measure of the amount of energy needed to raise the temperature of 1 kg of a substance by 1 °C

speed of light speed at which electromagnetic radiation travels through a vacuum – 300 000 000 metres per second

sperm male sex cell of an animal

stainless steel steel alloy containing chromium and nickel to resist corrosion

state symbol symbol used in equations to show whether something is solid, liquid, gas or in solution in water

states of matter substances can exist in three states of matter (solid, liquid or gas) – changes from one state to another are called changes of state

statin drug that reduces cholesterol level in the blood

steam cracking cracking hydrocarbons by mixing with steam and heating

steam distillation process of blowing steam through a mixture to vaporise volatile substances – used to extract essential oils from flowers

steel alloy of iron and steel, with other metals added depending on its intended use

step-down transformer transformer that changes alternating current to a lower voltage

step-up transformer transformer that changes alternating current to a higher voltage

sterile containing no living organisms

sterile technique handling apparatus and material to prevent microorganisms from entering them

stimulus a change in the environment that is detected by a receptor

sub-atomic particle particle that make up an atom – proton, neutron, electron

subduction zone area of ocean floor in which an oceanic plate is sinking beneath a continental plate

sugar sweet-tasting compound of carbon, hydrogen and oxygen such as glucose or sucrose

sulfur dioxide poisonous, acidic gas formed when sulfur or a sulfur compound is burned

surface area a measure of the area of an object that is in direct contact with its surroundings

sweat liquid secreted onto the surface skin that has a cooling effect as it evaporates

symbol (for an element) one or two letters used to represent a chemical element, for example C for carbon or Na for sodium

synapse gap between two neurones

synthetic made by people

target organ the part of the body affected by a hormone

tectonic plate section of Earth's crust that floats on the mantle and slowly moves across the surface

telecommunications communications over long distances using various types of electromagnetic radiation

thalidomide a drug that was originally prescribed to pregnant women but was found to cause deformities in fetuses

thermal decomposition chemical reaction in which a substance is broken down into simpler chemicals by heating it

tidal power station power station generating electricity using the energy transferred by moving tides

tissue group of cells that work together and carry out a similar task, such as lung tissue

toxin poisonous substance (pathogens make toxins that make us feel ill)

transfer (energy) energy transfers occur when energy moves from one place to another, or when there is a change in the way in which it is observed

transformer device by which alternating current of one voltage is changed to another voltage

transition metals group of metal elements in the middle block of the periodic table – includes many common metals

transmitter chemical chemical that transfers a nerve impulse across a synapse

transverse wave a wave in which the vibration of particles is at right angles to the direction in which the wave transfers energy

tropism response of a plant to a stimulus, by growing towards or away from it

tsunami huge waves caused by earthquakes – can be very destructive

turbine device for generating electricity – the turbine has coils of wire that rotate in a magnetic field to generate electricity

ultraviolet radiation electromagnetic radiation that can damage human skin

unsaturated fats liquid fats, containing C=C double bonds – usually from plants or fish

unsaturated hydrocarbon hydrocarbon containing one or more C=C double bonds

upright image image that is the same way up as the object

U-value a measure of how easily energy is transferred through a material as heat

vaccine killed microorganisms, or living but weakened microorganisms, that are given to produce immunity to a particular disease

vacuum a space in which there are no particles of any kind

vaporise change from liquid to gas (vapour)

variation differences between individuals belonging to the same species

vent crack or weak spot in the Earth's crust, through which magma reaches the surface

virtual image image that can be seen but cannot be projected onto a screen (a mirror forms a virtual image behind the mirror)

virus very small infectious organism that reproduces within the cells of living organisms and often causes disease

volcano landform (often a mountain) where molten rock erupts onto the surface of the planet

voltage a measure of the energy carried by an electric current (the old name for potential difference)

wasted energy energy that is transferred by a device or appliance in ways that are not wanted, or useful

watt unit of energy transfer – one watt is a rate of energy transfer of one joule per second

wave equation the speed of a wave is always equal to its frequency multiplied by its wavelength

wave power electricity generation using the energy transferred by water waves as the water surface moves up and down

wavelength distance between two wave peaks

wind turbine device generating electricity by using the energy in moving air to turn a turbine and a generator

X-rays ionising electromagnetic radiation – used in X-ray photography to generate pictures of bones

yeast single-celled fungus used in making bread and beer

zygote a diploid cell formed by the fusion of the nuclei of two gametes

Index

Internet research

The Internet is a great resource to use when you are working through your GCSE Science course.

Below are some tips to make the most of it.

1 Make sure that you get information at the right level for you by typing in the following words and phrases after your search: 'GCSE', 'KS4', 'KS3', 'for kids', 'easy', or 'simple'.

2 Use OR, AND, NOT and NEAR to narrow down your search.

- Use the word OR between two words to search for one or the other word.
- Use the word AND between two words to search for both words.
- Use the word NOT, for example, 'York NOT New York' to make sure that you do not get unwanted results (hits).
- Use the word NEAR, for example, 'London NEAR Art' to bring up pages where the two words appear very close to each other.

3 Be careful when you search for phrases. If you search for a whole phrase, for example, A Room with a View, you may get a lot of search results matching some or all of the words. If you put the phrase in quote marks, 'A Room with a View' it will only bring search results that have that whole phrase and so bring you more pages about the book or film and less about flats to rent!

4 For keyword searches, use several words and try to be specific. A search for 'asthma' will bring up thousands of results. But, a search for 'causes of asthma' or 'treatment of asthma' will bring more specific and fewer returns. Similarly, if you are looking for information on cats, for example, be as specific as you can by using the breed name.

5 Most search engines list their hits in a ranked order so that results that contain all your listed words (and so most closely match your request) will appear first. This means the first few pages of results will always be the most relevant.

6 Avoid using lots of smaller words such as A or THE unless it is particularly relevant to your search. Choose your words carefully and leave out any unnecessary extras.

7 If your request is country-specific, you can narrow your search by adding the country. For example, if you want to visit some historic houses and you live in the UK, search 'historic houses UK' otherwise it will search the world. With some search engines you can click on a 'web' or 'pages from the UK only' option.

8 Use a plus sign (+) before a word to force it into the search. That way only hits with that word will come up.

Group 1	Group 2											Group 3	Group 4	Group 5	Group 6	Group 7	0
			$^{1}_{1}$H hydrogen														$^{4}_{2}$He helium
$^{7}_{3}$Li lithium	$^{9}_{4}$Be beryllium											$^{11}_{5}$B boron	$^{12}_{6}$C carbon	$^{14}_{7}$N nitrogen	$^{16}_{8}$O oxygen	$^{19}_{9}$F fluorine	$^{20}_{10}$Ne neon
$^{23}_{11}$Na sodium	$^{24}_{12}$Mg magnesium											$^{27}_{13}$Al aluminium	$^{28}_{14}$Si silicon	$^{31}_{15}$P phosphorus	$^{32}_{16}$S sulfur	$^{35}_{17}$Cl chlorine	$^{40}_{18}$Ar argon
$^{39}_{19}$K potassium	$^{40}_{20}$Ca calcium	$^{45}_{21}$Sc scandium	$^{48}_{22}$Ti titanium	$^{51}_{23}$V vanadium	$^{52}_{24}$Cr chromium	$^{55}_{25}$Mn manganese	$^{56}_{26}$Fe iron	$^{59}_{27}$Co cobalt	$^{59}_{28}$Ni nickel	$^{64}_{29}$Cu copper	$^{65}_{30}$Zn zinc	$^{70}_{31}$Ga gallium	$^{73}_{32}$Ge germanium	$^{75}_{33}$As arsenic	$^{79}_{34}$Se selenium	$^{80}_{35}$Br bromine	$^{84}_{36}$Kr krypton
$^{85}_{37}$Rb rubidium	$^{88}_{38}$Sr strontium	$^{89}_{39}$Y yttrium	$^{91}_{40}$Zr zirconium	$^{93}_{41}$Nb niobium	$^{96}_{42}$Mo molybdenum	$^{99}_{43}$Tc technetium	$^{101}_{44}$Ru ruthenium	$^{103}_{45}$Rh rhodium	$^{106}_{46}$Pd palladium	$^{108}_{47}$Ag silver	$^{112}_{48}$Cd cadmium	$^{115}_{49}$In indium	$^{119}_{50}$Sn tin	$^{122}_{51}$Sb antimony	$^{128}_{52}$Te tellurium	$^{127}_{53}$I iodine	$^{131}_{54}$Xe xenon
$^{133}_{55}$Cs caesium	$^{137}_{56}$Ba barium	$^{139}_{57}$La lanthanum	$^{178}_{72}$Hf hafnium	$^{181}_{73}$Ta tantalum	$^{184}_{74}$W tungsten	$^{186}_{75}$Re rhenium	$^{190}_{76}$Os osmium	$^{192}_{77}$Ir iridium	$^{195}_{78}$Pt platinum	$^{197}_{79}$Au gold	$^{201}_{80}$Hg mercury	$^{204}_{81}$Tl thallium	$^{207}_{82}$Pb lead	$^{209}_{83}$Bi bismuth	$^{210}_{84}$Po polonium	$^{210}_{85}$At astatine	$^{222}_{86}$Rn radon
$^{223}_{87}$Fr francium	$^{226}_{88}$Ra radium	$^{227}_{89}$Ac actinium															

Modern periodic table. You need to remember the symbols for the highlighted elements.

Acknowledgements

The publishers wish to thank the following for permission to reproduce photographs. Every effort has been made to trace copyright holders and to obtain their permission for the use of copyright materials. The publishers will gladly receive any information enabling them to rectify any error or omission at the first opportunity.

p.8t Branislav Senic/iStockphoto, p.8c Pete Donofrio/Shutterstock, p.8b Gelpi/Shutterstock, p.9t Sean Nel/Shutterstock, p.9c iDesign/Shutterstock, p.9b Dmitry Naumov/iStockphoto, p.10 Tony McConnell/Science Photo Library, p.11t Branislav Senic/iStockphoto, p.11b Sean Nel/Shutterstock, p.12t Lauri Patterson/iStockphoto, p.12b Andresr/iStockphoto, p.13 Steve Stock/Alamy, p.14t Dr Jeremy Burgess/Science Photo Library, p.14b Scott Camazine/Alamy, p.15l The Print Collector/Alamy, p.15r Fstop/The Image Bank/Getty Images, p.16t Dr P Marazzi/Science Photo Library, p.16c National Cancer Research/Science Photo Library, p.17 Juergen Berger/Science Photo Library, p.18t Jean-Loup Charmet/Science Photo Library, p.18b Dr Jeremy Burgess/Science Photo Library, p.19 CNRI/Science Photo Library, p.20t Phototak Inc/Alamy, p.20c Dr Kari Lounatmaa/Science Photo Library, p.21 Ian Miles/Alamy, p.22t PHIL/Barbara Rice, p.22b Dmitry Naumov/iStockphoto, p.24 Satiris Zakeiris/Science Photo Library, p.25 Custom Medical Stock/Alamy, p.26 Andy Harmer/Alamy, p.28 AFP/Getty Images, p.30 Scott Camazine/Alamy, p.31 Juergen Berger/Science Photo Library, p.32 David Reed/Corbis, p.34 Ralf Herschbach/Shutterstock, p.35 Jim Feliciano/Shutterstock, p.36 Scott Camazine/Science Photo Library, p.38 Bubbles Photo Library/Alamy, p.39t David Gregs/Alamy, p.39c CC Studio/Science Photo Library, p.40l Scientifica Visuals/Science Photo Library, p.40r Scientifica Visuals/Science Photo Library, p.41 Nigel Cattlin/Alamy, p.44t Kris Hanke/iStockphoto, p.44b Bjorn Svensson/Alamy, p.45 Chuck Nacke/Alamy, p.46t Paul Fievez/Getty Images, p.46r Jim West/Alamy, p.48t forbis/Shutterstock, p.48c katsgraphicslv/iStockphoto, p.49 Anthony Collins/Alamy, p.54 kolvenbach/Alamy, p.56t Gelpi/Shutterstock, p.56c David Chapman/Alamy, p.56b Jubal Harshaw/Shutterstock, p.57t Antonio Jorge Nunes/Shutterstock, p.57u Roca/Shutterstock, p.57c Georgy Markov/Shutterstock, p.57l Adrian T Sumner/Science Photo Library, p.57b Jordan Tan/Shutterstock, p.58t Vlad61/Shutterstock, p.58r Chris Pancewicz/Alamy, p.59t Bruce MacQueen/Shutterstock, p.59l Roman Kobzarev/iStockphoto, p.59b Antonio Jorge Nunes/Shutterstock, p.60t Iggy1108/Shutterstock, p.60c Emiliano Rodriguez/Shutterstock, p.60b John Pitcher/iStockphoto, p.61t Seleznev Oleg/Shutterstock, p.61b YinYang/iStockphoto, p.62t Canadian Press/Rex Features, p.62c Andre Mueller/Shutterstock, p.62b David Chapman/Alamy, p.63t Rick Wylie/iStockphoto, p.63b Jakub Pavlinec/Shutterstock, p.64t Science Photo Library, p.64c Martin Shields/Alamy, p.64b Dave Bevan/Alamy, p.65 Elena Korenbaum/iStockphoto, p.66t Adriatic2Alps Photography Tours/iStockphoto, p.66b blickwinkel/Alamy, p.67 Roca/Shutterstock, p.68 DJ Mattaar/Shutterstock, p.69 Gert Ellstrom/Shutterstock, p.70 Mogens Trolle/Shutterstock, p.72t British Museum/Munoz-Yague/Science Photo Library, p.72c Georgy Markov/Shutterstock, p.73t Arti_Zav/Shutterstock, p.73c Iorga Studio/Shutterstock, p.73b Владимир Воронин/iStockphoto, p.74 Stefan Sollfors/Alamy, p.75 Chris Mattison/Alamy, p.76 sonya etchison/Shutterstock, p.78 Adrian T Sumner/Science Photo Library, p.79l Gelpi/Shutterstock, p.79b Valentina R./Shutterstock, p.80t Krystyna Szulecka Photography/Alamy, p.80c Nigel Cattlin/Alamy, p.80b Chantal de Bruijne/Shutterstock, p.81t Eye of Science/Science Photo Library, p.81c Dr Keith Wheeler/Science Photo Library, p.82t Andrey Arkusha/Shutterstock, p.82r Phototake Inc./Alamy, p.84 Ryan Carter/iStockphoto, p.85 chungking/Shutterstock, p.85l David White/Alamy, p.88 Penn State University/Science Photo Library, p.90t Joshua Rainey/iStockphoto, p.90l Eric Isselée/iStockphoto, p.90r Eric Isselée/iStockphoto, p.91t Michael W Tweedie/Science Photo Library, p.91b Michael W Tweedie/Science Photo Library, p.92 NASA, p.100t valdis torms/iStockphoto, p.100u Quayside/Shutterstock, p.100l Christina Richards/Shutterstock, p.100b James Davidson/iStockphoto, p.101t Martyn F Chillmaid/Photolibrary, p.101u Dreamframer/iStockphoto, p.101l Josef Bosak/Shutterstock, p.101b Alistair Scott/Shutterstock, p.102 Science Photo Library, p.103l 4Science, p.103cl 4Science, p.103cr 4Science, p.103r 4Science, p.103b Laguna Design/Science Photo Library, p.104 David Parker/Science Photo Library, p.105 Andrew Lambert Photography/Science Photo Library, p.106t vladoleg/iStockphoto, p.106r Martyn F Chillmaid/Science Photo Library, p.108t Greg Balfour Evans/Alamy, p.108r Martyn F Chillmaid/Photolibrary, p.110 Martyn F Chillmaid/Science Photo Library, p.112t Johnny Greig/iStockphoto, p.112l Dreamframer/iStockphoto, p.112r LifeJourneys/iStockphoto, p.113 Tashka/iStockphoto, p.114t LDWinsconsin/Wikimedia Commons, p.114r Ted Spiegel/Corbis, p.115 Andrew Lambert Photography/Science Photo Library, p.116 John Gollop/Alamy, p.118 David Monniaux/Wikimedia Commons, p.119t Dirk Wiersma/Science Photo Library, p.119l Wikimedia Commons, p.119r Science Photo Library, p.120t jordache/Shutterstock, p.120r Friedrich von Horsten/Alamy, p.122 AliquisNJ/iStockphoto, p.124t Marc Dozier/Photolibrary, p.124r LanceB/iStockphoto, p.125 discpicture/Shutterstock, p.126t Jan Zoetekouw/Shutterstock, p.126r parema/iStockphoto, p.128t NASA/KSC/Science Photo Library, p.128l Robert Fullerton/Shutterstock, p.129 Miriam Maslo/Science Photo Library, p.130t Richard Heinzen/Photolibrary, p.130l Gary Whitton/Shutterstock, p.130c Mike Tingle, p.130r Mike Tingle, p.131l Alexnn/iStockphoto, p.131r Bluerain/Shutterstock, p.134t wrangler/Shutterstock, p.134r Bartlomiej K. Kwieciszewski/Shutterstock, p.135l Wikimedia Commons, p.135c Karol Kozlowski/Shutterstock, p.135br Lesley Garland Picture Library/Alamy, p.136t NASA, p.136c Brian A Jackson/Shutterstock, p.137t Diego Cervo/Shutterstock, p.137l Alex Segre/Alamy, p.138t ribeiroantonio/Shutterstock, p.138c Michael J Thompson/Shutterstock , p.138b Rick Decker/Alamy, p.140 Kolvenbach/Alamy, p.148t Hiob/iStockphoto, p.148c Andrew Lambert Photography/Science Photo Library, p.148b Chapelle/Shutterstock, p.149t Joe Belanger/Shutterstock, p.149c Luis Carlos Jimenez del rio/Shutterstock, p.149b David Hodges/Alamy, p.150t Konstantin Sutyagin/Shutterstock, p.152 DWD-photo/Alamy, p.153t Mike Tingle, p.153b Andrew Lambert Photography/Science Photo Library, p.154 Chris Parypa/Alamy, p.155l Khoroshunova Olga/Shutterstock, p.155r Jim Parkin/Alamy, p.155b Mike Tingle, p.156t Monkey Business Images/Shutterstock, p.156r ultimathule/Shutterstock, p.157r kaband/Shutterstock, p.157b Joe Belanger/Shutterstock, p.158t Sipa Press/Rex Features, p.158l SIU/Science Photo Library, p.158r EpicStock/Shutterstock, p.159t mirounga/Shutterstock, p.159b Readius and Polymer Vision, p.160t Robert Galbraith/Reuters/Corbis, p.160l Lya_Cattel/iStockphoto, p.160r Lya_Cattel/iStockphoto, p.161t Stuart Kelly/Alamy, p.161b Green Stock Media/Alamy, p.162 John James/Alamy, p.164t Soyka/Shutterstock, p.164r AFP/Getty Images, p.165 Mike Tingle, p.166t Jim West/Alamy, p.166r Alvey & Towers Picture Library/Alamy, p.167 courtesy of www.Biodieselathome.biz, p.168tl ASP Food/Alamy, p.168tr Jeffrey Blackler/Alamy, p.168cr Lew Robertson/iStockphoto, p.169t Andrew Lambert Photography/Science Photo Library, p.169b mypix/Shutterstock, p.170t Monika Adamczyk/iStockphoto, p.170r Stuart Monk/iStockphoto, p.171 Martin Allinger/Shutterstock, p.172 Bon Appetit/Alamy, p.174 Planet Observer/Science Photo Library, p.175 Josemaria Toscano/iStockphoto, p.177t Wolfgang Steiner/iStockphoto, p.177b Charles & Josette Lenars/Corbis, p.178 Mana Photo/Shutterstock, p.179 David Hodges/Alamy, p.180 Rene Jansa/Shutterstock, p.181 Iain Sarjeant/iStockphoto, p.182 NASA, p.184 Warwick Lister-Kaye/iStockphoto, p.192t Rafal Olkis/Shutterstock, p.192c Scott Rothstein/Shutterstock, p.192b kak2s/Shutterstock, p.193t great_photos/Shutterstock, p.193c Ivonne Wierink/Shutterstock, p.193b Joey Boylan/iStockphoto, p.194t Andrew Holt/Alamy, p.194r s_oleg/Shutterstock, p.194l Monkey Business Images/Shutterstock, p.195 Ivonne Wierink/Shutterstock, p.196t Detlev van Ravenswaay/Science Photo Library, p.196r LianeM/Shutterstock, p.198 Jody Fairish/Alamy, p.199t Charles D Winter/Science Photo Library, p.199c Cann Balcioglu/Shutterstock, p.200 Kate Leigh/iStockphoto, p.201 iofoto/Shutterstock, p.202t Myrleen Pearson/Alamy, p.202r Shutterstock, p.204t Monkey Business Images/Shutterstock, p.204l mangostock/Shutterstock, p.204r Martyn F. Chillmaid/Science Photo Library, p.205 Anke van Wyk/Shutterstock, p.206t OxfordSquare/iStockphoto, p.206r Peter Alvey/Alamy, p.207 NIA, p.208 Mary Evans Picture Library/Alamy, p.209 The Print Collector/Alamy, p.210 JC Photography/Alamy, p.211 Oleksiy Maksymenko/Alamy, p.212 David Weintraub/Science Photo Library, p.214t Jonas Ahlman/Alamy, p.214l zoomstudio/iStockphoto, p.215 Thierry Maffeis/iStockphoto, p.218 NI Syndication, p.219t tr3gin/Shutterstock, p.219b Ivaschenko Roman/Shutterstock, p.220 Courtesy of Husqvarna UK, p.221t Baris Simsek/iStockphoto, p.221b Martyn F. Chillmaid/Science Photo Library, p.222t Paul M Thompson/Alamy, p.222b Joey Boylan/iStockphoto, p.223 Lev Lev/Photolibrary, p.230t Yegor Korzh/Shutterstock, p.230c Theo Gottwald/Alamy, p.230b Baris Simsek/iStockphoto, p.230t Apple's Eyes Studio/Shutterstock, p.230c James Doss/Shutterstock, p.230b NASA, p.232 Yves Grau/iStockphoto, p.233 ilbusca/iStockphoto, p.234t Igor Grochev/Shutterstock, p.234b Yegor Korzh/Shutterstock, p.235 Darren Baker/Shutterstock, p.236t Paul Glendell/Alamy, p.236b Rhoberazzi/iStockphoto, p.237 David Coultham/iStockphoto, p.238t Patrick Landmann/Science Photo Library, p.238b Luis Pedrosa/iStockphoto, p.239 Mikkel William Nielsen/iStockphoto, p.240t Karl-Friedrich Hohl/iStockphoto, p.240r Michael DeGasperis/Shutterstock, p.241t Apple's Eyes Studio/Shutterstock, p.241b indianstockimages/Shutterstock, p.242 NASA, p.243 NASA, p.244t mikeuk/iStockphoto, p.244b Dmitry Naumov/iStockphoto, p.245tr Tim Ridley Collection/Dorling Kindersley/Getty Images, p.245r sciencephotos/Alamy, p.245br Luis Sandoval Mandujano/iStockphoto, p.246 Ed Darack/Science Faction/Corbis, p.247tr Leslie Garland Picture Library/Alamy, p.247cr Jan van der Hoeven/Shutterstock, p.247bl Breadmaker/Shutterstock, p.248 NASA/Science Photo Library, p.250t Roger Bambe/Alamy, p.250b iofoto/Shutterstock, p.251 Tom Hahn/iStockphoto, p.252t Alan Williams/Alamy, p.252r Ted Kinsman/Science Photo Library, p.253tr Terry Williams/Rex Features, p.253cr Philippe Psaila/Science Photo Library, p.253bl Paul Gooney/Alamy, p.254 Stephen Meese/Shutterstock, p.256t Monkey Business Images/Shutterstock, p.256c View China Photo/Rex Features, p.256b Alberto Pomares/iStockphoto, p.257 Doncaster and Bassetlaw Hospitals/Science Photo Library, p.258t NASA, p.258b Corbis RF/Alamy, p.259 Tomas Bercic/iStockphoto, p.260 Goodluz/Shutterstock, p.261 Gastrolab/Science Photo Library, p.262l Friedrich Saurer/Science Photo Library, p.262c saiko3p/Shutterstock, p.262r Martyn F. Chillmaid/Science Photo Library, p.264t Wikimedia Commons, p.264r TexPhoto/iStockphoto, p.265 NASA, p.266 A. T. Willett/Alamy, p.268t sciencephotos/Alamy, p.268b Chris Butler/Science Photo Library, p.269 WMAP Science Team, NASA/Science Photo Library.